Effective Models in Heritage Science

Effective Models in Heritage Science

Editors

Peter Brimblecombe
Jenny Richards

MDPI

Basel • Beijing • Wuhan • Barcelona • Belgrade • Novi Sad • Cluj • Manchester

Editors
Peter Brimblecombe
National Sun Yat-sen University
Kaohsiung, Taiwan

Jenny Richards
University of Oxford
Oxford, UK

Editorial Office
MDPI
St. Alban-Anlage 66
4052 Basel, Switzerland

This is a reprint of articles from the Special Issue published online in the open access journal *Heritage* (ISSN 2571-9408) (available at: https://www.mdpi.com/journal/heritage/special_issues/emihs).

For citation purposes, cite each article independently as indicated on the article page online and as indicated below:

Lastname, A.A.; Lastname, B.B. Article Title. *Journal Name* **Year**, *Volume Number*, Page Range.

ISBN 978-3-0365-8654-0 (Hbk)
ISBN 978-3-0365-8655-7 (PDF)
doi.org/10.3390/books978-3-0365-8655-7

Cover image courtesy of Peter Brimblecombe

Contents

Contents

About the Editors

Peter Brimblecombe

Peter is an Honorary Chair Professor at National Sun Yat-Sen University in Taiwan. He was born in Australia, ultimately completing a PhD in Atmospheric Chemistry in Auckland, New Zealand. He spent four decades at the University of East Anglia studying long-term changes in urban air pollution and climate as it affects health and buildings, producing books such as The Big Smoke. In recent years, he has worked on air pollution and climate in Hong Kong and Taiwan.

Jenny Richards

Jenny holds a Career Development Fellowship at St John's College, Oxford, and is a Researcher at the School of Geography and the Environment, University of Oxford. She read Geography (BA Hons) at the University of Oxford before completing her MRes at UCL and DPhil at Oxford University. Her research investigates the interactions between the environment and heritage.

Preface

Heritage sites and objects are shaped by their environment. Modelling is a useful tool to investigate the interactions between heritage and its surroundings. Although process-based models are less commonly applied to the management of heritage than might be expected, they can be important in understanding the processes of change and informing management decisions.

This Special Issue, entitled "Effective Models in Heritage Science", explores the ways in which models can reveal the nature and effect of environmental pressures, thus contributing to better practice and management of heritage. Optimally modelling studies should encourage a two-way exchange of understanding between research and practice, such that feedback is likely to offer the potential for continuous improvement.

The papers collected here should be useful to those interested in environmental pressures on heritage. It is hoped that these contributions lead to further developments of effective models that address long-term change to heritage objects and sites. This Special Issue is likely to be useful to those concerned with the practical applications of models to the strategic and local management of heritage. The contributing authors come from both academic and heritage organisations, so they are well experienced at coupling theoretical understanding to practice.

We are especially grateful to the authors who have contributed to this Special Issue, and also to everyone who helped shape the content of the Special Issue by joining webinar discussions and online meetings. We were also fortunate to have had attentive and helpful referees, who deserve our thanks.

Peter Brimblecombe and Jenny Richards
Editors

heritage

MDPI

Editorial

Tuning and Effectiveness in Heritage Models

Jenny Richards [1,2] and Peter Brimblecombe [3,*]

[1] St John's College, Oxford University, Oxford OX1 3JP, UK; jennifer.richards@sjc.ox.ac.uk
[2] School of Geography and the Environment, Oxford University, Oxford OX1 3QY, UK
[3] Department of Marine Environment and Engineering, National Sun Yat-sen University, Kaohsiung 804201, Taiwan
* Correspondence: p.brimblecombe@uea.ac.uk

Abstract: Modelling can explore heritage responses to environmental pressures over wide spatial and temporal scales, testing both theory and process. However, compared to other fields, modelling approaches are not yet as common in heritage management. Some heritage models have become well known, though they struggle to have an impact beyond academia, with limited practical applications. Successful models appear to be adaptable to multiple sites or objects, intuitive to use, run using widely available software and produce output translatable into practical actions. Model tuning is also vital for the model to be effective. A specific purpose should be determined from the outset to enable tuning in the earliest design stages. Heritage models can be developed to explore theories or processes that affect or interact with heritage. Input should also be tuned to relevant temporal and spatial scales and consider duration and location. Additionally, it is important to account for materials and elements specific to heritage. Models need to be useful and usable if they are to be effective. User-friendly programs and interfaces help practical use. However, success can create problems, as input and output could become socially or commercially sensitive. The wider use of models may require broader discussion among heritage professionals and the provision of training.

Keywords: climate change; heritage climate; purpose; process; theory; scale; material; applicability

1. Introduction

Heritage sites and objects possess a uniqueness that makes them irreplaceable [1]. However, heritage faces many threats, which can result in a loss of artistic worth and evidential value [2,3], although others might see this as an acceptable process of change [4]. Heritage practitioners and scientists need novel approaches to investigate processes that pose a threat to heritage to manage the pressures at heritage sites [5]. The methods used in field experiments can be highly constrained due to the potential for causing damage to objects or sites, and laboratory experiments take place in highly controlled conditions. Modelling methods allow for a safe mode of experimentation that can incorporate multiple processes over a wide range of spatial and temporal scales and enable the testing of both theory and process regimes.

Within the heritage field, several models have become well known, and their findings are frequently cited. Examples include the model output of future heritage climate developed under EU-funded projects, such as NOAHs ARK [6] and Climate for Culture [7]. However, these models, along with ones published in the recent academic literature, have often struggled to have a substantial impact on practice, predominantly remaining within the academic sphere [8]. The use of models in heritage science has been relatively uncommon when compared with other practical fields [8].

When models are used in practice, such applications are often recorded in the grey literature, which can limit a widening readership. Models that have gained broader adoption in practical contexts and influenced heritage management have tended to be: (i) adaptable to multiple sites or objects, (ii) intuitive to use, with, for example, a graphical user interface, (iii) run using widely available software and (iv) produce outputs that are

Citation: Richards, J.; Brimblecombe, P. Tuning and Effectiveness in Heritage Models. *Heritage* **2023**, *6*, 5516–5523. https://doi.org/10.3390/heritage6070290

Received: 17 June 2023
Accepted: 11 July 2023
Published: 21 July 2023

translatable into practical actions [8]. Furthermore, for models to be effective in heritage science, "the modelled spatial and temporal scales need to be relevant to the impact of processes of change relevant to the heritage value and utility associated with sites and objects" [8].

As with any scientific approach, models and their underlying methods should be continually refined. This requires feedback between model developers and model users, "such that: (i) developers understand the challenges and benefits that users have with using the model, and (ii) users can share ideas and experience" [8]. There also needs to be considerable attention paid to the effect of uncertainty in model output. All models inherently contain uncertainty, but understanding how such uncertainties could result in errors being propagated through to actual decision making [9] requires further research in terms of heritage management. Recent work has suggested that while uncertainties can cause a range in output magnitude, when looking at change over time, the direction of change is an important indicator for developing management strategies [10,11].

Just as fieldwork or lab experiments are broad concepts that capture a range of methods, there are an array of models developed for a heritage context that we can use to explore the requirements and challenges in developing effective models. The Special Issue on *Effective Models in Heritage Science* aims to bring together modelling-based studies, and it acts as a basis to explore the idea of tuning heritage models that have emerged in developing this Special Issue.

2. Tuning Heritage Models

Models inherently require a system to be simplified. Therefore, model builders have to decide which elements and interactions within a given system to include. These decisions are context-dependent for the model to have the necessary specificity and precision to make the model and its outputs effective. In a heritage context, models should be tuned to: (i) the management requirements and purpose, (ii) relevant theory or process, (iii) issues of scale, spatial and temporal, and the location and time period under consideration and (iv) the relevance to specific heritage materials or elements of construction.

2.1. Purpose

When models are built to address a specific heritage question, ideas of model tuning can be included from the earliest design stages. This can be advantageous as the model can be readily tailored to research and management needs. When models are fit for purpose and have clear practical application, they can have far-reaching implications. For example, models developed for the National Trust were able to assess the economics of dust [12] and housekeeping [13]. Lloyd's [13] research demonstrates how an easy-to-use, adaptable model in Microsoft Excel can aid with housekeeping and conservation decisions across a portfolio of historic homes. However, many new models fail to gain traction in the wider heritage community, often due to a lack of awareness that the model exists and the type of information it could provide. Furthermore, specific technical expertise or computational resources likely required to run the model can act as a barrier to uptake [8].

Several heritage models, such as ECOS/RUNSALT [14], IMPACT [15] and HERIe [16], are used by both heritage academics and practitioners. These have a simple approach to input or graphical user interfaces. This enables researchers and practitioners to engage with a model developed to address a process that impacts heritage materials, such as salt crystallisation and changes in indoor environments [17,18]. It is highly important for experts to produce clear guidance for good practice, as well as avoiding common mistakes for these out-of-the-box models, as users can find the model limitations hard to identify, even though they were obvious to the developer. Godts et al. [19] provide an example of this for the widely used ECOS/RUNSALT model [20–23].

Model users must also grapple with model uncertainty. For example, in Lloyd's management model [13], errors may lead to a housekeeping allocation that is too small in some properties, such that individual property managers might feel under-resourced or, at other times, too large leading to over-resourcing. Although mathematical errors in models can be assessed, it is often difficult to understand how these errors might affect decisions. It is particularly challenging in models where the underlying processes are hidden, such as with multi-layered neural networks [24–27]. When using models to plan for future conservation needs, capturing the direction of change with high confidence will be important for informing management plans. The direction of change can often be established, even when there is uncertainty over the exact magnitude of change [11]. An additional source of error can arise from biases within the model. Although Lloyd [13] tries to minimise the bias in inputs, outputs might be skewed towards overprotection rather than underprotection. However, some have argued that biases may be helpful, especially where the costs of action and inaction are very different [9].

Finally, feedback is crucial to ensure that a model fulfils, and continues to fulfil, its desired purpose. However, in heritage research, there are few meaningful feedback pathways between academic researchers and practitioners. This means that the experience of using the model in practice is commonly not fed back to the researcher [8], constraining improvement, thus limiting its overall usefulness.

2.2. Theory and Process

Models in a heritage context can be developed to explore relevant processes or theories. Theory-focused models are valuable tools for advancing concepts and ideas that underpin our understanding of process, as well as having the potential to develop interdisciplinary conversations [8]. More often, models focus on assessing process. For example, Bretti and Ceseri [28] developed a model to translate the specifics of the carbonation process into a future changing climate, and Hart et al. [29] constructed a linear model to assess rainfall impacts on adobe blocks with a range of conservation treatments. Such models can capture key processes while retaining versatility in treating physical, chemical and biological threats (e.g., the role insects play in the degradation of wood [30] or other organic materials [31]). However, when models are built for a specific site or material, there is a danger that these can be difficult to transfer to other heritage sites or objects, limiting model effectiveness. Such situations highlight the problems of over-tuning a model to the point that it becomes so specific it cannot be used in other settings.

In addition to building new models tuned to a heritage context, existing non-heritage-specific model outputs can also be tuned to make them relevant to heritage. A common example of this is tuning parameters to meet the requirements of dose–response functions [32]. This often means reworking the inputs, which might be meteorological parameters or air pollutant concentrations. For example, outputs from global climate models were reprocessed to capture deterioration processes in timber [33], while Verney-Carron et al. [34] compared the ability of dose–response functions with kinetic laws to assess climate and pollution impacts on medieval stained glass.

Further work is needed to assess how models capture the effects of extreme events or the crossing of thresholds, as these are likely to become increasingly relevant in a future world experiencing extensive environmental changes, changing frequency of climate hazards and sea level rise [35], but they are infrequently explored in heritage models.

2.3. Scale: Time and Location

Heritage models also need to consider scale, with regard to both time and space. Modelling the effect of long-term pollution and climate change on sites and objects has been a consistent theme in heritage science. Declining levels of air pollution in recent years have been beneficial to heritage [36], while concern over climate change impacts has increased e.g., the Noah's Ark Project [6]. Past exposure conditions for sites and objects can be difficult to determine, and although some have tried, e.g., assessment of historical weathering [37], examining the deposits on buildings [38–40], archive photographs [41,42] and economic records of repair [43], it can be difficult. Thus, there has been considerable interest in trying to model conditions in the past and relate them to observed damage [34,44,45].

There are usually questions about the relevance of time periods under investigation. A 75-year span is useful to model significant changes, which might occur by the end of the 21st century and is frequently used in long-term assessments [46–48]. However, planning for heritage management is usually on a shorter time scale. Even the 30-year periods adopted as part of the notion of climate normals can seem long to heritage planners [49]. Budgets may be set over three-year timescales, or models might need to be tuned to the lifetime of maintenance and repair [13]. Thus, shorter planning horizons or recommendations for immediately implementable actions [17] will see an increasing need for effective models to consider reduced time periods.

Time is also important with regard to the resolution of input and output data. Heritage science sometimes uses annual or monthly averages (e.g., [50,51]), but for models to capture processes that drive deterioration, much shorter time scales are often needed, commonly requiring daily [33] or sub-daily data [17,18]. In some cases, it can be possible to extract the higher-resolution data from monthly averages, but this is not always possible [52]. Thus, datasets at higher resolution can be of importance in a heritage context.

Heritage models are developed for a wide range of spatial scales from individual materials [28,34], objects and sites [17,29,30], countries and regions [53,54], as well as global-scale analyses [33]. The spatial scale of a heritage model will be determined by the purpose of the model, with strategic models tending to have larger spatial scales, while, for example, chemical changes within heritage materials will require smaller-scale models. Challenges can arise when transferring between spatial scales as the dominant processes can be dependent on the scale [30]. This can be particularly difficult when relating global climate to indoor conditions [55]. However, as heritage is influenced by both global and local environmental processes [56], effective modelling and heritage management are dependent on being able to combine findings across multiple scales.

Models should also be able to be tuned to a specific location. This might include modelling heritage in the context of its surrounding landscape or environmental conditions [57–60]. Flexibility in the set-up parameters enables models to represent local conditions, which is important in determining processes for specific objects or sites. It is important that these set-up parameters remain adjustable so that the model is not constrained to a fixed set of starting conditions.

2.4. Materials and Heritage Elements

Models assessing physical, chemical or biological change to heritage materials are predominantly developed for a specific material. For example, Hart et al.'s [29] model was developed for adobe blocks, while Bretti and Ceseri's [28] model was for concrete. This means that models can consider the relevant processes that cause change in a material, as well as accounting for the varying rates at which different materials undergo change [18]. Other models have focused on capturing the conditions that drive the change, rather than the change itself [33], e.g., ECOS/RUNSALT [14,19]. Using a model to capture conditions might make it more easily alterable to fit a different material or context. For example, understanding the number of freeze–thaw cycles a site is exposed to will be relevant to many materials, including stone, earth and brick [54,61,62]. Some models, such as those that consider earthquakes, sea level rise, wind-driven sand or rain or flooding, frequently need to account for multiple materials that form heritage buildings or sites, more than focussing on isolated elements [59,60,63,64].

A clear aim is vital for models to be effective; models with a site- or object-specific purpose will require a much closer alignment to specific materials than more strategic models looking at country- and regional-scale patterns of change.

3. Effective Models

For a model to represent a system in a useful manner, the purpose of a heritage model needs to be clearly defined from the outset. By understanding the purpose, this can dictate other tuning factors, such as time, place and materials, and allows the parameterisation to be undertaken more efficiently.

These models also need to be useful and usable if they are to be effective. Usefulness requires that models can be used in practice, but engaging with models can be daunting due to: (i) complex underlying theory and (ii) specialised computer programming languages. It seems that approachable models include those written in Excel (as in Lloyd [13]) or as Java applets (e.g., IMPACT [15]). More complex models have to be written in more formal programming languages. Here, they benefit from Graphical User Interfaces that guide the user through the process of data entry and provide output that is easy to interpret. For models to remain applicable to other sites or objects, the underlying code should be flexible to enable such a transfer to be possible.

Elements of successful models include being able to: (i) address a given challenge and (ii) be used by relevant stakeholder groups. In academia, a proxy for success is typically measured using citations to a model, but such citations may not reflect practical use. The practical use of models may be much discussed among professionals at conferences and training events, but quantifiable metrics are harder to establish. For example, Lloyd's [13] model of housekeeping has been used as a part of masters' courses in Collections Care and Conservation Management [65], but such success is not captured by standard metrics.

When a model is used in practice, both the input and output may become sensitive as they can influence decision making regarding finance, staffing or management. Thus, success for a model might also mean that it becomes a valuable asset, which can consequently limit sharing it as an open resource.

The long-term useability of a model requires documentation and ongoing training if knowledge inherent to ensure the model is not confined to one person (or a small team of people). Thus, groups with a good understanding of previous model iterations are required to update a model.

4. Conclusions

Models in heritage science have the potential to be a useful tool for research and management but face limited uptake in practice. This may arise from a belief that they are theoretical constructs and do not confront the complex reality inherent to heritage, or that such models may introduce bias or errors in decision making. Tuning models with respect to purpose, process and theory, spatial and temporal scale and materials will help improve the effectiveness of heritage models. Many process-based models used in a heritage context engage with climate. While understanding the impacts of climate change on heritage is of great importance, heritage models need to widen their scope to include other aspects of the heritage environment. To improve the understanding of models, it is likely that engagement with professional societies and the development of training courses are needed to support a widening understanding of the contributions that can be made as a result of effective modelling in heritage science.

Author Contributions: Conceptualization, methodology, analysis, drafting and editing: P.B. and J.R. All authors have read and agreed to the published version of the manuscript.

Data Availability Statement: This paper predominantly draws on findings from the Special Issue "Effective Models in Heritage Science", available open access: https://www.mdpi.com/journal/heritage/special_issues/emihs (accessed on 16 June 2023).

Acknowledgments: We thank all the authors who have contributed to the Special Issue.

Conflicts of Interest: The authors declare no conflict of interest.

References

1. Taboroff, J. Cultural Heritage and Natural Disasters: Incentives for Risk Management and Mitigation. In *Managing Disaster Risk in Emerging Economies*; Kreimer, A., Arnold, M., Eds.; World Bank: Washington, DC, USA, 2000; pp. 71–79.
2. Zammit, N.; Bianco, L. Computation of Heritage Values: Towards a Holistic Method to Assess Built Heritage. *Herit. Sustain. Dev.* **2022**, *4*, 101–110. [CrossRef]

3. Clark, K. Values in Cultural Resource Management. In *Heritage Values in Contemporary Society*; Smith, G.S., Messenger, P., Soderland, H., Eds.; Left Coast Press: Walnut Creek, CA, USA, 2010; pp. 89–100.
4. Holtorf, C. Averting Loss Aversion in Cultural Heritage. *Int. J. Herit. Stud.* **2015**, *21*, 405–421. [CrossRef]
5. Kennedy, C.J. The Role of Heritage Science in Conservation Philosophy and Practice. *Hist. Environ. Policy Pract.* **2015**, *6*, 214–228. [CrossRef]
6. Sabbioni, C.; Cassar, M.; Brimblecombe, P.; Tidblad, J.; Kozlowski, R.; Drdácký, M.; Sáiz-Jiménez, C.; Grøntoft, T.; Wainwright, I.; Ariño, X. Global Climate Change Impact on Built Heritage and Cultural Landscapes. In Proceedings of the International Conference on Heritage, Weathering and Conservation, HWC, Madrid, Spain, 21–24 June 2006; pp. 395–401.
7. Leissner, J.; Kilian, R.; Kotova, L.; Jacob, D.; Mikolajewicz, U.; Broström, T.; Ashley-Smith, J.; Schellen, H.L.; Martens, M.; van Schijndel, J.; et al. Climate for Culture: Assessing the Impact of Climate Change on the Future Indoor Climate in Historic Buildings Using Simulations. *Herit. Sci.* **2015**, *3*, 38. [CrossRef]
8. Richards, J.; Brimblecombe, P. The Transfer of Heritage Modelling from Research to Practice. *Herit. Sci.* **2022**, *10*, 17. [CrossRef]
9. Johnson, D.D.P.; Blumstein, D.T.; Fowler, J.H.; Haselton, M.G. The Evolution of Error: Error Management, Cognitive Constraints, and Adaptive Decision-Making Biases. *Trends Ecol. Evol.* **2013**, *28*, 474–481. [CrossRef]
10. Richards, J.; Brimblecombe, P.; Engelstaedter, S. Modelling Temperature-Precipitation Pressures on African Timber Heritage. *Int. J. Climatol.* (under review).
11. Brimblecombe, P.; Richards, J. Köppen Climates and Scheffer Index as Indicators of Timber Risk in Europe. *Herit Sci.* **2023**, *11*, 148. [CrossRef]
12. Lloyd, H.; Brimblecombe, P.; Lithgow, K. Economics of Dust. *Stud. Conserv.* **2007**, *52*, 135–146. [CrossRef]
13. Lloyd, H. Quantifying Housekeeping Challenge and Conservation Need. *Heritage* **2023**, *6*, 3757–3776. [CrossRef]
14. Bionda, D. RUNSALT—A Graphical User Interface to the ECOS Thermodynamic Model for the Prediction of the Behaviour of Salt Mixtures under Changing Climate Conditions. 2005. Available online: http://science.sdf-eu.org/runsalt/ (accessed on 16 June 2023).
15. Blades, N.; Kruppa, D.; Cassar, M. Development of a Web-Based Software Tool for Predicting the Occurrence and Effect of Air Pollutants inside Museum Buildings. In Proceedings of the ICOM Committee for Conservation 13th Triennial Meeting, Rio de Janeiro, Brazil, 20–27 September 2002; ICOM, Ed.; James & James: London, UK, 2002.
16. Kupczak, A.; Jędrychowski, M.; Strojecki, M.; Krzemień, L.; Bratasz, Ł.; Łukomski, M.; Kozłowski, R. HERIe: A Web-Based Decision-Supporting Tool for Assessing Risk of Physical Damage Using Various Failure Criteria. *Stud. Conserv.* **2018**, *63*, 151–155. [CrossRef]
17. Grøntoft, T.; Stoveland, L.P. Painted Wood Climate Risk Analysis by the HERIe Model of Building Protection and Conservation Heating Scenarios in Norwegian Medieval Stone Churches. *Heritage* **2023**, *6*, 3089–3112. [CrossRef]
18. Thickett, D. Practical Use of Damage Functions for Environmental Preventive Conservation and Sustainability—Examples from Naturally Ventilated Buildings. *Heritage* **2023**, *6*, 2633–2649. [CrossRef]
19. Godts, S.; Steiger, M.; Orr, S.A.; Stahlbuhk, A.; Desarnaud, J.; De Clercq, H.; Cnudde, V.; De Kock, T. Modeling Salt Behavior with ECOS/RUNSALT: Terminology, Methodology, Limitations, and Solutions. *Heritage* **2022**, *5*, 3648–3663. [CrossRef]
20. Chabas, A.; Kloppmann, W.; Sizun, J.P.; Wille, G.; Coman, A.; Petitmangin, A.; Nowak, S.; Martin, E.; Jurgens, M.A. Sources and Chronology of Soluble Salt Formation in a Medieval Dovecote Caught up in Urbanisation: A Resilience Story? *Env. Earth Sci.* **2022**, *81*, 550. [CrossRef]
21. Wang, Y.; Zhang, H. The Synergic Impacts of Salt Mixture and Frost Damage on Rock Decay: Implications for the Deterioration of Rock-Hewn Heritages. *Herit. Sci.* **2023**. [CrossRef]
22. Hu, T.; Brimblecombe, P.; Zhang, Z.; Song, Y.; Liu, S.; Zhu, Y.; Duan, J.; Cao, J.; Zhang, D. Capillary Rise Induced Salt Deterioration on Ancient Wall Paintings at the Mogao Grottoes. *Sci. Total Environ.* **2023**, *881*, 163476. [CrossRef]
23. Pintér, F. The Combined Use of Ion Chromatography and Scanning Electron Microscopy to Assess Salt-Affected Mineral Materials in Cultural Heritage. *J. Am. Inst. Conserv.* **2021**, *61*, 85–99. [CrossRef]
24. Perez, H.; Tah, J.H.M.; Mosavi, A. Deep Learning for Detecting Building Defects Using Convolutional Neural Networks. *Sensors* **2019**, *19*, 3556. [CrossRef]
25. Hatır, M.E.; İnce, İ.; Korkanç, M. Intelligent Detection of Deterioration in Cultural Stone Heritage. *J. Build. Eng.* **2021**, *44*, 102690. [CrossRef]
26. Hart, A.; Wyatt, J. Evaluating Black-Boxes as Medical Decision Aids: Issues Arising from a Study of Neural Networks. *Med. Inf.* **1990**, *15*, 229–236. [CrossRef] [PubMed]
27. Huang, L.; Song, Y. Intangible Cultural Heritage Management Using Machine Learning Model: A Case Study of Northwest Folk Song Huaer. *Sci. Program.* **2022**, *2022*, 1383520. [CrossRef]
28. Bretti, G.; Ceseri, M. Climate Change Effects on Carbonation Process: A Scenario-Based Study. *Heritage* **2023**, *6*, 236–257. [CrossRef]
29. Hart, S.; Raymond, K.; Williams, C.J.; Rutherford, W.A.; DeGayner, J. Modeling Earthen Treatments for Climate Change Effects. *Heritage* **2023**, *6*, 4214–4226. [CrossRef]
30. Brimblecombe, P.; Richards, J. Moisture as a Driver of Long-Term Threats to Timber Heritage—Part II: Risks Imposed on Structures at Local Sites. *Heritage* **2022**, *5*, 2966–2986. [CrossRef]
31. Manachini, B. Alien Insect Impact on Cultural Heritage and Landscape: An Underestimated Problem. *Conserv. Sci. Cult. Herit.* **2015**, *15*, 61–72. [CrossRef]

32. Strlič, M.; Thickett, D.; Taylor, J.; Cassar, M. Damage Functions in Heritage Science. *Stud. Conserv.* **2013**, *58*, 80–87. [CrossRef]
33. Richards, J.; Brimblecombe, P. Moisture as a Driver of Long-Term Threats to Timber Heritage—Part I: Changing Heritage Climatology. *Heritage* **2022**, *5*, 1929–1946. [CrossRef]
34. Verney-Carron, A.; Sessegolo, L.; Lefèvre, R.-A.; Brimblecombe, P. Modelling the Alteration of Medieval Stained Glass as a Function of Climate and Pollution: Comparison between Different Methodologies. *Heritage* **2023**, *6*, 3074–3088. [CrossRef]
35. Sesana, E.; Gagnon, A.S.; Ciantelli, C.; Cassar, J.; Hughes, J.J. Climate Change Impacts on Cultural Heritage: A Literature Review. *WIREs Clim. Change* **2021**, *12*, e710. [CrossRef]
36. Vidal, F.; Vicente, R.; Mendes Silva, J. Review of Environmental and Air Pollution Impacts on Built Heritage: 10 Questions on Corrosion and Soiling Effects for Urban Intervention. *J. Cult. Herit.* **2019**, *37*, 273–295. [CrossRef]
37. Inkpen, R.J.; Viles, H.A.; Moses, C.; Baily, B.; Collier, P.; Trudgill, S.T.; Cooke, R.U. Thirty Years of Erosion and Declining Atmospheric Pollution at St Paul's Cathedral, London. *Atmos. Environ.* **2012**, *62*, 521–529. [CrossRef]
38. Bonazza, A.; Brimblecombe, P.; Grossi, C.M.; Sabbioni, C. Carbon in Black Crusts from the Tower of London. *Environ. Sci. Technol.* **2007**, *41*, 4199–4204. [CrossRef]
39. Del Monte, M.; Ausset, P.; Lefèvre, R.A.; Thiébault, S. Evidence of Pre-Industrial Air Pollution from the Heads of the Kings of Juda Statues from Notre Dame Cathedral in Paris. *Sci. Total Env.* **2001**, *273*, 101–109. [CrossRef]
40. Wilhelm, K.; Longman, J.; Orr, S.A.; Viles, H. Stone-Built Heritage as a Proxy Archive for Long-Term Historical Air Quality: A Study of Weathering Crusts on Three Generations of Stone Sculptures on Broad Street, Oxford. *Sci. Total Environ.* **2021**, *759*, 143916. [CrossRef]
41. Thornbush, M.; Viles, H. The Changing Façade of Magdalen College, Oxford: Reconstructing Long-Term Soiling Patterns from Archival Photographs and Traffic Records. *J. Arch. Conserv.* **2005**, *11*, 40–57. [CrossRef]
42. Grossi, C.M.; Brimblecombe, P. Past and Future Colouring Patterns of Historic Stone Buildings. *Mater. Construcción* **2008**, *58*, 143–160. [CrossRef]
43. Brimblecombe, P. Air Pollution and Architecture: Past, Present and Future. *J. Arch. Conserv.* **2000**, *6*, 30–46. [CrossRef]
44. Brimblecombe, P.; Grossi, C.M. Millennium-Long Damage to Building Materials in London. *Sci. Total Env.* **2009**, *407*, 1354–1361. [CrossRef]
45. Grøntoft, T. Historical Dry Deposition of Air Pollution in the Urban Background in Oslo, Norway, Compared to Western European Data. *Atmos. Environ.* **2021**, *267*, 118777. [CrossRef]
46. Bonazza, A.; Sardella, A. Climate Change and Cultural Heritage: Methods and Approaches for Damage and Risk Assessment Addressed to a Practical Application. *Heritage* **2023**, *6*, 3578–3589. [CrossRef]
47. Kapsomenakis, J.; Douvis, C.; Poupkou, A.; Zerefos, S.; Solomos, S.; Stavraka, T.; Melis, N.S.; Kyriakidis, E.; Kremlis, G.; Zerefos, C. Climate Change Threats to Cultural and Natural Heritage UNESCO Sites in the Mediterranean. *Environ. Dev. Sustain.* **2022**, 1–26. [CrossRef]
48. Saha, A.; Pal, S.C.; Santosh, M.; Janizadeh, S.; Chowdhuri, I.; Norouzi, A.; Roy, P.; Chakrabortty, R. Modelling Multi-Hazard Threats to Cultural Heritage Sites and Environmental Sustainability: The Present and Future Scenarios. *J. Clean. Prod.* **2021**, *320*, 128713. [CrossRef]
49. Daly, C.; Engel Purcell, C.; Donnelly, J.; Chan, C.; MacDonagh, M.; Cox, P. Climate Change Adaptation Planning for Cultural Heritage, a National Scale Methodology. *J. Cult. Her. Manag. Sustain. Dev.* **2021**, *11*, 313–329. [CrossRef]
50. Tola, H.G.; Brimblecombe, P. Environmental Pressures at Dirre Sheikh Hussein Sanctuary. *Heritage* **2022**, *5*, 2661–2672. [CrossRef]
51. Hernández-Montes, E.; Hdz-Gil, L.; Coletti, C.; Dilaria, S.; Germinario, L.; Mazzoli, C. Prediction Model for the Evolution of the Deterioration of Bricks in Heritage Buildings in Venice Caused by Climate Change. *Heritage* **2023**, *6*, 483–491. [CrossRef]
52. Brimblecombe, P.; Richards, J. Temporal Resolution of Climate Pressures on Façades in Oxford 1815–2021. *Theor. Appl. Clim.* **2023**, *153*, 561–572. [CrossRef]
53. Bonazza, A.; Messina, P.; Sabbioni, C.; Grossi, C.M.; Brimblecombe, P. Mapping the Impact of Climate Change on Surface Recession of Carbonate Buildings in Europe. *Sci. Total Environ.* **2009**, *407*, 2039–2050. [CrossRef]
54. Vyshkvarkova, E.; Sukhonos, O. Climate Change Impact on the Cultural Heritage Sites in the European Part of Russia over the Past 60 Years. *Climate* **2023**, *11*, 50. [CrossRef]
55. Ryhl-Svendsen, M.; Smedemark, S.H. Mass-Transfer Air Pollution Modeling in Heritage Buildings. *Heritage* **2023**, *6*, 4768–4786. [CrossRef]
56. Richards, J.; Orr, S.A.; Viles, H. Reconceptualising the Relationships between Heritage and Environment within an Earth System Science Framework. *J. Cult. Herit. Manag. Sustain. Dev.* **2019**, *10*, 122–129. [CrossRef]
57. Al-Maiyah, S.; Elkadi, H. The Role of Daylight in Preserving Identities in Heritage Context. *Renew. Sustain. Energy Rev.* **2007**, *11*, 1544–1557. [CrossRef]
58. Matthews, T.; Grant-Smith, D. Managing Ensemble Scale Heritage Conservation in the Shandon Architectural Conservation Area in Cork, Ireland. *Cities* **2017**, *62*, 152–158. [CrossRef]
59. Pineda, P.; Iranzo, A. Analysis of Sand-Loaded Air Flow Erosion in Heritage Sites by Computational Fluid Dynamics: Method and Damage Prediction. *J. Cult. Herit.* **2017**, *25*, 75–86. [CrossRef]
60. Richards, J.; Mayaud, J.; Zhan, H.; Wu, F.; Bailey, R.; Viles, H. Modelling the Risk of Deterioration at Earthen Heritage Sites in Drylands. *Earth Surf. Process Landf.* **2020**, *45*, 2401–2416. [CrossRef]
61. Cui, K.; Wu, G.; Du, Y.; An, X.; Wang, Z. The Coupling Effects of Freeze-Thaw Cycles and Salinization Due to Snowfall on the Rammed Earth Used in Historical Freeze-Thaw Cycles Relics in Northwest China. *Cold Reg. Sci. Technol.* **2019**, *160*, 288–299. [CrossRef]

62. Deprez, M.; De Kock, T.; De Schutter, G.; Cnudde, V. A Review on Freeze-Thaw Action and Weathering of Rocks. *Earth Sci. Rev.* **2020**, *203*, 103143. [CrossRef]
63. Hanazato, T.; Niitsu, Y.; Morii, M.; Minowa, C.; Nitto, K.; Yokoo, T. Seismic and Wind Performance of Five-Storied Pagoda of Timber Heritage Structure Affected by Great East Japan Earthquake of 2011 and Typhoon Jelawat of 2012. In Proceedings of the Structural Analysis of Historical Constructions: Anamnesis, diagnosis, therapy, controls—Proceedings of the 10th International Conference on Structural Analysis of Historical Constructions, SAHC, Leuven, Belgium, 13–15 September 2016; CRC Press/Balkema: London, UK, 2016; pp. 1343–1348.
64. Ravanelli, R.; Riguzzi, F.; Anzidei, M.; Vecchio, A.; Nigro, L.; Spagnoli, F.; Crespi, M. Sea Level Rise Scenario for 2100 A.D. for the Archaeological Site of Motya. *Rend. Lincei* **2019**, *30*, 747–757. [CrossRef]
65. West Dean College MA Collections Care & Conservation Management. Available online: https://www.westdean.ac.uk/study/degrees-and-diplomas/courses/ma-collections-care-and-conservation-management (accessed on 13 June 2023).

heritage

MDPI

Article

Modeling Salt Behavior with ECOS/RUNSALT: Terminology, Methodology, Limitations, and Solutions

Sebastiaan Godts [1,2,3,*], Michael Steiger [4], Scott Allan Orr [5], Amelie Stahlbuhk [4], Julie Desarnaud [1,6], Hilde De Clercq [1], Veerle Cnudde [3,7] and Tim De Kock [2]

1 Monuments Lab, Royal Institute for Cultural Heritage (KIK-IRPA), 1000 Brussels, Belgium
2 Faculty of Design Sciences, Antwerp Cultural Heritage Sciences (ARCHES), University of Antwerp, 2000 Antwerpen, Belgium
3 Department of Geology, PProGRess, Ghent University, 9000 Ghent, Belgium
4 Department of Chemistry, University of Hamburg, 20146 Hamburg, Germany
5 Institute for Sustainable Heritage, University College London (UCL), London WC1E 6BT, UK
6 Renovation & Heritage Lab, Belgium Building Research Institute (BBRI), 1060 Saint-Gilles, Belgium
7 Department of Earth Sciences, Utrecht University, 3584 Utrecht, The Netherlands
* Correspondence: sebastiaan.godts@kikirpa.be

Abstract: Damage to porous materials in heritage buildings caused by salt mixture crystallization is driven by the surrounding environmental conditions. To understand the crystallization behavior of a mixed salt solution as a function of changing climatic conditions (i.e., relative humidity and temperature), excluding factors such as the internal pore structure, the thermodynamic model ECOS/RUNSALT is the only freeware available that requires simple input and includes the most relevant ions for heritage buildings and solids. We suggest the use of specific terminology and describe how to use the model and how to interpret the output, with emphasis on key limitations for which solutions are provided. When used correctly, the model output can be trusted, specifically when it is used to inform preventive conservation (e.g., environmental conditions in which salt crystallization cycles should not occur). However, salt mixture kinetics and the internal pore structure remain crucial parameters that are not considered in the model. These aspects need further attention to develop a better understanding and correctly model salt damage in relation to climatic changes.

Keywords: salt mixtures; thermodynamic modeling; crystallization behavior; climate; built environment; conservation; masonry

Citation: Godts, S.; Steiger, M.; Orr, S.A.; Stahlbuhk, A.; Desarnaud, J.; De Clercq, H.; Cnudde, V.; De Kock, T. Modeling Salt Behavior with ECOS/RUNSALT: Terminology, Methodology, Limitations, and Solutions. *Heritage* **2022**, *5*, 3648–3663. https://doi.org/10.3390/heritage5040190

Academic Editor: Peter Brimblecombe

Received: 25 October 2022
Accepted: 21 November 2022
Published: 23 November 2022

1. Introduction

Salt deterioration is a common issue when dealing with the conservation of porous materials in built heritage [1–6]. However, understanding salt behavior is a complex subject due to the presence of a wide variety of ions [7], which are often the result of groundwater infiltration by capillary rising damp, rainwater infiltration, and atmospheric, biological, or internal material contamination. Over time, these result in an accumulation of salts in the first few millimeters or centimeters of a material's drying front. Damage to the material occurs when salts fill a porous material and crystallization cycles are provoked by changing environmental conditions. The individual mixture composition found in the material determines the crystallization behavior of each possible solid that can occur, as described by Price and Brimblecombe [8] in the context of porous materials. This behavior is further influenced by a wide range of internal and external factors of the salt-bearing porous material and salt solution properties, such as supersaturation, viscosity, pore characteristics, inner pore processes [9–15], and ambient environment. The sheer scope of all the parameters involved limits our current understanding of the real-world processes that underpin the damage potential of salts over time. However, the outcomes of specific scientific projects contribute to the understanding of complex salt behavior, as described in Price (Ed.) [16]. This has resulted in a thermodynamic model, Environmental Control of Salts (ECOS), used to determine the environmental conditions needed to reduce salt damage in porous materials. Based on the ECOS model, Bionda [17] developed RUNSALT, which is a graphical user interface (GUI) that takes care of the data pre-/postprocessing and the visualization of outputs. The freeware and published work [18–31] of the first use cases can be found on the RUNSALT website [32].

Since its development, the model has been used extensively to aid management decisions for the preservation of built heritage worldwide. Although a limited amount of literature is available [33–50], it is important to note that its use is mostly undocumented in peer-reviewed literature; for example, in Belgium the software has been used for over 300 heritage sites [51]. Like any model, ECOS and RUNSALT have limitations and pitfalls, with ineffective or non-existent transfer from research to practice, a challenge faced more widely in heritage science, as recently explored by Richards and Brimblecombe [52]. Some of these limitations are linked to data processing and issues with the input parameters and outputs. Several of the limitations are reported on and addressed in this paper with the aim to advance the conservation field when considered or adjusted in future versions of the software. Additionally, when dealing with salt crystallization, an ambiguous use of terminology and discrepancies can be found in the literature; thus, specific terminology and abbreviations are suggested for salt mixtures.

2. Models and Theory

When dealing with salt mixtures, there are several models available that output, amongst others, specific saturation, crystallization, dissolution, and transition relative humidity points, that allow a deeper understanding of the mixture behavior under changing climatic conditions. However, most programs or models are designed for specific purposes, such as atmospheric, industrial (brines) or (sea) water chemistry, e.g., FREEZCHEM [53], E-AIM [54,55], and [56–61]. Since these programs and models have been developed primarily to address specific applications, they exclude relevant ions and data for salts typically found in building materials [7]. The computer program PHREEQC [62] and the ECOS/RUNSALT model [16,17] are the most cited in literature for this purpose. PHREEQC includes a variety of options and incorporations of ions, such as those described by, e.g., Benavente et al. [63] and Pérez-Diez et al. [64]. However, an important limitation in the aqueous model is the lack of internal consistency in the databases [62]. PHREEQC has a limited pre-installed dataset of solids and non-validated parameters, which the user needs to update. Moreover, experimental data are often lacking or contain inconsistencies in the literature [65]; it is thus a complicated and tedious task to complete the datasets and derive reliable results, particularly for systems containing nitrate. However, the program has potential in stone conservation, as it permits the implementation of, e.g., kinetics and in-pore situations.

ECOS/RUNSALT is currently the only model with simple inputs that include the most relevant salt phases found in the built environment, and can handle more complex systems when compared to PHREEQC. ECOS (Environmental Control of Salts) is a chemical equilibrium model initially developed on the molality-based thermodynamic approach of the ion interaction model of Pitzer [66]. There, the solubilities of the included mineral phases, as well as the water vapor–salt solution equilibrium, are considered. It is based on a molality-dependent expression for excess Gibbs energy which includes empirically determined interaction parameters and, by its minimization, allows the iterative determination of the activity and osmotic coefficient. While the latter coefficient is related to the water activity of the electrolyte solution, the activity coefficient corrects for the non-ideal behavior of ions in the solution. During the development of ECOS, it turned out that the parameterization of the model was well suited for the calculation of solubilities in mixed electrolyte solutions in the desired range, but in combination with the algorithm used for the calculation of the amounts of crystallized salts and solution, partly incorrect results were obtained. Especially in cases of high concentrations, the algorithm based on the original Pitzer molality-based model passed unrealistic conditions in ranges where the model was already invalid. It was possible to overcome this hurdle by using equations in terms of mole fractions [55,67–69], which was implemented in Fortran [70]. Having equivalent principles to the molality-based model, the mole fraction one, generally known as the Pitzer–Simonson–Clegg model, includes ion concentrations expressed as mole fractions [71]. Nevertheless, the original molality-based approach was later improved, as described by Steiger et al. [72], and should be considered as a valid alternative. In either case, it must be noted that ECOS always considers equilibrium conditions, neglecting kinetic aspects. Thus, certain metastable pathways are possible, and salt damage is linked to the kinetically driven supersaturation of the salt solution.

To further understand the ECOS calculations, we refer to the literature, e.g., [16,73–78], including data related to activity coefficients and (solubility) phase diagrams (solution concentration as molality (number of moles of dissolved salt per kilogram water) (m(salt)/mol·kg^{-1}), volume (V), relative humidity (RH), or water activity (a_w) over temperature (T)). Phase diagrams are best suited for binary or ternary systems to illustrate the crystallization pathways at given concentrations. However, the graphical representation of quaternary or higher systems becomes more complicated. If all data from such phase diagrams should be derived, x-y-z plots are required with, for example, x as RH, y as the number of moles of crystalline salt (n) or volume (V), and z as temperature (T). Such a plot is, in principle, the same as a combination of RUNSALT plots derived from calculations at different T or RH and presented with three axes, as shown by Menéndez [33].

Equilibrium conditions also mean that at each RH, ECOS considers an equilibration with the surroundings. In reality, RH changes are generally faster where non-equilibrium dynamic RH changes occur, so there are larger gradients between the vapor pressure of the solution and the surrounding air influencing evaporation.

When looking at phase diagrams in the context of ions found in building materials, at least senary or septenary systems should be considered that include the most important ions (CO_3^{2-}, Cl^-, NO_3^-, SO_4^{2-}, Na^+, K^+, Mg^{2+}, and Ca^{2+}) while excluding less common ions such as fluoride, phosphate, oxalate, ammonium, acetate, or formate, as previously described [79,80]. Since the least soluble salts will rapidly crystallize from a mixed salt system, carbonates and gypsum can be excluded in most cases. Thus, a senary system of more soluble salts will remain, including Cl^-, NO_3^-, Na^+, K^+, and Mg^{2+} with either SO_4^{2-} or Ca^{2+}, as further described in [7], which is implemented in ECOS.

3. Terminology for Mixed Salt Systems and Methodology for Using RUNSALT

Understanding salt mixture behavior in porous media under changing climatic conditions is not a straightforward task and requires in-depth knowledge of the mixture composition and material characteristics, as well as internal and external factors. A first step, however, is knowing the correct salt mixture present, which requires data input preparation for the model, as described further in Steiger and Heritage [79] and recently verified by applying the method to a large dataset including several additional steps in Godts et al. [7].

Before moving forward with the methodology for using RUNSALT, Table 1 is given to overcome the ambiguous use of terminology found in the literature and to clarify the crystallization pathways of mixed salt systems shown in RUNSALT plots (see example Figure 1). Specific RH points of interest are linked to the suggested symbols presented in the table, and the letters A to F are further detailed in the legend of Figure 1. Note how RH points of interest overlap depending on how a plot is read from a humid to a dry environment or vice versa. The term mutual (m) is chosen as the behavior of each solid is influenced by the mixture composition, and m is removed when dealing with single salts. The symbols are recommended for future use to make scientific information comparable.

Table 1. Overview of terminology and suggested symbols to describe RUNSALT plots showing the crystallization pathway of mixed salt systems under changing *RH*. Refer also to Figure 1, the legend of Figure 1, and Table 2.

	Meaning	Base Symbol	Species-Specific Symbol [1]
	Explanation following RUNSALT plots (example Figure 1)		
1.	Mutual crystallization relative humidity *RH* point at the onset of any line shown in a plot corresponding to the start of crystallization; the number shown in the specific symbol refers to the species/solid in order of appearance from a humid to a dry environment. The use of the number (e.g., 1) in relation to the solids can be useful to understand the sequence of crystallization. The solution at this point is saturated with respect to a specific solid. When available, the first letters of the mineral name or chemical formula can used to replace the number, e.g., $RH^{m}_{cry_1}$ is aphthitalite = aph and thus $RH^{m}_{cry_{aph}}$ (letter A in Figure 1). Aphthitalite is the first that crystallizes in the mixture and the same base symbol is used for the mutual crystallization relative humidity of all solids that crystallize (indicated with the letter B in Figure 1). This is only relevant when solution is still available before crystallization takes place (reactions in solution in Table 2).	RH^{m}_{cry}	$RH^{m}_{cry_1}$
2.	Mutual dissolution relative humidity *RH* point at the end of a horizontal line in a plot, looking from a dry to a humid environment, equals the start of dissolution; e.g., in Figure 1 this is illustrated by the RH points indicated with the letter C, and thus when solution becomes available.	RH^{m}_{dis}	$RH^{m}_{dis_1}$
3.	Mutual deliquescence relative humidity *RH* point at the end of a horizontal line in a plot when no more solution is available, looking from a dry to a humid environment, e.g., indicated as letter D in Figure 1. Here, the last solid that crystallizes is darapskite, and afterwards no more solution is available. Thus, $RH^{m}_{del\ dar}$, as further illustrated by reaction number 2 shown in Table 2.	RH^{m}_{del}	$RH^{m}_{del_1}$
4.	Mutual transition relative humidity *RH* point at which salt transitions occur. The numbers refer to the solids involved in the transition, starting with solids before the dash (e.g., 3 in [3–5]) at more humid conditions transitioning to solids after the dash (e.g., 5 in [3–5]) at dryer conditions. Either a phase change (hydration, dehydration), decomposition, or the formation (addition) of solids occur under both wetting and drying conditions. For example, the transition of mirabilite to thenardite is $RH^{m}_{tra_{\|mir-the\|}}$, or is more complicated, as shown by reaction 3 in Table 2 (letter E in Figure 1).	RH^{m}_{tra}	$RH^{m}_{tra_{\|3-5\|}}$
	Additional terms that are useful when calculating water activities or concentrations. Values that are not included in the RUNSALT output data yet could be derived from the ECOS calculations.		
5.	Mutual equilibrium relative humidity Any *RH* point at which a solution is in equilibrium with its environment = water activity at any concentration if solution is available, e.g., in Figure 1 any *RH* point above D, and thus $RH^{m}_{del_{dar}}$.	RH^{m}_{eq}	
6.	Mutual saturation relative humidity Any *RH* point at which a solution is saturated (points on the curves, e.g., in Figure 1, all *RH* points between A and C on the curve of aphthitalite crystallization), equal to the RH^{m}_{eq} points during crystallization (when solid and solution are available).	RH^{m}_{sat}	

[1] For practical considerations, double subscripts can be replaced by a comma between subscripts, e.g., $RH^{m}{}_{cry,aph}$.

Table 2. Summary of the reactions under drying conditions shown in the RUNSALT plot (Figure 1, noted as # 1, 2 and 3 above the figure).

Start Composition of the Solution (mol): $2Na^+ + 2K^+ + 1Cl^- + 1NO_3^- + 1SO_4^{2-}$	
#	Reactions in solution
1.	$2Na^+ + 6K^+ + 4SO_4^{2-} \rightarrow Na_2SO_4 \cdot 3K_2SO_4$ (cr)
2.	$Na_2SO_4 \cdot 3K_2SO_4$ (cr) + $11Na^+ + Cl^- + 10NO_3^- + 4H_2O \rightarrow$ NaCl (cr) + $6KNO_3$ (cr) + $4NaNO_3 \cdot Na_2SO_4 \cdot H_2O$ (cr)
	Solid-state reactions
3.	$6NaNO_3 \cdot Na_2SO_4 \cdot H_2O$ (cr) + $Na_2K\ SO_4)_4$ (cr) $\rightarrow 6KNO_3$ (cr) + $10Na_2SO_4$ (cr)

The reactions presented in the RUNSALT plot shown in Figure 1 are further detailed in Table 2.

Legend by Figure 1, with the letters A to F indicating specific *RH* points of interest:

A. The first mutual crystallization relative humidity of the mixture ($RH^{m}_{cry_1}$) represents the *RH* at which crystallization initiates for the first solid that appears under drying conditions (aphthitalite here, and thus $RH^{m}_{cry_{aph}}$). The solution is saturated with respect to aphthitalite, and above this *RH* all solids are dissolved.

B. The mutual crystallization relative humidity of all the following solids that crystallize from the solution in the mixture, and thus is the *RH* at which crystallization first begins for $RH^{m}_{cry_{hal}}$, $RH^{m}_{cry_{nit}}$ and $RH^{m}_{cry_{dar}}$.

C. The mutual dissolution relative humidity of all solids in the mixture when solution becomes available is equal to the *RH* points when a crystal starts to dissolve for $RH^{m}_{dis_{aph}}$, $RH^{m}_{dis_{hal}}$, $RH^{m}_{dis_{nit}}$ and $RH^{m}_{dis_{dar}}$. B and C are often at the same *RH*; here, the resolution of the plot distorts the position for halite and niter, a phenomenon explained further on.

D. The mutual deliquescence relative humidity of the mixture is the *RH* determined by the solids in the mixture at which the first dissolution starts to occur and solution becomes available; here, $RH^{m}_{del_{dar}}$ also equals the dissolution relative humidity of $RH^{m}_{dis_{aph}}$, $RH^{m}_{dis_{hal}}$ and $RH^{m}_{dis_{nit}}$.

E. The mutual transition relative humidity $RH^{m}_{tra_{[dar/aph]-[nit/the]}}$. Here, under drying conditions, thenardite is formed and the amount of niter increases from the decomposition of darapskite and aphthitalite in a solid-state reaction.

F. Plot stacking artifact caused by transition reactions, herein identified by chloride that is not available in other solids.

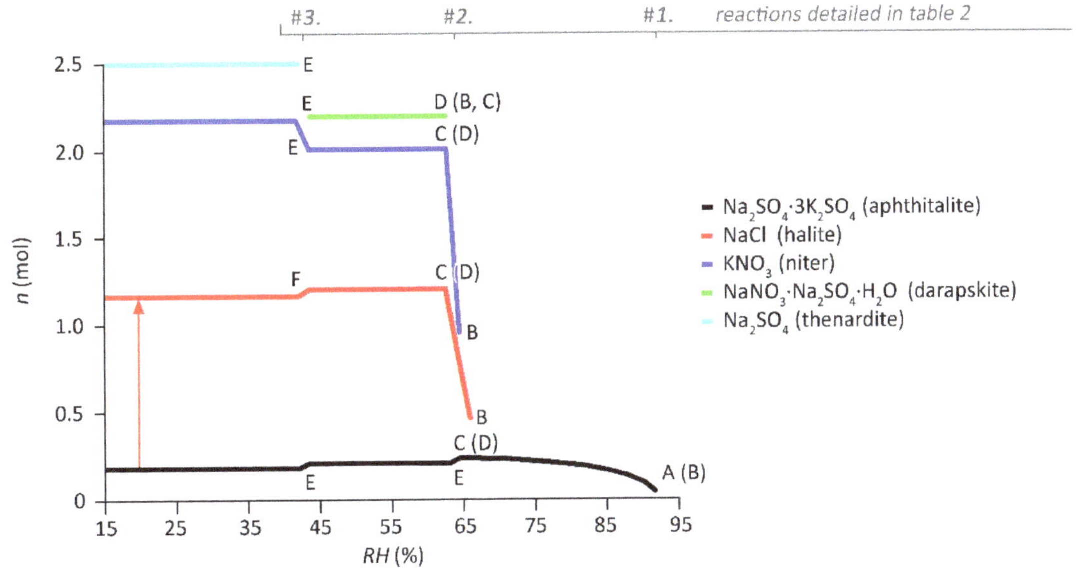

Figure 1. Example of an ECOS/RUNSALT output with 2 mol Na^+ and K^+, and 1 mol Cl^-, NO_3^-, and SO_4^{2-}, calculated at 20 °C with *RH* (%) from 15% to 95% (resolution 1.6% *RH* points), with the latter on the *x*-axis and the amount of crystalline salt, *n* (mol), stacked on the *y*-axis. The red arrow illustrates the absolute amount of the solid NaCl (halite) at the specific *RH*, here from 0.16 to 1.16 mol, thus with an absolute amount of 1 mol. The letters A to F indicate specific *RH* points of interest and are explained in the legend below. The numbers 1 to 3 indicate the *RH* points at which reactions take place when looking from a humid to a dry environment, as further detailed in Table 2.

As described in the previous section, the ions used for the model input are Cl^-, NO_3^-, Na^+, K^+, and Mg^{2+}, with either SO_4^{2-} or Ca^{2+}. This excludes, amongst others, carbonates and the equimolar contents of calcium and sulfate, with the latter considered as the gypsum content. Hence, the model primarily calculates a maximum of six ions.

The system composition is entered into the RUNSALT interface as mol or weight, with the selection of either a *RH* range between 15% and 98%, or a *T* range between −30 and +50 °C, after which either the *T* or *RH* value is fixed. The RUNSALT interface generates a temporary .DAT file that includes the inputs required by the model. These inputs are read in by the batch executable .EXE file which then initiates the model. In another temporary file, ECOS outputs the equilibrium composition at 50 equally spaced intervals for either the specific temperature or humidity range, and is then read by RUNSALT to produce a graphical representation of the crystallization behavior. The output can be exported in graphical and textual (.CSV) formats for further analysis through RUNSALT.

The crystallization behavior of the mixture is graphically represented by RUNSALT with the specified relative humidity (*RH*) or temperature (*T*) range on the *x*-axis, while the *y*-axis returns the amount of substance (mol) (Figure 1). After the plot is generated, one can choose to show the *y*-axis as volume (*V*) in cm^3 (molar volume of salt, that is, equilibrium crystal volume) (Figure 2), which gives a more realistic visualization of the salt content in the pores. For example, aphthitalite is present in approximately one tenth of the total mol content (Figure 1), and at least a third of the total solid volume (Figure 2). The latter is thus more indicative of risk in the pore structure and will determine the overall interpretation and conservation advice. The use of volume in the outputs was recently illustrated in relation to climatic conditions by Costa et al. [81]. Expressing the results as volume is additionally useful to estimate pore filling, as the molar volume (V_m) can be used as an input value in the calculation, as described in Gulotta et al. [82]. All values on the *y*-axis are cumulative (stacked), meaning that the amount of the first solid should be deducted from the second to know the absolute value of each individual solid. The individual amount of salt is illustrated for halite (NaCl) at the given *RH* with the arrow in Figure 1.

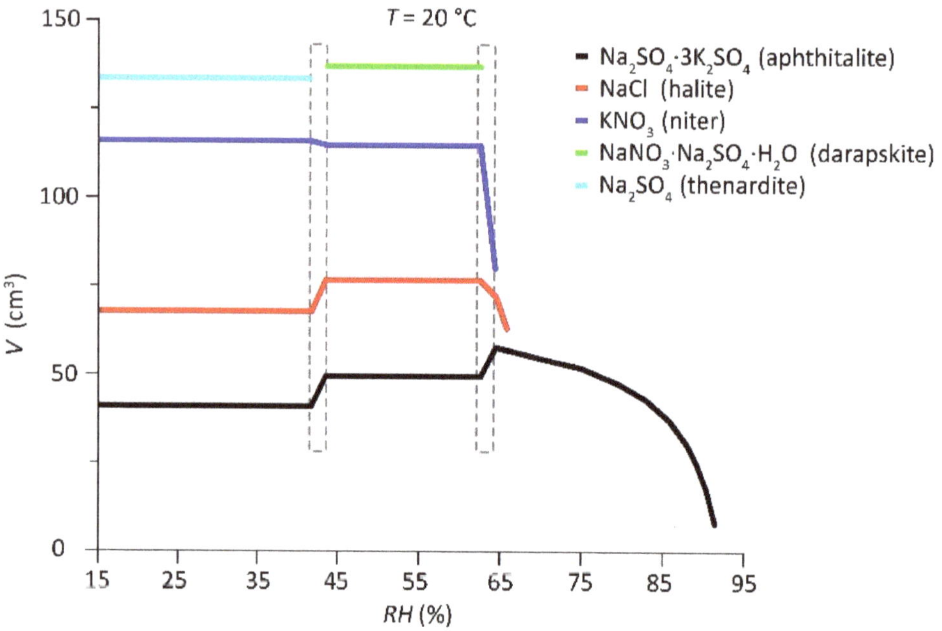

Figure 2. Example of an ECOS/RUNSALT output of 2 mol Na^+ and K^+, and 1 mol Cl^-, NO_3^- and SO_4^{2-}, calculated at 20 °C with *RH* (%) from 15% to 95% (resolution 1.6% *RH*), with the latter on the *x*-axis and the amount of crystalline salt in volume, *V* (cm^3), stacked on the *y*-axis. The dashed rectangles illustrate plot artifacts caused by the *RH* resolution; the non-vertical lines are to be read as vertical ones and the gap between darapskite and thenardite is closed.

As the data points are systematically calculated for 50 points within the chosen range of environmental parameters (*RH* or *T*), the smaller the range the higher the resolution. Hence, data related to smaller successive intervals can be stitched together to obtain a more detailed output. However, changes in stacking order in the detailed plots can occur at higher resolution. This process can be an important step to correct certain artifacts caused by the resolution of the chosen environmental parameters. For example, in Figure 3, if the *RH* range between 60 and 70% is entered into RUNSALT, the resolution of the plot increases to 0.2% *RH*, as opposed to 1.6% *RH* when generating a plot from 15 to 95% *RH* (Figure 1). Thus, the thermodynamically calculated mutual relative humidity points of interest, for example, $RH^{m}_{cry_{nit}}$ and $RH^{m}_{cry_{hal}}$, are more accurate. The plot also shows that $RH^{m}_{cry_{dar}}$ or $RH^{m}_{del_{dar}}$ is equal to $RH^{m}_{dis_{nit}}$, $RH^{m}_{dis_{hal}}$, and $RH^{m}_{tra_{[aph]-[hal/nit/dar]}}$. The resolution makes little to no difference to the final conservation advice given to the field in terms of the risk assessment of *RH* ranges of crystallization/dissolution. However, selecting a narrower range in the environmental parameters can clarify certain artifacts and uncertainties in the plot. In particular, the slightly non-vertical lines of aphthitalite, halite, and niter are caused by the resolution 0.2% *RH*, and these lines are in fact to be read as vertical (location shown by the dashed rectangles in Figures 2 and 3). The same is true for the gap (transition $RH^{m}_{tra\ [dar/aph]-[nit/the]}$) and all non-vertical lines at lower *RH* at approximately 43% (shown in Figure 2).

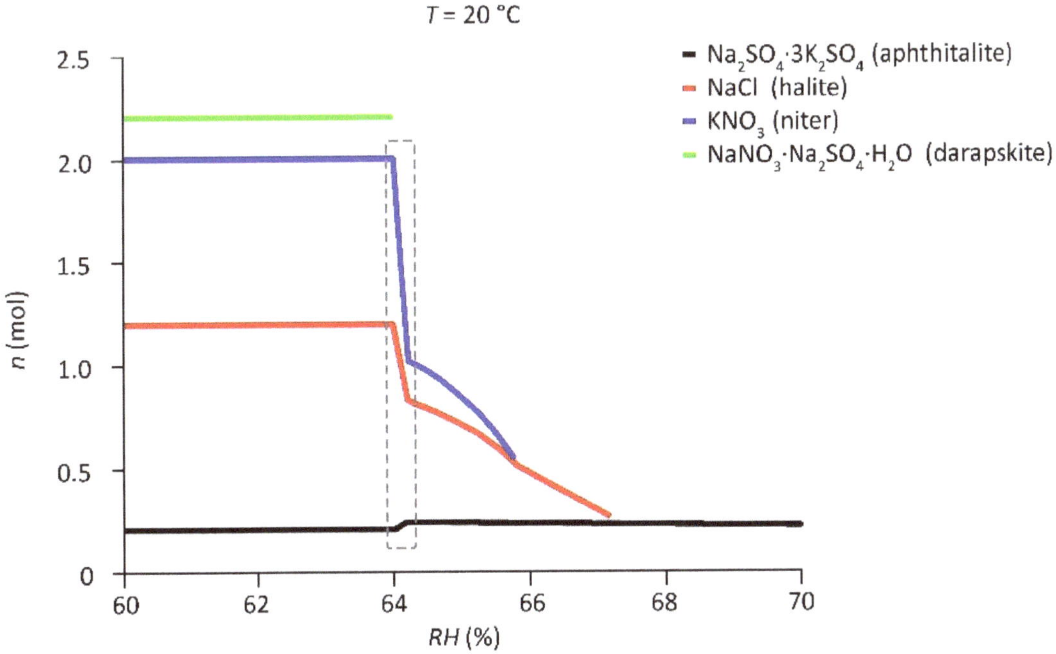

Figure 3. Detail of the ECOS/RUNSALT output shown in Figure 1, calculated at 20 °C with *RH* (%) from 60% to 70% (resolution 0.2% *RH* points), with the latter on the *x*-axis and the amount moles of crystalline salt, *n* (mol), stacked on the *y*-axis. The dashed rectangle illustrates plot artifacts caused by the *RH* resolution; the non-vertical lines are to be read as vertical.

When reading the complete RUNSALT plot, we can start looking at the *x*-axis from a more humid environment on the far right (95% *RH*) to a dry environment on the far left (15% *RH*). Under more humid conditions, and before the first line appears, all salts are in solution. The solution has a certain concentration corresponding to the given *RH*; the further away from the first solid, the more the solution is diluted, which is theoretically infinite until pure water is reached at 100% *RH*. Just before the RH at which the first solid crystallizes (for example, in Figure 1, aphthitalite ($Na_2SO_4{\cdot}3K_2SO_4$)), the solution is at its highest degree of saturation. The solution is saturated and in equilibrium with the environment, and is referred to as mutual saturation relative humidity (RH^{m}_{sat}). In a mixture, this can be defined at any *RH* point where curved lines are visible in the RUNSALT outputs. The lines indicate the crystallization of a solid. Looking further at Figure 1, aphthitalite starts to crystallize at approximately 92% (20 °C); this point is the mutual crystallization relative humidity of the solution (RH^{m}_{cry}). Following the crystallization of aphthitalite to dryer conditions, the remaining ions are still in solution and become more concentrated until halite starts to crystallize, followed by niter and darapskite. Each solid has a crystallization relative humidity (RH^{m}_{cry}) of 67, 66, and 64% (±0.1), respectively. Similar to aphthitalite, for both halite and niter, the solid amount increases over a *RH* range, while all darapskite crystallizes at a specific *RH* point, as shown in Figure 3. This *RH* of 64% is also the point at which no more solution is available. This is the mutual deliquescence relative humidity (RH^{m}_{del}) of the mixture.

Looking further down the remaining crystallization pathway at dryer conditions below $RH^{m}_{cry_{dar}}$ (=RH^{m}_{del} of the mixture), while keeping in mind that no solution is available, the following solid-state reactions can be observed. First, a small amount of aphthitalite decomposes at the same *RH* as $RH^{m}_{cry_{dar}}$, and from 64% to 43% *RH* all solids remain crystallized. For this mixture, the *RH* of 64% is the most important to avoid crystallization cycles and damage to porous materials. In practice, it would be common to advise a stable *RH* between 50% and 60% *RH* at 20 °C; that is, if all water sources are eliminated, other artifacts in the area remain well preserved under these conditions, the mixture compositions are representative for the entire salt risk assessment, and the location allows such a narrow range of *RH* to be maintained. However, some flexibility should be considered, specifically toward the lower *RH* range as the solid-state reactions might have limited effect on the substrates. Additionally, an *RH* increase over a limited time should be acceptable due to the kinetics considering dissolution/crystallization rates [43]. However, more research is needed to further understand these processes.

As shown in Table 2, the formation of thenardite is the result of the decomposition of aphthitalite and darapskite, which also explains the increase in niter at the same *RH* of approximately 43%. The decrease in halite at the same *RH* is an artifact of plot stacking (amount of substance or volume) on the *y*-axis. This can be derived from the fact that no other salt is formed with Cl^-. In drier conditions (below 43%), all salts remain crystallized.

Looking back at higher *RH* in the figures, it is important to understand that, for example, aphthitalite in the system will only start to dissolve if all other salts have gone into solution and the solution has reached the specific dilution above 64% *RH*. Specifically, the dissolution of aphthitalite is dependent on the concentration of the surrounding solution. The solution will accumulate moisture, which can cause discoloration, moisture stains, the peeling of paint layers, and attract biological growth. At $RH^{m}_{cry_{aph}}$ of 92%, the solution is saturated with respect to aphthitalite. Above this *RH* point, the solution becomes further diluted (until infinity at 100%). The amount of water vapor absorbed by the solution at a given *RH* can be calculated with ECOS; however, the data are currently not given in RUNSALT. Details on the backend calculations of the model are extensively described in Price et al. [16].

The above example shows the value of the model to derive specific advice for environmental salt risk assessment. The model has proven extremely valuable for the field and certain aspects have been verified with four ion mixtures by Rörig-Dalgaard [83]; however, several limitations and issues should be taken into consideration before application. In the following, the most common limitations and solutions are provided, while we abstain from considering deviations in the crystallization pathways if a solid becomes isolated from the remaining solution.

4. ECOS/RUNSALT Limitations and Solutions

Comparable to any model, ECOS/RUNSALT has limitations and pitfalls. An important obstacle in the calculations is caused when there are extremely high concentrations or supersaturation in hygroscopic mixtures, including calcium nitrate and calcium chloride, resulting in water activities higher than expected from thermodynamics. This obstacle was overcome by the incorporation of certain assumptions and non-verified solids, such as $MgCa(NO_3)_4 \cdot 10H_2O$, which rarely appears in the outputs. However, it is as yet unclear whether extreme hygroscopic solids crystallize in these conditions.

Concerning the input data, an issue occurs with the autobalance option in RUNSALT. When using theoretical charge input data with integer numbers (e.g., $1Cl^-$ and $1Na^+$), the autobalance works correctly. However, with experimental output of ion analyses, values with several decimal places are common and the autobalance only corrects the chloride content, rendering the output incorrect. It is thus recommended to abstain from using the 'autobalance' feature and consider the use of charge balance calculations as described in [7], including the downloadable calculation sheet and R script (.R and .xlsx) at [51]. Furthermore, an error message can occur due to rounding issues produced by the thermodynamic calculations, which is caused by the number of decimals of each ion value entered. This can be resolved by changing the place values of all ions equally to ones, tens, hundreds, or thousands, depending on the initial concentration; although the total amount on the *y*-axis (mol or volume) varies, the output remains identical.

In certain cases, an error message appears when either entering the full *RH* range from 15 to 98% as input in the environmental parameters, or under certain temperatures, depending on the mixture composition. Both errors are easily overcome by limiting the *RH* to 95% or increasing or decreasing the temperature by one to five degrees. In either case, the results obtained from the model for the limitations of output values are considered more than sufficient. Another issue in the environmental parameters is the possibility to use values below 0 °C, although the formation of ice is not incorporated in the outputs. It is thus advised not to use a temperature input values below 0 °C in RUNSALT.

Moving forward to complications specifically related to single salts in the ECOS calculations and RUNSALT outputs, inconsistencies are seen with more recent studies related to $Ca(NO_3)_2$, K_2SO_4, and $MgSO_4$ hydrates [84]. The critical *RH* values calculated by the ECOS of $Ca(NO_3)_2$, K_2SO_4, and $MgSO_4$ are presented in the RUNSALT output in Figures 4 and 5, respectively. The crystallization behaviors of Ca^{2+} and NO_3^- show two critical *RH* values (Figure 4, left), the first at 51.82% for tetrahydrate (nitrocalcite) and the second at 37.98% for anhydrous calcium nitrate (at 20 °C *RH* resolution = 0.02%; resolution not represented in the figure). However, it is known that dehydration of the tetrahydrate only occurs over extended periods of time and at extreme low *RH* with transition values between the anhydrous di-, tri-, and tetrahydrate at 8.3%, 12.4%, and 20.5%, respectively [75]. Thus, considering that the dehydration of nitrocalcite is kinetically hindered, one can expect that the crystallization *RH* of the tetrahydrate is the only one to be considered, keeping in mind that more research is needed to understand the in-pore effects under realistic climatic conditions.

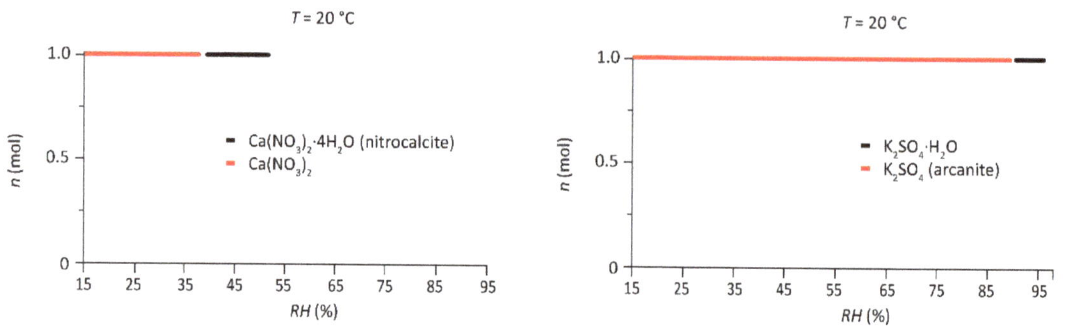

Figure 4. ECOS/RUNSALT outputs, 1 mol Ca^{2+} and 2 mol NO_3^- (**left**), and 2 mol K^+ and 1 mol SO_4^{2-} (**right**). Calculated at 20 °C for *RH* ranging from 15% to 95% or to 98% (**right**) (resolution 1.6% *RH*), with the latter on the *x*-axis and the amount of crystalline salt, *n* (mol), on the *y*-axis.

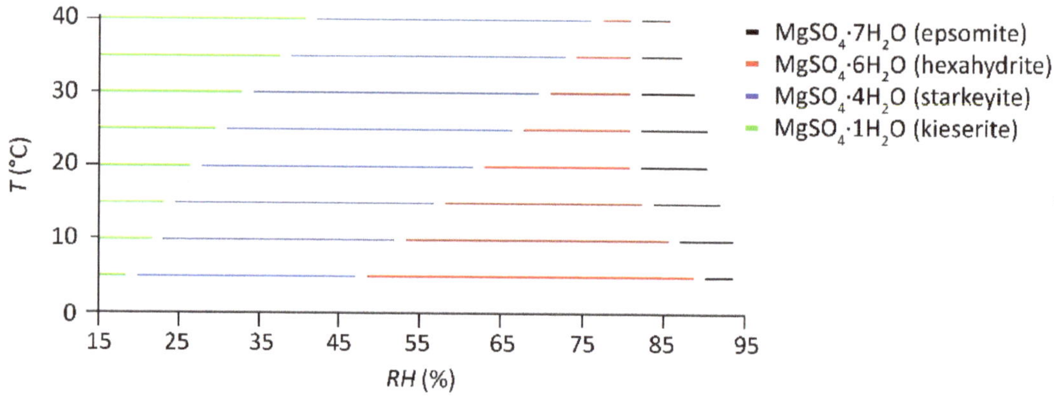

Figure 5. Calculated $MgSO_4$ phases at different temperatures derived from ECOS/RUNSALT plots (with equimolar contents of Mg^{2+} and SO_4^{2-}); relative humidity from 15 to 95% on the *x*-axis (resolution 1.6% *RH*) and temperature, *T* (°C), on the *y*-axis.

The RUNSALT outputs of K^+ and SO_4^{2-} (calculated at 20 °C) show two critical *RH* values (Figure 4, right) with a relative humidity crystallization of potassium sulfate monohydrate starting at 97.7%, followed by dehydration to form arcanite at 89.7%. However, as detailed further by Archer and Kirklin [85], it has been found in several studies that the monohydrate phase does not occur below 9 °C, and if stable at all this is less probable at higher temperatures. Furthermore, the RH_{cry} should decrease with decreasing temperature; currently, the outputs show the opposite at lower temperatures, with a decreasing amount of the monohydrate (not illustrated). Thus, caution should be taken when looking at the critical *RH* values of potassium sulfate. Here, at 20 °C, K_2SO_4 is likely to start at 97.7% and the monohydrate can be ignored.

For magnesium sulfate (Figure 5), four critical *RH* values are shown at 20 °C, starting with the crystallization relative humidity, $RH_{cry_{eps}}$ at 91.54% for $MgSO_4{\cdot}7H_2O$ (epsomite), followed by the transition to $MgSO_4{\cdot}6H_2O$ (hexahydrite), $RH_{tra_{[eps-hex]}}$ at 81.94%, and to $MgSO_4{\cdot}4H_2O$ (starkeyite) and $MgSO_4{\cdot}1H_2O$ (kieserite), with $RH_{tra_{[hex-sta]}}$ and $RH_{tra_{[sta-kie]}}$ at 62.3% and 27.16%, respectively. However, from experimental results and improved thermodynamic calculations (see [72,84]), important deviations specifically concerning starkeyite are derived. The data show that the values used in ECOS for this phase are inaccurate and no change in the mixtures from one to the other hydrate should be taken into consideration within the range of 5–40 °C. The original data from the ECOS/RUNSALT outputs are shown in Figure 5, while the corrected data are given in Figure 6. Here, at 20 °C, the RH_{del} of $MgSO_4{\cdot}7H_2O$ (epsomite) is 91.2% *RH* with a transition to $MgSO_4{\cdot}1H_2O$ (kieserite) at 46.6% *RH*; the hexahydrate only occurs at higher temperatures. The latter figure illustrates how ECOS/RUNSALT outputs with a wide *RH* range calculated in a variety of temperatures will correspond to the phase diagram, following the transition between kieserite, hexahydrite, and epsomite, and the RH_{del} at different temperatures. In addition to the described issues that should be considered in the model outputs, it remains important to understand that certain phases can be metastable and kinetically hindered, as further described for magnesium sulfates in Steiger et al. [84].

Several solids are missing in ECOS/RUNSALT, although they can play a role in the crystallization pathways and deterioration processes, such as Ca-K-NO_3 double salts [86]. The efflorescence found on monuments [80,87–93] reveals salts that are currently not available or not consistently incorporated in the model outputs, for example, humberstonite ($Na_7K_3Mg_2(SO_4)_6(NO_3)_2{\cdot}6H_2O$). Moreover, the removal of equimolar contents of calcium and sulfate (gypsum) can alter the RUNSALT outputs, including double salts such as glauberite ($Na_2Ca(SO_4)_2$), gorgeyite ($K_2Ca_5(SO_4)_6{\cdot}H_2O$), and syngenite ($K_2Ca(SO_4)_2{\cdot}H_2O$). However, the formation of the latter three might be kinetically hindered or occur as solid-state reactions over longer periods of time.

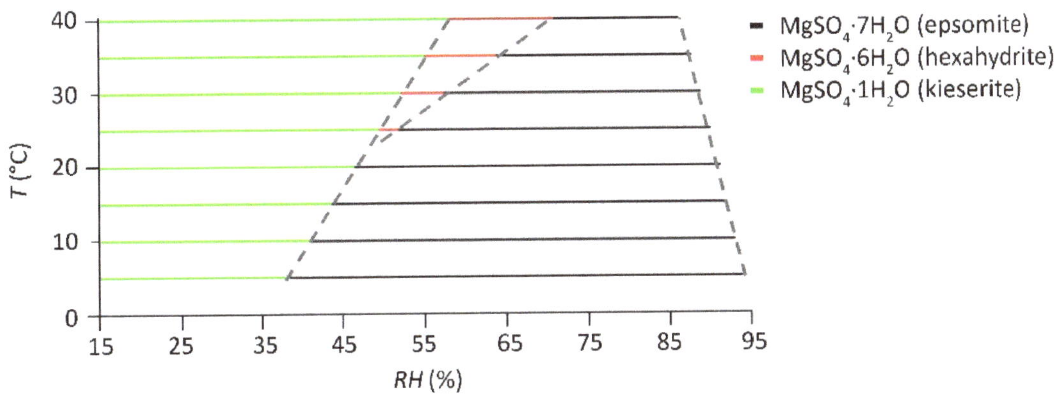

Figure 6. Calculated and experimental $MgSO_4$ phases at different temperatures derived from Steiger et al. [84] (the dashed lines correspond to data of the phase diagram); the relative humidity from 15 to 95% (resolution 1.6% *RH*) is on the *x*-axis, and temperature, *T* (°C), is on the *y*-axis.

As detailed in [35], the model is not capable of systematically integrating the presence of an equimolar amount of calcium and sulfate ions. In rare cases, when the model allows calculations with calcium and sulfate, the crystallization pathway can change. In these cases, it is often observed that the common salt darapskite ($NaNO_3{\cdot}Na_2SO_4{\cdot}H_2O$) is replaced by glauberite ($Na_2SO_4{\cdot}CaSO_4$), as illustrated in Figure 7. As mentioned earlier, several issues are visible in the plot on the right, such as the kinetically hindered solid-state phase change between gypsum and anhydrite. Additionally, due to the stacking of the solids and the *RH* resolution, the vertical lines for glauberite and nitratine should remain horizontal; thus, the latter is simply an artifact of the gap between the transition $RH^{m}_{tra_{[gyp-anh]}}$. Further research is needed to understand the formation of double salts containing calcium and sulfate.

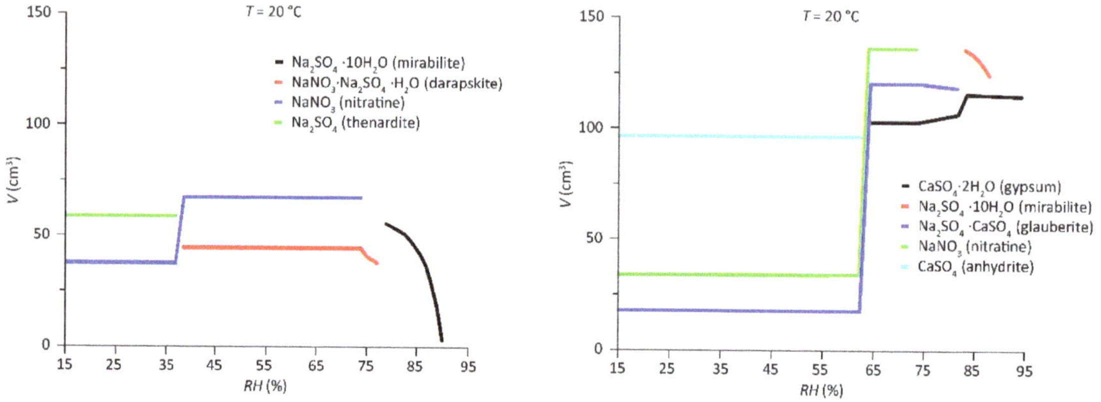

Figure 7. ECOS/RUNSALT outputs to illustrate the influence of Ca^{2+} and $SO_4{}^{2-}$ (considered as the gypsum content) on the crystallization behavior of a salt mixture. **Left:** excluding equimolar contents of Ca^{2+} and $SO_4{}^{2-}$, input parameters in mol are Na^+: 1.8, Ca^{2+}: 0, $NO_3{}^-$: 1, and $SO_4{}^{2-}$: 0.4. **Right:** including equimolar contents of Ca^{2+} and $SO_4{}^{2-}$, thus Na^+: 1.8, Ca^{2+}: 3.6, $NO_3{}^-$: 1, and $SO_4{}^{2-}$: 4. Both outputs show the *RH* (%) on the *x*-axis and the amount of crystalline salt in volume, *V* (cm^3), on the *y*-axis.

Furthermore, relevant carbonate salts such as thermonatrite ($Na_2CO_3{\cdot}H_2O$), natron ($Na_2CO_3{\cdot}10H_2O$), trona ($Na_3CO_3HCO_3{\cdot}2H_2O$), and kalicinite ($KHCO_3$) are absent from the model outputs. However, these soluble carbonates are rare and only seen in approximately 2% of samples taken in Belgium heritage [7]. Other rare anions not included but contributing to the total charge balance and crystallization pathway are fluoride, phosphate, oxalate, acetate, ammonium, and formate [79].

5. Discussion and Conclusions

ECOS/RUNSALT is currently the only model that includes the most relevant salt phases found in built heritage. We provide an overview to further the understanding of its use and suggest specific terminology for salt mixture behavior to simplify and clarify the model, and make scientific information comparable. Furthermore, as several limitations and pitfalls exist when using the model, possible incorrect interpretations of the derived outputs can occur. However, when the presented issues and solutions are taken into consideration, the RUNSALT outputs can be considered highly accurate. The most important issues described concern calcium nitrate, potassium sulfate, and magnesium sulfate phases (hydrates), and the possible influence of calcium sulfate on the formation of different solids under specific conditions.

It also remains important to examine the variety of factors that can cause deviations from the modeled crystallization behavior. Some of these factors include the in-pore situation, the material characteristics, impurities in the system, and salt kinetics. The latter is specifically relevant to environmental conditions, the separation of solids from the solution, gradients in solution concentrations, kinetically hindered salt crystallization, and rates of crystallization/dissolution, including long-term solid-state reactions. In any case, if a specific range of *RH* is considered safe when interpreting RUNSALT outputs, meaning salt crystallization/transition cycles are less likely to occur in the specific environment, the model outputs can be trusted, keeping in mind that different crystallization pathways are possible when certain salts are separated from the solution. Overall, more research is needed considering salt mixture kinetics in the pore system. An important aspect to focus on is the rate at which phase transitions occur, as this can guide conservation scientists towards a better prediction of salt damage in relation to climatic changes over time. Thereby, updates of the current model and wider accessibility of the source codes are important for the future.

Author Contributions: S.G., M.S., S.A.O. and A.S.: Conceptualization, Methodology, Validation, Investigation; S.G.: Data curation, Writing—original draft, Visualization, Funding acquisition; M.S., S.A.O., A.S., J.D., H.D.C., V.C. and T.D.K.: Writing—review and editing. All authors have read and agreed to the published version of the manuscript.

Funding: This research is funded by the Belgium Science Policy (Belspo) within the framework of BRAIN-be 2.0, Belgian Research Action through Interdisciplinary Networks: project B2/191/P1/PREDICT (Research action B2); joint PhD project PREDICT, Phase Transitions of Salts under Changing Climatic Conditions.

Conflicts of Interest: The authors declare no conflict of interest.

References

1. Evans, I.S. Salt Crystallization and Rock Weathering: A Review. *Rev. Geomorphol. Dyn.* **1970**, *19*, 153–177.
2. Goudie, A.; Viles, H. *Salt Weathering Hazards*; John Wiley & Sons: Chichester, UK, 1997; ISBN 0-471-95842-5.
3. Doehne, E. Salt Weathering: A Selective Review. *Geol. Soc. Spec. Publ.* **2002**, *205*, 51–64. [CrossRef]
4. Doehne, E.; Price, C.A. *Stone Conservation: An Overview of Current Research*, 2nd ed.; Research in Conservation; Getty Conservation Institute: Los Angeles, CA, USA, 2010; ISBN 978-1-60606-046-9.
5. Snethlage, R.; Sterflinger, K. Stone Conservation. In *Stone in Architecture*; Siegesmund, S., Snethlage, R., Eds.; Springer Berlin Heidelberg: Berlin/Heidelberg, Germany, 2011; pp. 411–544. ISBN 978-3-642-14474-5.
6. Oguchi, C.T.; Yu, S. A Review of Theoretical Salt Weathering Studies for Stone Heritage. *Prog. Earth Planet. Sci.* **2021**, *8*, 1–23. [CrossRef]
7. Godts, S.; Steiger, M.; Orr, S.A.; De Kock, T.; Desarnaud, J.; De Clercq, H.; Cnudde, V. Charge Balance Calculations for Mixed Salt Systems Applied to a Large Dataset from the Built Environment. *Sci. Data* **2022**, *9*, 324. [CrossRef]
8. Price, C.; Brimblecombe, P. Preventing Salt Damage in Porous Materials. *Stud. Conserv.* **1994**, *39*, 90–93. [CrossRef]
9. Espinosa, R.M.; Franke, L.; Deckelmann, G. Model for the Mechanical Stress Due to the Salt Crystallization in Porous Materials. *Constr. Build. Mater.* **2008**, *22*, 1350–1367. [CrossRef]
10. Espinosa, R.M.; Franke, L.; Deckelmann, G. Phase Changes of Salts in Porous Materials: Crystallization, Hydration and Deliquescence. *Constr. Build. Mater.* **2008**, *22*, 1758–1773. [CrossRef]
11. Naillon, A.; Joseph, P.; Prat, M. Ion Transport and Precipitation Kinetics as Key Aspects of Stress Generation on Pore Walls Induced by Salt Crystallization. *Phys. Rev. Lett.* **2018**, *120*, 034502. [CrossRef]
12. Li, L.; Kohler, F.; Dziadkowiec, J.; Røyne, A.; Espinosa Marzal, R.M.; Bresme, F.; Jettestuen, E.; Dysthe, D.K. Limits to Crystallization Pressure. *Langmuir* **2022**, *38*, 11265–11273. [CrossRef]
13. Espinosa-Marzal, R.M.; Scherer, G.W. Advances in Understanding Damage by Salt Crystallization. *Acc. Chem. Res.* **2010**, *43*, 897–905. [CrossRef]
14. Flatt, R.; Aly Mohamed, N.; Caruso, F.; Derluyn, H.; Desarnaud, J.; Lubelli, B.; Espinosa Marzal, R.M.; Pel, L.; Rodriguez-Navarro, C.; Scherer, G.W.; et al. Predicting Salt Damage in Practice: A Theoretical Insight into Laboratory Tests. *RILEM Tech. Lett.* **2017**, *2*, 108–118. [CrossRef]
15. Espinosa-Marzal, R.M.; Scherer, G.W. Impact of In-Pore Salt Crystallization on Transport Properties. *Environ. Earth Sci.* **2013**, *69*, 2657–2669. [CrossRef]
16. Price, C. *An Expert Chemical Model for Determining the Environmental Conditions Needed to Prevent Salt Damage in Porous Materials: Protection and Conservation of the European Cultural Heritage*; Project ENV4-CT95-0135 (1996–2000) Final Report; Protection and conservation of the European Cultural Heritage; Archetype: London, UK, 2000; ISBN 978-1-873132-52-4.
17. Bionda, D. RUNSALT—A Graphical User Interface to the ECOS Thermodynamic Model for the Prediction of the Behaviour of Salt Mixtures under Changing Climate Conditions (2005). Available online: Http://Science.Sdf-Eu.Org/Runsalt/ (accessed on 9 October 2022).
18. Bionda, D.; Storemyr, P. Modelling the Behaviour of Salts Mixtures in Walls: A Case Study from Tenaille von Fersen Building, Suomenlinna, Finland. In *Proceedings of the Study of Salt Deterioration Mechanisms. Decay of Brick Walls Influenced by Interior Climate Changes*; von Konow, T., Ed.; Suomenlinnan Hoitokunta: Suomenlinna, Finland, 2002; pp. 95–101.
19. Bionda, D. Methodology for the Preventive Conservation of Sensitive Monuments: Microclimate and Salt Activity in a Church. In Proceedings of the 10th International Congress on Deterioration and Conservation of Stone, Stockholm, Sweden, 27 June–2 July 2004; Kwiatkowski, D., Löfvendahl, R., Eds.; ICOMOS Sweden: Stockholm, Sweden, 2004; pp. 627–634.
20. Klenz Larsen, P. The Salt Decay of Medieval Bricks at a Vault in Brarup Church, Denmark. *Environ. Geol.* **2007**, *52*, 375–383. [CrossRef]
21. Prokos, P. Equilibrium Conditions of Marine Originated Salt Mixtures: An ECOS Application at the Archaeological Site of Delos, Greece. In Proceedings of the SWBSS, Salt Weathering of Buildings and Stone Sculptures, Copenhagen, Denmark, 22–24 October 2008; Ottosen, L.M., Ed.; Technical University of Denmark: Lyngby, Denmark, 2008.
22. Eklund, S. *Stone Weathering in the Monastic Building Complex on Mountain of St Aaron in Petra, Jordan*; University of Helsinki: Helsinki, Jordan, 2008.
23. Franzen, C.; Mirwald, P.W. Moisture Sorption Behaviour of Salt Mixtures in Porous Stone. *Geochemistry* **2009**, *69*, 91–98. [CrossRef]

24. Maguregui, M.; Sarmiento, A.; Martínez-Arkarazo, I.; Angulo, M.; Castro, K.; Arana, G.; Etxebarria, N.; Madariaga, J.M. Analytical Diagnosis Methodology to Evaluate Nitrate Impact on Historical Building Materials. *Anal. Bioanal. Chem.* **2008**, *391*, 1361–1370. [CrossRef] [PubMed]
25. Maguregui, M.; Knuutinen, U.; Castro, K.; Madariaga, J.M. Raman Spectroscopy as a Tool to Diagnose the Impact and Conservation State of Pompeian Second and Fourth Style Wall Paintings Exposed to Diverse Environments (House of Marcus Lucretius). *J. Raman Spectrosc.* **2010**, *41*, 1400–1409. [CrossRef]
26. Price, C.A. Predicting Environmental Conditions to Minimise Salt Damage at the Tower of London: A Comparison of Two Approaches. *Environ. Geol.* **2007**, *52*, 369–374. [CrossRef]
27. Sawdy, A.; Heritage, A. Evaluating the Influence of Mixture Composition on the Kinetics of Salt Damage in Wall Paintings Using Time Lapse Video Imaging with Direct Data Annotation. *Environ. Geol.* **2007**, *52*, 303–315. [CrossRef]
28. Sawdy, A.; Price, C. Salt Damage at Cleeve Abbey, England. *J. Cult. Herit.* **2005**, *6*, 125–135. [CrossRef]
29. Steiger, M. Modellierung von Phasengleichgewichten. In *Proceedings of the DBU Workshops im Februar 2008 in Osnabrück, Salzschäden an Kulturgütern Stand des Wissens und Forschungsdefizite*; Schwarz, H.-J., Steiger, M., Eds.; Deutsche Bundesstiftung Umwelt: Hannover, Germany, 2009; pp. 80–99.
30. Zehnder, K.; Schoch, O. Efflorescence of Mirabilite, Epsomite and Gypsum Traced by Automated Monitoring on-Site. *J. Cult. Herit.* **2009**, *10*, 319–330. [CrossRef]
31. Sawdy, A. *The Kinetics of Salt Weathering of Porous Materials*; Institute of Archaeology University College London: London, UK, 2001.
32. Bionda, D. RUNSALT. Available online: http://science.sdf-eu.org/runsalt/ (accessed on 6 January 2022).
33. Menéndez, B. Estimation of Salt Mixture Damage on Built Cultural Heritage from Environmental Conditions Using ECOS-RUNSALT Model. *J. Cult. Herit.* **2017**, *24*, 22–30. [CrossRef]
34. Scatigno, C.; Prieto-Taboada, N.; Festa, G.; Madariaga, J.M. Soluble Salts Quantitative Characterization and Thermodynamic Modeling on Roman Bricks to Assess the Origin of Their Formation. *Molecules* **2021**, *26*, 2866. [CrossRef] [PubMed]
35. Godts, S.; Hayen, R.; De Clercq, H. Investigating Salt Decay of Stone Materials Related to the Environment, a Case Study in the St. James Church in Liège, Belgium. *Stud. Conserv.* **2017**, *62*, 329–342. [CrossRef]
36. Gómez-Laserna, O.; Cardiano, P.; Diez-Garcia, M.; Prieto-Taboada, N.; Kortazar, L.; Olazabal, M.Á.; Madariaga, J.M. Multi-Analytical Methodology to Diagnose the Environmental Impact Suffered by Building Materials in Coastal Areas. *Environ. Sci. Pollut. Res.* **2018**, *25*, 4371–4386. [CrossRef]
37. Gibeaux, S.; Thomachot-Schneider, C.; Eyssautier-Chuine, S.; Marin, B.; Vazquez, P. Simulation of Acid Weathering on Natural and Artificial Building Stones According to the Current Atmospheric SO2/NOx Rate. *Environ. Earth Sci.* **2018**, *77*, 327. [CrossRef]
38. Heinrichs, K.; Azzam, R. Quantitative Analysis of Salt Crystallization–Dissolution Processes on Rock-Cut Monuments in Petra/Jordan. In *Engineering Geology for Society and Territory—Volume 8*; Lollino, G., Giordan, D., Marunteanu, C., Christaras, B., Yoshinori, I., Margottini, C., Eds.; Springer International Publishing: Cham, Sweden, 2015; pp. 507–510. ISBN 978-3-319-09407-6.
39. Crevals, V.; Godts, S.; Desarnaud, J. Salt Problems and Climate Control in the Case of the Church of Sint Aldegondis in Mespelare, Belgium, an ECOS/Runsalt Approach. In Proceedings of the Fifth International Conference on Salt Weathering of Buildings and Stone Sculptures, Delft, The Netherlands, 22–24 September 2021; Lubelli, B., Kamat, A.A., Quist, W.J., Eds.; TU Delft Open: Delft, The Netherlands, 2021; pp. 13–20.
40. Godts, S.; Hayen, R.; Clercq, H.D. Common Salt Mixtures Database: A Tool to Identify Research Needs. In Proceedings of the 3rd International Conference on Salt Weathering of Buildings and Stone Sculptures, Brussels, Belgium, 14–16 October 2014; KIK-IRPA: Brussels, Belgium, 2014; p. 14.
41. Godts, S.; Hayen, R. Onderzoek en preventie van vocht- en zoutschade gerelateerd aan het klimaat in de crypte van de Sint-Baafskathedraal te Gent. In Proceedings of the WTA-PRECOM^{3}OS Studiedag: Preventieve Conservatie: Van klimaat- en schademonitoring naar een geïntegreerde benadering, Leuven, Belgium, 5 April 2019; Verstynge, E., van Bommel, B., Vernimme, N., van Hees, R., Eds.; TUDelft—KULeuven: Leuven, Belgium, 2019.
42. Godts, S.; Clercq, C.D. Analysis of the Salt Content during Water Bath Desalination of a Polychrome Limestone Relief. In Proceedings of the 14th International Congress on the Deterioration and Conservation of Stone, Monument Future: Decay and Conservation of Stone, Göttingen, Germany, 7–12 September 2020; Siegesmund, S., Middendorf, B., Eds.; Mitteldeutscher Verlag: Göttingen, Germany, 2020; pp. 853–858.
43. Godts, S.; Orr, S.A.; Desarnaud, J.; Steiger, M.; Wilhelm, K.; De Clercq, H.; Cnudde, V.; De Kock, T. NaCl-Related Weathering of Stone: The Importance of Kinetics and Salt Mixtures in Environmental Risk Assessment. *Herit. Sci.* **2021**, *9*, 1–13. [CrossRef]
44. Godts, S.; Clercq, H.D.; Hayen, R.; Roy, J.D. Risk Assessment and Conservation Strategy of a Salt Laden Limestone Mausoleum and the Surrounding Funeral Chapel in Boussu, Belgium. In Proceedings of the 12th International Congress on the Deterioration and Conservation of Stone, New York, NY, USA, 22–26 October 2012; Columbia University: New York, NY, USA, 2012.
45. De Clercq, H.; Godts, S. Rehabilitation of Farmhouses and Barns: Limits of Salt Content. *Appl. Phys. A* **2016**, *122*, 831. [CrossRef]
46. De Clercq, H.; Godts, S.; Hayen, R. Impact of the Indoor Climate on the Performance of Building Materials Contaminated with Salt Mixtures. *Conserv. Manag. Archaeol. Sites* **2014**, *16*, 39–55. [CrossRef]
47. Menéndez, B. Estimators of the Impact of Climate Change in Salt Weathering of Cultural Heritage. *Geosciences* **2018**, *8*, 401. [CrossRef]

48. Laue, S.; Poerschke, D.; Hübner, B. Investigation and Conservation of Salt Damaged Epitaphs in the Church of Werben (Saxony-Anhalt, Germany). SWBSS 2017 Fourth Int Conf Salt Weather Build Stone Sculpt. In Proceedings of the Fourth Int Conf salt Weather Build Stone Sculpt, Potsdam, Germany, 20–22 September 2017; Laue, S., Ed.; Verlag der Fachhochschule: Potsdam, Germany, 2017.
49. Chabas, A.; Kloppmann, W.; Sizun, J.-P.; Wille, G.; Coman, A.; Petitmangin, A.; Nowak, S.; Martin, E.; Jurgens, M.-A. Sources and Chronology of Soluble Salt Formation in a Medieval Dovecote Caught up in Urbanization: A Resilience Story? *Environ. Sci. Pollut. Res.* **2022**. *preprint*. [CrossRef]
50. Pintér, F. The Combined Use of Ion Chromatography and Scanning Electron Microscopy to Assess Salt-Affected Mineral Materials in Cultural Heritage. *J. Am. Inst. Conserv.* **2022**, *61*, 85–99. [CrossRef]
51. Godts, S.; Orr, S.A.; Steiger, M.; De Kock, T. Mixed Salt Systems in the Built Environment—Charge Balance Calculations [Data Set]. Available online: https://zenodo.org/record/6280617#.Y32dBX1BxPY (accessed on 22 September 2022).
52. Richards, J.; Brimblecombe, P. The Transfer of Heritage Modelling from Research to Practice. *Herit. Sci.* **2022**, *10*. [CrossRef]
53. Marion, G.M.; Mironenko, M.V.; Roberts, M.W. FREZCHEM: A Geochemical Model for Cold Aqueous Solutions. *Comput. Geosci.* **2010**, *36*, 10–15. [CrossRef]
54. Wexler, A.S. Atmospheric Aerosol Models for Systems Including the Ions H^+, NH_4^+, Na^+, SO_4^{2-}, NO_3^-, Cl^-, Br^-, and H_2O. *J. Geophys. Res.* **2002**, *107*, 4207. [CrossRef]
55. Clegg, S.L.; Pitzer, K.S.; Brimbiecombet, P. Thermodynamics of Multicomponent, Miscible, Ionic Solutions. 2. Mixtures Includlng Unsymmetrical Electrolytes. *J Phys Chem* **1992**, *96*, 9470–9479. [CrossRef]
56. Gruszkiewicz, M.S.; Palmer, D.A.; Springer, R.D.; Wang, P.; Anderko, A. Phase Behavior of Aqueous Na–K–Mg–Ca–Cl–NO_3 Mixtures: Isopiestic Measurements and Thermodynamic Modeling. *J. Solut. Chem.* **2007**, *36*, 723–765. [CrossRef]
57. Thomsen, K.; Rasmussen, P.; Gani, R. Correlation and Prediction of Thermal Properties and Phase Behaviour for a Class of Aqueous Electrolyte Systems. *Chem. Eng. Sci.* **1996**, *51*, 3675–3683. [CrossRef]
58. Wang, P.; Anderko, A.; Young, R.D. A Speciation-Based Model for Mixed-Solvent Electrolyte Systems. *Fluid Phase Equilibria* **2002**, *203*, 141–176. [CrossRef]
59. Wang, P.; Springer, R.D.; Anderko, A.; Young, R.D. Modeling Phase Equilibria and Speciation in Mixed-Solvent Electrolyte Systems. *Fluid Phase Equilibria* **2004**, *222*, 11–17. [CrossRef]
60. Wang, P.; Anderko, A.; Springer, R.D.; Young, R.D. Modeling Phase Equilibria and Speciation in Mixed-Solvent Electrolyte Systems: II. Liquid–Liquid Equilibria and Properties of Associating Electrolyte Solutions. *J. Mol. Liq.* **2006**, *125*, 37–44. [CrossRef]
61. Hingerl, F.F.; Wagner, T.; Kulik, D.A.; Kosakowski, G.; Driesner, T.; Institut, P.S.; Hingerl, F. Development of a New Activity Model for Complex Mixed-Salt Solutions from Ambient to Geothermal Conditions. In Proceedings of the Geophysical Research Abstracts, Vienna, Austria, 22–27 April 2012; Volume 14, p. 5332.
62. Parkhurst, D.L.; Appelo, C.A.J. *Description of Input and Examples for PHREEQC Version 3: A Computer Program for Speciation, Batch-Reaction, One-Dimensional Transport, and Inverse Geochemical Calculations*; Techniques and Methods; U.S. Geological Survey: Reston, VA, USA, 2013; Volume 6-A43, p. 519.
63. Benavente, D.; Brimblecombe, P.; Grossi, C.M. Thermodynamic Calculations for the Salt Crystallisation Damage in Porous Built Heritage Using PHREEQC. *Environ. Earth Sci.* **2015**, *74*, 2297–2313. [CrossRef]
64. Pérez-Diez, S.; Fernández-Menéndez, L.J.; Morillas, H.; Martellone, A.; De Nigris, B.; Osanna, M.; Bordel, N.; Caruso, F.; Madariaga, J.M.; Maguregui, M. Elucidation of the Chemical Role of the Pyroclastic Materials on the State of Conservation of Mural Paintings from Pompeii. *Angew. Chem. Int. Ed.* **2021**, *60*, 3028–3036. [CrossRef] [PubMed]
65. Marcilla, A.; Reyes-Labarta, J.A.; Olaya, M.M. Should We Trust All the Published LLE Correlation Parameters in Phase Equilibria? Necessity of Their Assessment Prior to Publication. *Fluid Phase Equilibria* **2017**, *433*, 243–252. [CrossRef]
66. Pitzer, K.S. Ion Interaction Approach: Theory and Data Correlation. In *Activity coefficients in electrolyte solutions;* CRC Press: Boca Raton, FL, USA, 1991; pp. 75–153.
67. Clegg, S.L.; Pitzer, K.S. Thermodynamics of Multicomponent, Miscible, Ionic Solutions: Generalized Equations for Symmetrical Electrolytes. *J. Phys. Chem.* **1992**, *96*, 3513–3520. [CrossRef]
68. Pitzer, K.S.; Simonson, J.M. Thermodynamics of Multicomponent, Miscible, Ionic Systems: Theory and Equations. *J. Phys. Chem.* **1986**, *90*, 3005–3009. [CrossRef]
69. Pitzer, K.S. The Treatment of Ionic Solutions over the Entire Miscibility Range. *Berichte Bunsenges. Für Phys. Chem.* **1981**, *85*, 952–959. [CrossRef]
70. Clegg, S.L. *Personal Communication about the ECOS/Runsalt Model and Backend Calculations;* University of East Anglia: Norwich, UK, 2021.
71. Clegg, S.L.; Brimblecombe, P. Pitzer Model of Electrolyte Solutions. In *An Expert Chemical Model for Determining the Environmental Conditions Needed to Prevent Salt Damage in Porous Materials: Project ENV4-CT95-0135 (1996-2000) Final Report;* Protection and conservation of the European Cultural Heritage; Price, C.A., Ed.; Archetype: London, UK, 2000; pp. 13–18. ISBN 978-1-873132-52-4.
72. Steiger, M.; Kiekbusch, J.; Nicolai, A. An Improved Model Incorporating Pitzer's Equations for Calculation of Thermodynamic Properties of Pore Solutions Implemented into an Efficient Program Code. *Constr. Build. Mater.* **2008**, *22*, 1841–1850. [CrossRef]
73. Pitzer, K.S.; Mayorga, G. Thermodynamics of Electrolytes. III. Activity and Osmotic Coefficients for 2–2 Electrolytes. *J. Solut. Chem.* **1974**, *3*, 539–546. [CrossRef]

74. Steiger, M.; Asmussen, S. Crystallization of Sodium Sulfate Phases in Porous Materials: The Phase Diagram Na_2SO_4–H_2O and the Generation of Stress. *Geochim. Cosmochim. Acta* **2008**, *72*, 4291–4306. [CrossRef]
75. Steiger, M. Salts in Porous Materials: Thermodynamics of Phase Transitions, Modeling and Preventive Conservation. *Restor. Build. Monum.* **2005**, *11*, 419–432. [CrossRef]
76. Lindström, N.; Talreja, T.; Linnow, K.; Stahlbuhk, A.; Steiger, M. Crystallization Behavior of Na_2SO_4–$MgSO_4$ Salt Mixtures in Sandstone and Comparison to Single Salt Behavior. *Appl. Geochem.* **2016**, *69*, 50–70. [CrossRef]
77. Shen, Y.; Linnow, K.; Steiger, M. Crystallization Behavior and Damage Potential of Na_2SO_4–NaCl Mixtures in Porous Building Materials. *Cryst. Growth Des.* **2020**, *20*, 5974–5985. [CrossRef]
78. Steiger, M.; Charola, A.E.; Sterflinger, K. Weathering and Deterioration. In *Stone in Architecture*; Siegesmund, S., Snethlage, R., Eds.; Springer Berlin Heidelberg: Berlin/Heidelberg, Germany, 2014; pp. 225–316. ISBN 978-3-642-45154-6.
79. Steiger, M.; Heritage, A. Modelling the Crystallization Behaviour of Mixed Salt Systems: Input Data Requirements. In Proceedings of the 12th International Congress on the Deterioration and Conservation of Stone, New York, NY, USA, 22–26 October 2012.
80. Arnold, A.; Zehnder, K. *Monitoring Wall Paintings Affected by Soluble Salts*; Cather, S., Ed.; The Courtauld Institute of Fine Art and The Getty Conservation Institute: London, UK, 1987; pp. 103–135.
81. Costa, F.M.d.C.; Henriques Rosa, M.A.N.; Canetto, M.; Sobral da Fonseca, M.J. The Degradation of the "Study Room" (Convent of Christ, Tomar, Portugal), from a Preliminary Analysis towards a Sustainable Maintenance. *Ge-conservacion* **2022**, *21*, 95–107. [CrossRef]
82. Gulotta, D.; Godts, S.; De Kock, T.; Steiger, M. Comparative Estimation of the Pore Filling of Single Salts in Natural Stone. In Proceedings of the Fifth International Conference on Salt Weathering of Buildings and Stone Sculptures, Delft, The Netherlands, 22–24 September 2021; Lubelli, B., Kamat, A.A., Quist, W.J., Eds.; TU Delft Open: Delft, The Netherlands, 2021.
83. Rörig-Dalgaard, I. Direct Measurements of the Deliquescence Relative Humidity in Salt Mixtures Including the Contribution from Metastable Phases. *ACS Omega* **2021**, *6*, 16297–16306. [CrossRef] [PubMed]
84. Steiger, M.; Linnow, K.; Ehrhardt, D.; Rohde, M. Decomposition Reactions of Magnesium Sulfate Hydrates and Phase Equilibria in the $MgSO_4$–H_2O and Na^+–Mg^{2+}–Cl^-–$SO_4{}^{2-}$–H_2O Systems with Implications for Mars. *Geochim. Cosmochim. Acta* **2011**, *75*, 3600–3626. [CrossRef]
85. Archer, D.G.; Kirklin, D.R. Enthalpies of Solution of Sodium Chloride and Potassium Sulfate in Water. Thermodynamic Properties of the Potassium Sulfate + Water System. *J. Chem. Eng. Data* **2002**, *47*, 33–46. [CrossRef]
86. Flatt, R.; Bocherens, P. Sur Le Système Ternaire Ca^{++}_K^+_NO_3^-_H_2O. *Helv Chim Acta* **1962**, *45*, 187–195. [CrossRef]
87. Siedel, H. Salt Efflorescence as Indicator for Sources of Damaging Salts on Historic Buildings and Monuments: A Statistical Approach. *Environ. Earth Sci.* **2018**, *77*, 572. [CrossRef]
88. Morillas, H.; Maguregui, M.; Paris, C.; Bellot-Gurlet, L.; Colomban, P.; Madariaga, J.M. The Role of Marine Aerosol in the Formation of (Double) Sulfate/Nitrate Salts in Plasters. *Microchem. J.* **2015**, *123*, 148–157. [CrossRef]
89. Benavente, D.; de Jongh, M.; Cañaveras, J.C. Weathering Processes and Mechanisms Caused by Capillary Waters and Pigeon Droppings on Porous Limestones. *Minerals* **2020**, *11*, 18. [CrossRef]
90. Von Konow, T. *Test Results. The Study of Salt Deterioration Mechanisms. Decay of Brick Walls Influenced by Interior Climate Changes*; von Konow, T., Ed.; Suomenlinnan Hoitokunta: Helsinki, Finland, 2002.
91. Charola, A.; Lewin, S. Efflorescence on Building Stones—SEM in the Characterization and Elucidation of the Mechanism of Formation. *Scan Electron Microsc* **1979**, *79*, 379–387.
92. Arnold, A.; Küng, A. Crystallization and Habit of Salt Efflorescences on Walls I. In Proceedings of the 5th International Congress on Deterioration and Conservation of Stone, Lausanne, Switzerland, 25–27 September 1985; Félix, G., Ed.; Presses Romandes: Lausanne, Switzerland, 1985; pp. 255–267.
93. Bionda, D. *Modelling Indoor Climate and Salt Behaviour in Historical Buildings*; Swiss Federal Institute of Technology: Zurich, Switzerland, 2006.

heritage

MDPI

Article

Modeling Earthen Treatments for Climate Change Effects

Sharlot Hart [1,*], Kara Raymond [1], C. Jason Williams [2], William A. Rutherford [2] and Jacob DeGayner [1]

[1] US National Park Service, Southern Arizona Office, Phoenix, AZ 85712, USA; kara_raymond@nps.gov (K.R.); jacob_degayner@nps.gov (J.D.)
[2] US Department of Agriculture, Agricultural Research Service, Southwest Watershed Research Center, Tucson, AZ 85719, USA; jason.williams@usda.gov (C.J.W.); austin.rutherford@usda.gov (W.A.R.)
* Correspondence: sharlot_hart@nps.gov; Tel.: +1-602-882-7223

Abstract: Adobe has been used globally for millennia. In the US Southwest, cultural heritage sites made of adobe materials have lasted hundreds of years in an arid/semi-arid environment. A common prediction across multiple climate change models, however, is that rainfall intensity will increase in the US Southwest. This increased erosivity threatens the long-term protection and preservation of these sites, and thus resource managers are faced with selecting effective conservation practices. For this reason, modeling tools are needed to predict climate change impacts and plan for adaptation strategies. Many existing strategies, including patching damaged areas, building protective caps and shelter coating walls are already commonly utilized. In this study, we modeled adobe block construction, subjected extant walls to a local 100-year return interval rainfall intensity, and tested earthen-coat-based strategies to minimize the deterioration of earthen fabric. Findings from the resultant linear models indicate that the patching of earthen architecture alone will not prevent substantial damage, while un-amended encapsulation coats and caps provide similar, and significantly greater protection than patching. The use of this model will enable local heritage resource managers to better target preservation methods for a return on investment of the material and labor costs, resulting in better preservation overall and the retention of culturally valuable resources.

Keywords: cultural heritage; deterioration; adobe; earthen architecture; climate change; heritage preservation; erosion

Citation: Hart, S.; Raymond, K.; Williams, C.J.; Rutherford, W.A.; DeGayner, J. Modeling Earthen Treatments for Climate Change Effects. *Heritage* **2023**, 6, 4214–4226. https://doi.org/10.3390/heritage6050222

Academic Editors: Peter Brimblecombe and Jenny Richards

Received: 16 April 2023
Revised: 4 May 2023
Accepted: 7 May 2023
Published: 9 May 2023

1. Introduction

Moisture is a major driver of damage to adobe architecture. The intensity and duration of rain events contribute to the severity of erosion and the occurrence of catastrophic collapse of earthen structures. Climate projections for the US Southwest have heightened the need to understand the relationship between rain intensity and resource damage, and to identify effective preservation methods for withstanding a range of climate futures. Climate change models for the US Southwest indicate that the frequency of high intensity rainfall is likely to increase over the next century [1–5]. Other studies have found that the frequency of extreme precipitation is already increasing in the US Southwest [6–8]. These changing precipitation patterns are likely to increase the deterioration of earthen structures [9].

The US National Park Service (NPS) preserves historic period adobe resources in a state that conveys their appearance as ruins. In general, ruins are defined as resources that no longer have their basic structural components. For adobe resources on NPS lands, many adobe ruins no longer have roofs [10]. Repairs such as patches, caps, and encapsulations are common strategies to protect unroofed and otherwise unprotected walls from damage caused by precipitation. The constituency of stabilization and repair materials is, therefore, critical for providing long-term protection at these sites. Fort Selden in the US state of New Mexico, the location of the Getty Institute's long-term study on the productivity of earthen shelter coat amendments [11], is a prime example of attempts to test materials on adobe ruins. While many NPS sites in the US have test walls, they are built to test treatment efficacy in ambient weather conditions. This is the case at both Pecos National Historical Park in northern New Mexico [12] and on site at Tumacácori National Historical Park (TUMA) [13] in southern Arizona.

The experiment and results presented here grew out of a pilot study modeled on extant adobe walls at the site of Mission Los Santos Angeles de Guevavi, a part of TUMA [9], but was physically conducted in Tucson, Arizona, where the logistics of modeling could be streamlined. While the current work is centered primarily at sites in central and southern Arizona, US, damage from increasing precipitation intensity is expected to impact adobe resources throughout the US Southwest. To systematically investigate the impact of precipitation intensity and duration on earthen architecture, 20 adobe test walls were constructed using materials and methods consistent with the historic fabric comprising adobe buildings at TUMA. Test wall materials and construction are described in detail in the aforementioned Hart et al. [9] companion study. The bricks and walls were constructed during two training sessions for NPS cultural resource personnel and their cooperators in August and October 2018. The training sessions were led by three instructors experienced in both masonry and adobe construction and in preservation in southern Arizona.

No preservation treatments or amendments were applied to the test walls. In November 2018, a rainfall simulator was used to apply one of four high-intensity rain treatments to each test wall, based on the return interval for 1-year, 25-year, and 100-year storms as estimated for the TUMA weather station USC000028865 [9,14]. The 30 min rainfall treatments were (1) control: no rainfall, (2) 1-year storm: 3.6 cm h^{-1}, (3) 25-year storm: 8.5 cm h^{-1}, and (4) 100-year storm: 10.6 cm h^{-1}. This study found that the 30 min, 100-year storm caused a mean wall material loss of 5.64% and affected a mean area of 8790 cm^2 of the wall surface area. This earlier study found that an increasing rainfall intensity, as predicted in the climate change models mentioned above, will cause increased rates of erosion in unprotected adobe block construction.

This valuable data on the impact of the rainfall amount and intensity on bare adobe will allow cultural resource managers to anticipate and prepare for how a range of rain intensities can impact bare adobe. Since many adobe resources on NPS land already have protection strategies in place to mitigate the erosive effects of increasing rainfall intensity; the purpose of the current study is to evaluate three common protection measures using unamended earthen treatment options—patches, caps, and encapsulation/shelter coats. To address the goal, we (1) randomly applied patch, cap, and encapsulation/shelter coats to individual adobe test walls, (2) applied a local 100-year rainfall intensity for 30 min to each wall using rainfall simulation methods, and (3) used terrestrial laser scanning methods to quantify the wall deterioration.

2. Materials and Methods

Since their original construction, the test walls have experienced variable amounts of erosion in response to both ambient weather and the applied rainfall events [9]. As detailed above, in November 2018, rainfall simulators were used to apply one of four high-intensity rain treatments to each test wall. Additionally in that study, subsequent additional low-intensity rainfall simulation experiments were conducted in May–June 2019 on a subset of walls applying 0.97 cm h^{-1} of rainfall in either a single event (1 event, 240 min) or two events (2 event, 80 min and 160 min events separated by a 48 h hiatus) to assess the effects of prolonged low-intensity rainfall on wall degradation. Untreated walls (control treatment) for the low-intensity simulations utilized the same original five control test walls from the previous high-intensity rainfall simulations. The walls received no preservation treatments or amendments after the high- or low-intensity experiments. After these two experiments, all the walls were primarily exposed to ambient weather conditions. By Autumn of 2022, the walls' erosional state had visually appeared as "melted adobe ruins" (Figure 1).

Figure 1. Wall O when newly constructed in 2018 (**left**) and before rain simulation in 2022 (**right**).

While the previous study [9] used rainfall intensities to correlate the precipitation and material loss, this study sought to investigate the effectiveness of common treatments under one future rainfall intensity. In this case, all the walls, including the test locations originally classified as control walls, were subjected to 100-year rainfall intensity (10.6 cm h^{-1}) for 30 min. This scenario was intended to test the effectiveness of three common types of treatments. Each treatment used unamended earthen material, consisting of the same locally-sourced soil with clay content and sand as the previous study [9]. Unamended earthen material is defined as having no chemical additives such as polymer adhesives to help with water shedding. As each locally-procured soil will have distinct characteristics, there is no one formula for mixing the clay, sand, and water, and studies and guides on adobe give a range of 0 to 30% clay content [15–18]. However, a certain consistency is desirable and was reached, namely, one where the wet material stayed adhered to a trowel when inverted and spread easily on the wall surfaces. All the treatments were hand applied with trowels. The treatments were applied to 5 walls each in the following configurations (see also Figure 2):

1. Control: no unamended earthen material added.
2. Patch: earthen material used to fill in voids and cracks, in some cases cobbling/rajuelas were applied.
3. Cap: earthen material in a wet-plaster consistency applied over the tops of the walls and vertically down the four faces of the walls to 10 cm; depth/thickness of plaster did not exceed 1.25 cm, excluding locations where patching was performed. Patching was performed prior to capping.
4. Encapsulate (encapsulation/shelter coat): earthen material in a plaster consistency applied to the entire wall surface; that is, a cap plus coating the walls to the ground surface. Depth/thickness did not exceed 1.25 cm, excluding locations where patching was performed. Patching was performed prior to encapsulation.

Walls were randomly assigned to treatments. Because the walls had variable amounts of erosion from previous rain simulations and impacts from ambient rainfall, a one-way analysis of variance (ANOVA) was performed to identify any differences in the overall degradation between the randomized treatment groups. This analysis found that the treatment groups chosen immediately prior to the treatment application in 2022 were not significantly different from each other relative to their total volume loss (p = 0.79) and affected surface area (p = 0.95).

Figure 2. Lidar scan/wall models of treated walls prior to 100-year (10.6 cm h^{-1}) rainfall event application.

The mean amount of unamended material added per treatment is summarized in Table 1. While the control treatment did not receive any added material, the LiDAR scans indicate the control walls lost a mean of 577 cm^3 of original fabric prior to this study. This is primarily due to the physical disturbance from covering the walls with tarps. Tarps were applied to the walls one week prior to the rainfall simulations to protect them from forecasted rain immediately prior to the experiments. Some gravel-sized rocks were dislodged from the top of the walls where the surrounding adobe material had previously eroded. Wind likely caused the secured tarps to abrade the walls, eroding fines and causing material loss. All the walls were tarped, and similar losses likely occurred at the non-control walls but could not be quantified. The high standard deviation for the patch and encapsulate methods is due to the variable degradation of the walls prior to applying the treatment (Table 1). Comparatively, the cap application was much more uniform.

Table 1. Mean (standard deviation) volume (cm^3) of material added for each treatment and percent added relative to pre-treatment.

Treatment	Mean Added Material (cm^3)		Mean Added Material Relative to the Pre-Treatment Volume (%)	
Control	−577	(SD 228)	−0.253	(SD 0.107)
Patch	8043	(SD 4265)	4.06	(SD 2.00)
Cap	14,015	(SD 1760)	6.61	(SD 0.515)
Encapsulate	27,157	(SD 17,287)	14.3	(SD 4.59)

2.1. Rainfall Simulator

Rainfall simulations were conducted over the period of 25 October–1 November 2022 with consistent daily ambient weather conditions without natural rainfall. Rainfall simulations employed the same portable single-nozzle, Meyer and Harmon-type, oscillating-arm rainfall simulator [19–21] used in earlier high- and low-intensity rainfall studies on the test walls [9]. The rainfall simulator (Figure 3) was fitted with a VeeJet 80–100 nozzle raised 3 m above the ground surface and supplied with water pressurized at 41 N m^{-2}. The raindrop size and kinetic energy of the simulated rainfall was within 1 mm and

70 kJ ha^{-1} mm^{-1}, respectively, of the values reported for natural convective rainfall [19,22,23]. Tarps were applied to the rainfall simulator prior to the experiments to prevent wind effects on rainfall and to ensure consistency in the rainfall application rate. Specifically, the simulator configuration described above applies rainfall with a kinetic energy at the ground surface of approximately 200 kJ ha^{-1} mm^{-1} and about 2 mm average drop size [19]. The simulator was calibrated multiple times daily for the target intensity (10.6 cm h^{-1}) by simulating rainfall over a calibration pan for five minutes [20,21]. The calibrations resulted in an average application rate of 10.57 cm h^{-1} (applied for a 30 min duration) with a standard deviation of 0.01 cm h^{-1} across all the simulations. All the calibration runs and wall simulations were controlled for wind by tarping around the simulator and respective test wall.

Figure 3. Rainfall simulator with tarps removed, October 2022.

2.2. Lidar Application

A Surphaser® Model 10 was used to measure the test walls in 3D space, with a margin of error likely to remain under one millimeter for distances within 15 m of the scanner [9,24]. In addition, the same two methods of registration (i.e., spherical targets and permanent control points) used in Hart et al. [9] were again employed to improve accuracy.

We recorded test walls in three scanning epochs, each consisting of approximately 30 scans, over a 42-day period. The epochs consist of:

- Epoch IX: immediately prior to treatment application; 6 September 2022,
- Epoch X: after treatment application and before the rain simulation; 28 September 2022 (Figure 2),
- Epoch XI: after the rain simulation; 3 November 2022.

Measured and adjusted locations of the control points from each scanning epoch were compared to the global mean location derived from all three epochs, resulting in a mean target residual distance of 0.54 mm.

Post-processing methodologies remained the same as in previous experiments [9], including processing and registration in FARO Scene software (SCENE, version 2021.1), and then co-registering with the other epochs using the five permanent control points. We again defined the reference planes above mass wasting build-up levels and removed all the surfaces below.

Wall-degradation metrics (i.e., material loss for treatment and original fabric, and material loss, affected surface area, and maximum recession distance for original fabric only) were calculated for each wall using the before and after lidar scans. The material (%) and volume losses (cm^3) were defined as the relative difference between pre- and post-treatment wall volumes. Scanning epochs were transformed in single point clouds and converted to polygonal meshes in the same way as the previous study, but with Geomagic Wrap (v.2021). A hole-filling operation was again used and the volumes calculated.

The affected surface area was defined as wall surfaces that exhibited a deviation greater than or equal to 2 mm from the previous epoch. We calculated this error threshold in previous experiments by measuring deviations on non-treated control walls, yielding a confidence interval of approximately 99%. We re-used this value due to the identical methodology used in this study. Distances were computed using the Cloud-to-Mesh Distance tool in CloudCompare (v.2.11.2) to generate a scalar field, calculating a signed distance value to each mesh facet. Negative values indicate surfaces on a compared model below the corresponding area of the reference model. Conversely, positive values indicate areas above the reference model. Model facets representing positive and negative values in excess of 2 mm in either direction were isolated for each wall model (Figure 4). Thus, surface areas exhibiting negative values on Epoch IX compared to Epoch XI represent the areas where wall surfaces have eroded past treatment material into the "original" fabric (Figure 5). Positive values indicate areas where preservation material was still present. The maximum recession distance records the greatest negative value present for each test wall.

Material losses for the treatment and original fabric were derived as the relative difference between the Epoch X and XI wall volumes. Because of the material addition between scanning Epochs IX and X, direct measurements of the wall model volume could not be used to assess the amount of original fabric material loss. Thus, the material loss of original fabric only was calculated as the affected surface area measurements referenced above, multiplied by the signed distances to generate an indirect measurement of loss.

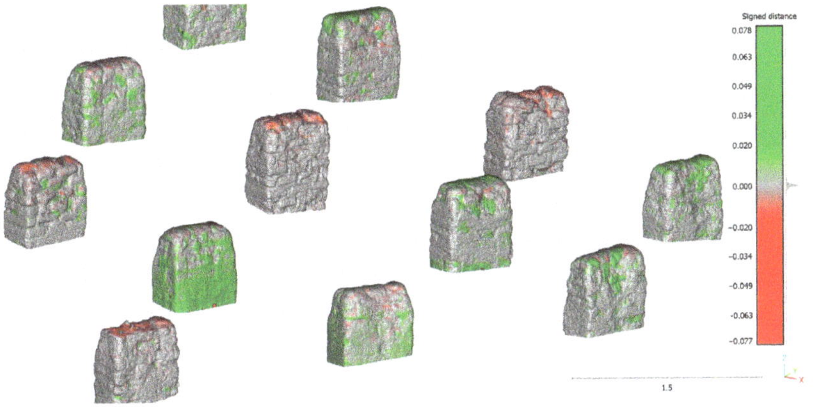

Figure 4. Epoch XI walls compared to Epoch IX showing positive signed areas representing additions (green) and negative signed areas representing loss (red).

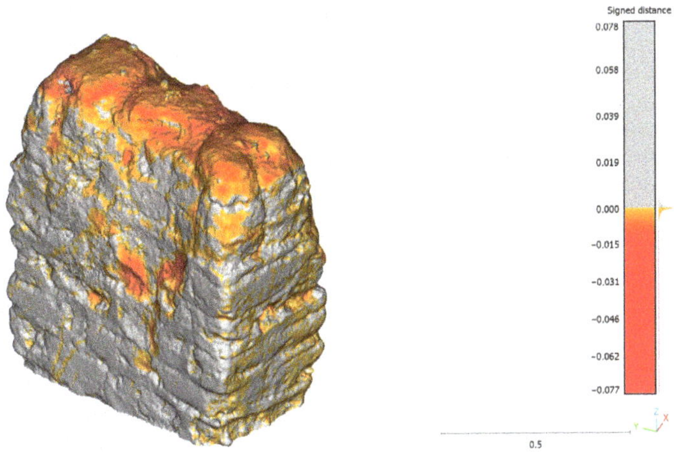

Figure 5. Affected area and negative signed values (in orange-red) used to calculate original material loss on Wall C.

2.3. *Data Analysis*

We computed a linear model by treatment (i.e., control, patch, cap, and encapsulate) for the four metrics: loss of original and treatment material, surface area receded past original material, impact on original material, and maximum deviation from original material. The metrics were analyzed using a one-way analysis of ANOVA and Tukey's post-hoc test.

We used the R programming language with RStudio (v.4.2.0) for the analyses: stats (v.4.2.0) for linear models, emmeans (v.1.8.3) to estimate the marginal means, car (v.3.1-1) for the type-II ANOVA (F-tests for linear models), and multcomp (v.1.4-20) to generate group letters of Tukey pairwise comparisons.

3. Results

The simulated rain event caused the patch, cap and encapsulate treatments to lose a mean of 3.37–3.75% of their total volume, including the original fabric and added treatment material (Table 2, Figure 6A). There was no significant difference between the three preservation treatments. Significantly less material eroded away in the control treatment, with a mean of 0.77% of the total volume, which was only comprised of original fabric.

Table 2. Wall treatments and erosion metrics.

Wall	Treatment	Material Loss of Original and Treatment Fabric (%) (cm^3)	Affected Surface Area of Original Material (%) (cm^2)	Material Loss of Original Fabric (cm^3)	Maximum Recession Distance from Original Material (cm)
C	Control	1.03 (2432)	15.85 (3718)	2880	4.11
M	Control	0.70 (1681)	13.76 (3250)	2109	3.26
O	Control	0.61 (1059)	12.93 (2427)	1653	2.70
P	Control	0.73 (1726)	13.93 (3177)	2114	3.92
R	Control	0.79 (1753)	15.84 (3539)	2298	3.27
A	Patch	2.56 (5240)	18.32 (3801)	3555	4.43
B	Patch	3.24 (6330)	13.16 (2517)	1416	3.73
H	Patch	4.08 (8106)	13.32 (2814)	2055	4.33
I	Patch	3.34 (6154)	13.52 (2590)	2193	3.25
T	Patch	3.60 (7779)	6.00 (1256)	796	2.97
D	Cap	4.35 (9171)	2.88 (564)	829	6.11
K	Cap	3.35 (6970)	5.00 (1055)	852	2.70
L	Cap	2.61 (6198)	3.04 (677)	328	2.68
Q	Cap	3.53 (6946)	4.24 (833)	482	3.04
S	Cap	3.15 (7514)	6.20 (1412)	865	3.61
E	Encapsulate	3.16 (6666)	2.56 (487)	286	2.95
F	Encapsulate	3.95 (7594)	11.97 (2210)	1264	2.95
G	Encapsulate	4.27 (9191)	4.19 (842)	546	3.11
J	Encapsulate	3.30 (7317)	3.45 (682)	658	8.79
N	Encapsulate	4.07 (8316)	7.41 (1524)	936	2.65

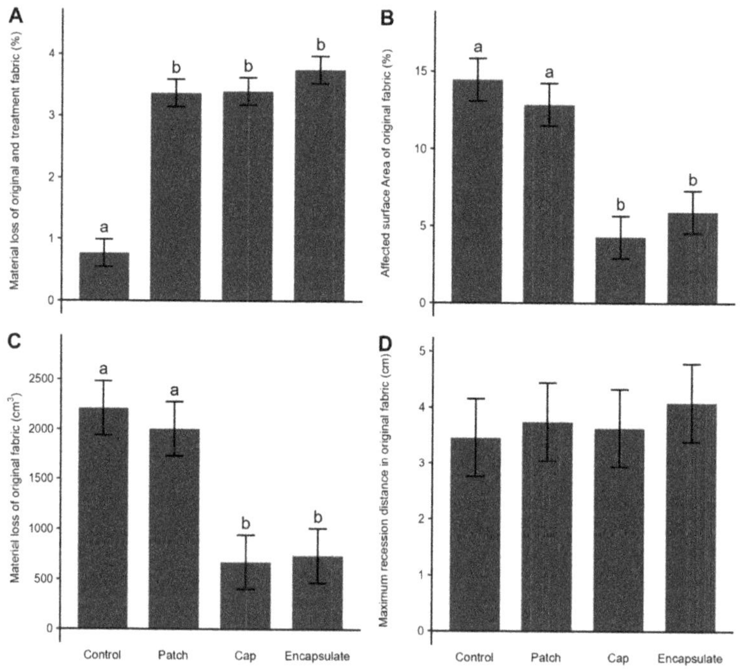

Figure 6. Estimated marginal mean (±SE) for erosion and moisture metrics by preservation treatments. Lowercase letters across treatments within a metric (**A–D**) indicate statistically different groups based on post-hoc pairwise comparisons, Tukey method ($p \leq 0.05$).

The affected surface area and material loss metrics related to the original walls show where the erosion went past the treatment material and into the original wall fabric. These metrics indicate that the control and patch treatments resulted in significantly greater degradation of the original fabric than the cap and encapsulate treatments (Table 2, Figure 6B,C). The simulated rain event caused a mean of 14.5% of the surface area of the pretreatment control and 12.9% of the patch walls to recede $\geq$2 mm; however, the two treatments were statistically the same. Erosion of the surface area was significantly less severe in the cap and encapsulate treatments, with a mean of 4.27% and 5.92%, respectively. The material loss of original fabric indicates mean losses of 2210 and 2000 cm^3 for the control and patch treatments, respectively. The cap and encapsulate treatments showed significantly less material loss on the original fabric, with approximately one-third of that for the control and patch treatments.

The type of preservation treatment used was not a strong predictor for the maximum recession distance into the original fabric (Figure 6D). The mean maximum recession distance was lowest in the control treatment and highest in the encapsulate treatment. However, the F-test (for variability between the group means) indicated that none of the treatments were significantly different, and thus, post-hoc tests were not conducted (Table 3).

Table 3. Analysis of variance results for erosion metrics.

Parameter	Material Loss of Original and Treatment Fabric (%) (cm^3)	Affected Surface Area of Original Material (%) (cm^2)	Material Loss of Original Fabric (cm^3)	Maximum Recession Distance from Original Material (cm)
F-test	39.0 (41.4)	14.9 (13.4)	9.07	0.147
p-value	<0.001 (<0.001)	<0.001 (<0.001)	<0.001	0.93

4. Discussion

4.1. Return on Investment for Sacrificial Materials

The preservation treatments in this study are all designed to be sacrificial treatments on top of the historic adobe, providing a protective layer that will erode before the historic adobe underneath. The earthen mixture used is similar to mortar and has a higher erodibility due to its higher content of clay and other fine particles compared to the adobe block fabric [9]. This study supports what historic preservationists have observed in the field, namely, that "the cap is the life of the wall" [25]. Statistically, there is no difference in the material affected, across all metrics, between a cap and an encapsulation coat treatment. There is, however, a statistical difference regarding the affected surface area and material loss on the original fabric affected between solely patching and either placing a sacrificial cap or encapsulation coat.

Each agency or entity that maintains historic architecture somehow defines its maintenance cycle, and the associated costs. Since those terminologies and definitions differ, here we borrow the concept of "return on investment" (ROI) from the business world, where ROI answers the question if there is a net benefit to investing in one of these treatments, e.g., [26]. In a general sense, it is easy to evaluate the staff time and materials needed for each of the three treatments: patching at the lower end of investment, capping in the middle, and full encapsulation at the upper end of the needed investment. For example, during the workshop used to apply the treatments to the walls in this study, a skilled practitioner was able to patch two walls, in the time it took another skilled practitioner to fully encapsulate one wall. This model not only validates the idea that an applied cap is vital to the preservation of original earthen fabric, but also provides heritage practitioners and managers with data to decide when and where to focus staff energy and funding.

For example, given the projected increased rainfall intensity discussed earlier, if a heritage area or other adobe site has full-time preservation practitioners on staff, our models show that their time investment to re-cap or re-encapsulate multiple times within a year may be a positive ROI. A site without full-time practitioners on staff, however, likely does not have the capacity to make the same choice and will need to look for other treatment options (e.g., amended earthen coats) to extend their cyclic preservation maintenance. This is certainly the case at another NPS heritage site in southern Arizona, Fort Bowie National Historical Park, where Porter [27] (p. 13) reported: "Sheltering with unamended earthen plaster requires a substantial commitment to maintenance of the shelters following each heavy rainfall; this is far beyond the current capacity of Fort Bowie, where one staff person is responsible for maintenance of park facilities as well as preservation maintenance of the ruins".

4.2. Armoring of Original Fabric

The results of the present study and high-intensity rainfall treatments on these test walls in 2018 suggest the walls became more armored, increasing their resistance to erosion [9]. In the 2018 study, a treatment group of the newly constructed bare adobe walls received a 30 min applied rainfall with an intensity of 10.6 cm h^{-1}, the same as in the study presented here; however, the control walls (n = 5) in this present study were not new, as they had been exposed to four years of ambient weather and then simulated rain events. We found that the erosion results were substantially different for these weathered walls, compared to the new walls that received the same simulated rainfall event (n = 5) (Table 4). The mean material loss on the new walls was over eight times that of the degraded walls. The mean affected surface area of the walls in the 2018 study was twice as much as the affected surface area of the weathered walls. This suggests that the original fabric became more resistant to intense rainfall over the period of four years.

Table 4. Mean (and standard deviation) erosion metrics (cm^3 and %) for new and degraded walls, both exposed to a 30 min, 10.6 h^{-1} simulated rain event.

Wall Age	Material Loss cm^3	%	Affected Surface Area cm^2	%
New walls (2018 study)	14,400 (SD 2560)	5.64 (SD 1.00)	6440 (SD 432)	28.9 (SD 3.29)
Weathered walls (present study)	1730 (SD 435)	0.764 (SD 0.139)	3220 (SD 443)	14.5 (SD 1.18)

Progressive armoring of the weathered adobe walls may account for some of the differences in the rate of erosion. Soil armoring may be accomplished through natural or anthropogenic means [28,29], but in this case we refer to the ability of larger particles including gravel and rock to reduce erosion [30,31]. The newly constructed walls had a smooth finish and flat top (Figure 1). Through successive simulated and ambient rain events, wind erosion, and abrasion from periodic tarping (for short periods immediately prior to experiments), the surface texture and shape of the walls changed drastically; fine particles were transported off the walls leaving the larger sand and gravel particles, the tops became more convex, and the mortar eroded faster than the bricks. Protruding sand and gravel can absorb the impact of raindrops and slow runoff; thus, reducing erosion. The convex top prevents the pooling of water, which may reduce the concentrated pour over and rilling on wall faces. Some runoff may route along the incised mortar lines, which may slow the runoff at the horizontal sections. Generally, these results support the observation that the condition of the top of a wall is a key determinant to the preservation of the wall, and even a natural armored adobe wall can provide some protection. Protection is relative, however; while the weathered walls here lost a fraction of the original fabric compared to the new walls under the same storm conditions, 1730 cm^3 and 14,400 cm^3, respectively, that smaller loss still constitutes a heritage conservation issue (Table 4).

4.3. Wind-Driven Rain

This study omitted the effects of wind on rainfall and wall degradation and material loss. This was intended to focus more specifically on the effects of rainfall intensity and to ensure consistency in the rainfall application rate across treatments. As such, the wall degradation and losses as quantified here may be conservative relative to responses under wind-driven rain [32,33]. Wind can strongly influence rainfall intensity and the impact velocity and direction of raindrops [33]. The magnitude of these effects varies with the horizontal wind velocity [34]. The respective relationships are complex [32–35] and are outside the scope of the current study. Substantive discussion on the effects of wind-driven rain on the degradation of historical building materials is provided by Blocken and Carmeliet [36] and Erkal et al. [35].

5. Conclusions

This study attests to the real-world application of models in heritage work. There is often a disconnect between the offices and staff who have the capacity to perform modeling experiments and those who are on the ground, faced with everything from hairline cracks to imminent collapse. This study seeks to test common strategies used to minimize deterioration to adobe ruins—and one that comes from the field. While all the treatments tested provided a level of protection, capping walls with unamended earthen material will provide some protection in light of increasing storm intensity. It is also less time and material intensive than an application of an encapsulation coat.

However, the data provided by the models do not conclude that either the cap or the encapsulation treatments "saved" the walls from increasing rainfall intensity. Indeed, while these treatments were statistically different from the control and patch treatments, they still had measurable impacts to the original fabric, after only a 30 min storm at the 100-year intensity. The application cycles for unamended earthen caps would need to be far shorter than they currently are at many NPS sites.

As such, we intend to use these same models to continue the experiments, testing different additives to earthen material and other sacrificial coat options. Finally, due to the unexpected findings regarding differences in the rate of erosion of new walls to weathered walls, further study is needed to refine our models, and to investigate soil armoring on three-dimensional objects (walls) to maximize the return on investment in preserving these nationally important heritage sites.

Author Contributions: Conceptualization, S.H.; methodology, S.H., K.R., J.D. and C.J.W.; formal analysis, J.D. and K.R.; investigation, validation, resources, data curation, writing—original draft preparation, writing—review and editing, S.H., K.R., C.J.W., W.A.R. and J.D. All authors have read and agreed to the published version of the manuscript.

Funding: This research received no external funding.

Data Availability Statement: The data are available as figures in the manuscript. Underlying data files can be obtained from the corresponding author.

Acknowledgments: The authors wish to thank all the adobe practitioners who came out to train on the test walls while also participating in our study, especially Chris Schrager, US Forest Service, and Ramon Madril, NPS—Tumacácori National Historic Park, for leading the training. Iraida Rodriguez, NPS—Southern Arizona office spent much of her time with in-field scanning, and Matthew Guebard, from the same office, gave invaluable advice and comments on the draft. The authors also thank Fred Pierson of the USDA Agricultural Research Service (ARS), Northwest Watershed Research Center, Boise, Idaho, for use of rainfall simulators applied in this study. We also thank Erin Phelps, co-affiliated with University of Arizona (School of Natural Resources and the Environment) and USDA-ARS Southwest Watershed Research Center, for assistance in completing rainfall simulation experiments. Resources for rainfall simulations were provided by the USDA-ARS. The USDA is an equal opportunity provider and employer. Mention of a proprietary product does not constitute endorsement by USDA and does not imply its approval to the exclusion of the other products that may also be suitable).

Conflicts of Interest: The authors declare no conflict of interest.

References

1. Easterling, D.R.; Kunkel, K.E.; Arnold, J.R.; Knutson, T.; LeGrande, A.N.; Leung, L.R.; Vose, R.S.; Waliser, D.E.; Wehner, M.F. Precipitation change in the United States. In *Climate Science Special Report: Fourth National Climate Assessment*; Wuebbles, D.J., Fahey, D.W., Hibbard, K.A., Dokken, D.J., Stewart, B.C., Maycock, T.K., Eds.; U.S. Global Change Research Program: Washington, DC, USA, 2017; Volume 1. [CrossRef]
2. Wang, B.; Biasutti, M.; Byrne, M.P.; Castro, C.L.; Chang, C.; Cook, K.H.; Fu, R.; Grimm, A.M.; Ha, K.; Hendon, H.; et al. Monsoons Climate Change Assessment. *Bull. Am. Meteorol. Soc.* **2020**, *102*, E1–E19. [CrossRef]
3. Pascale, S.; Carvalho, L.M.V.; Adams, D.K.; Castro, C.L.; Cavalcanti, I.F.A. Current and future variations of the monsoons of the Americas in a warming climate. *Curr. Clim. Chang. Rep.* **2019**, *5*, 125–144. [CrossRef]
4. Luong, T.M.; Castro, C.L.; Chang, H.; Lahmers, T.; Adams, D.K.; Ochoa-Moya, C.A. The more extreme nature of north American monsoon precipitation in the southwestern United States as revealed by a historical climatology of simulated severe weather events. *J. Appl. Meteor. Clim.* **2017**, *56*, 2509–2529. [CrossRef]
5. Bukovsky, M.S.; Carrillo, C.M.; Gochis, D.J.; Hammerling, D.M.; McCrary, R.R.; Mearns, L.O. Toward assessing NARCCAP regional climate model credibility for the North American monsoon: Future climate simulations. *J. Clim.* **2015**, *28*, 6707–6728. [CrossRef]
6. Kunkel, K.E.; Karl, T.R.; Brooks, H.; Kossin, J.; Lawrimore, J.H.; Arndt, D.; Bosart, L.; Changnon, D.; Cutter, S.L.; Doesken, N.; et al. Monitoring and understanding trends in extreme storms: State of knowledge. *Bull. Am. Meteor. Soc.* **2013**, *94*, 499–514. [CrossRef]
7. Demaria, E.M.C.; Hazenberg, P.; Scott, R.L.; Meles, M.B.; Nichols, M.; Goodrich, D. Intensification of the North American Monsoon rainfall as observed from a long-term high-density gauge network. *Geophys. Res. Lett.* **2019**, *46*, 6839–6847. [CrossRef]
8. Wright, D.B.; Bosma, C.D.; Lopez-Cantu, T.U.S. hydrologic design standards insufficient due to large increases in frequency of rainfall extremes. *Geophys. Res. Lett.* **2019**, *46*, 8144–8153. [CrossRef]
9. Hart, S.; Raymond, K.; Williams, C.J.; Johnson, J.; DeGayner, J.; Guebard, M.C. Precipitation impacts on earthen architecture for better implementation of cultural resource management in the US Southwest. *Herit. Sci.* **2021**, *9*, 143. [CrossRef]
10. National Park Service. *How to Apply the National Register Criteria for Evaluation*; US Department of Interior: Washington, DC, USA, 1997.
11. Oliver, A. *Fort Selden Adobe Test. Wall Project: Phase I: Final Report*; Getty Adobe Project, Getty Conservation Institute, and Museum of New Mexico: Los Angeles, CA, USA, 2000. Available online: http://hdl.handle.net/10020/gci_pubs/fort_selden_project (accessed on 1 May 2023).
12. Scott, K.; Moss, J. *Historic Preservation 2017: Pecos National Historical Park*; National Park Service: Washington, DC, USA, 2017.
13. Woodham, D.; Citto, C.; Porter, D.W.; Bass, A. Investigation, Analysis, and Treatment Testing for the Mission Church, Tumacácori National Historical Park. *J. Archit. Eng.* **2020**, *26*, 05020001. [CrossRef]
14. Bonnin, G.M.; Martin, D.; Lin, B.; Parzybok, T.; Yekta, M.; Riley, D. *Precipitation-Frequency Estimates Atlas of the United States NOAA Atlas 14:1*; Version 5.0.; National Weather Service: Silver Spring, MD, USA, 2011. Available online: https://www.weather.gov/media/owp/hdsc_documents/Atlas14_Volume1.pdf (accessed on 1 May 2023).
15. Cornerstones Community Partnership. *Adobe Conservation: A Preservation Handbook*; Sunstone Press: Santa Fe, NM, USA, 2006.
16. Cavicchio, A. An Evaluation of Shelter Coating as a Preventive Conservation Method for Earthen Sites. Master's Thesis, University of Pennsylvania, Philadelphia, PA, USA, 2022.
17. Rezende, M.A.P.; Do Vale, J.L.R. Adobe with 1% clay and 2.11 MPa resistance: A case study. In *Terra Lyon. 2016*; Joffroy, T., Guillaud, H., Sadozaï, C., Eds.; CRAterra: Villefontaine, France, 2018.
18. Hohn, C.M. *Guide G-521 ABCs of Making Adobe Bricks*; New Mexico State University: Las Cruces, NM, USA, 2011.
19. Meyer, L.D.; Harmon, W.C. Multiple-intensity rainfall simulator for erosion research on row sideslopes. *Trans. ASAE* **1979**, *22*, 100–103. [CrossRef]
20. Pierson, F.B.; Moffet, C.A.; Williams, C.J.; Hardegree, S.P.; Clark, P.E. Prescribed-fire effects on rill and interrill runoff and erosion in a mountainous sagebrush landscape. *Earth Surf. Process. Landforms.* **2009**, *34*, 193–203. [CrossRef]
21. Williams, C.J.; Pierson, F.B.; Kormos, P.R.; Al-Hamdan, O.Z.; Hardegree, S.P.; Clark, P.E. Ecohydrologic response and recovery of a semi-arid shrubland over a five year period following burning. *Catena* **2016**, *144*, 163–176. [CrossRef]
22. Wischmeier, W.H.; Smith, D.D. Rainfall energy and its relationship to soil loss. *Eos Trans. Am. Geophys. Union* **1958**, *39*, 285–291. [CrossRef]
23. Carter, C.E.; Greer, J.D.; Braud, H.J.; Floyd, J.M. Raindrop characteristics in south central United States. *Trans. ASAE* **1974**, *17*, 1033–1037. [CrossRef]
24. Surphaser. *Surphaser 3D Laser Scanners*; Basis Software Inc.: Redmond, WA, USA, 2016.
25. Yubeta, D. (Retired National Park Service, Tumacacori, AX, USA). Personal communication, 2017.
26. Phillips, J.J.; Phillips, P.P. *Return on Investment (ROI) Basics*; ASTD Press: Alexandria, VA, USA, 2005.
27. Porter, D. *Summary Report: Condition Assessment of First and Second Fort, Fort Bowie National Historic Site*; National Park Service: Washington, DC, USA, 2017; unpublished work.
28. Soil Health: Principle 1 of 4—Soil Armor. Available online: https://www.nrcs.usda.gov/conservation-basics/conservation-by-state/north-dakota/soil-health-principle-1-of-4-soil-armor (accessed on 5 April 2023).

29. Lightfoot, D.R.; Eddy, F.W. The agricultural utility of lithic-mulch gardens: Past and present. *GeoJournal* **1994**, *34*, 425–437. [CrossRef]
30. Dorn, R.I. Impact of consecutive extreme rainstorm events on particle transport: Case study in a Sonoran Desert range, western USA. *Geomorphology* **2015**, *250*, 53–62. [CrossRef]
31. Larson, P.H. Desert Fluvial Terraces and Their Relationship with Basin Development in the Sonoran Desert, Basin and Range: Case Studies from South-Central Arizona. Ph.D. Thesis, Arizona State University, Phoenix, AZ, USA, 2013.
32. Erpul, G.; Norton, L.D.; Gabriels, D. Raindrop-induced and wind-driven soil particle transport. *CATENA* **2002**, *47*, 227–243. [CrossRef]
33. Erpul, G.; Norton, L.D.; Gabriels, D. The effect of wind on raindrop impact and rainsplash detachment. *Trans. ASAE* **2003**, *46*, 51–62. [CrossRef]
34. Erpul, G.; Gabriels, D.; Norton, L.D.; Flanagan, D.C.; Huang, C.H.; Visser, S.M. Mechanics of interrill erosion with wind-driven rain. *Earth Surf. Process. Landf.* **2013**, *38*, 160–168. [CrossRef]
35. Erkal, A.; D'Ayala, D.; Sequeira, L. Assessment of wind-driven rain impact, related surface erosion and surface strength reduction of historic building materials. *Build. Environ.* **2012**, *57*, 336–348. [CrossRef]
36. Blocken, B.; Carmeliet, J. A review of wind-driven rain research in building science. *J. Wind. Eng. Ind. Aerodyn.* **2004**, *92*, 1079–1130. [CrossRef]

heritage

MDPI

Article

Climate Change Effects on Carbonation Process: A Scenario-Based Study

Gabriella Bretti * and Maurizio Ceseri

Istituto per le Applicazioni del Calcolo–CNR, Via dei Taurini 19, 00185 Rome, Italy
* Correspondence: gabriella.bretti@cnr.it

Abstract: Using a mathematical model of concrete carbonation that describes the variation in porosity as a consequence of the involved chemical reactions, we both validated and calibrated the related numerical algorithm of degradation. Once calibrated, a simulation algorithm was used as a forecasting tool for predicting the effects on the porosity of concrete exposed to increasing levels of CO_2 emissions, as well as to rising temperatures. Taking into account future projections of environmental modifications deriving from climate changes, some scenarios were produced numerically by the mathematical algorithm that showed the effects of different pollution levels and global warming on the porosity of Portland cement in a time window of years. Finally, a theoretical study on the effects of pollution levels on the carbonation constant determining the advancement in the carbonation front was carried out for the analyzed scenarios.

Keywords: concrete carbonation; reaction and diffusion models; climate changes; model parameter estimation; mathematical algorithms

Citation: Bretti, G.; Ceseri, M. Climate Change Effects on Carbonation Process: A Scenario-Based Study. *Heritage* **2023**, *6*, 236–257. https://doi.org/10.3390/heritage6010012

Academic Editors: Peter Brimblecombe and Jenny Richards

Received: 9 November 2022
Revised: 15 December 2022
Accepted: 22 December 2022
Published: 27 December 2022

1. Introduction

The interactions between environment and building heritage, i.e., monuments, archaeological sites, and historical and modern buildings have always been, and will continue to be, crucial for conservation issues. Moreover, the deterioration and damage of materials caused by weathering processes is still not completely understood since it is a highly complex phenomenon resulting from the interaction of both chemical and mechanical processes.

As natural stones are exposed to the modification of the environment and surrounding landscapes, concrete is also subjected to attack by multiple damaging factors, such as weathering, chemical aggression and abrasion, that may cause its deterioration in terms of a modification of the original form, quality and serviceability. Such weathering processes are always associated with water flow within the material, determined by wetting or infiltration, caused by meteoric precipitation or groundwater capillary rise, respectively.

An increased rate of extreme conditions due to climate change also constitutes a further threat, increasing the decaying rates and contributing to the occurrence of new degradation mechanisms. This happens because climatic changes can not only influence the frequency and intensity of hazardous events but can also worsen the physical, chemical and biological mechanisms causing the degradation of the structure and its materials. In recent years, research communities have started to discuss the impact of climate change on cultural heritage (CH). Since 2003, the number of papers on this subject increased significantly; see [1,2]. The changes in the climate system have been studied by climate scientists; see, for instance, [3–5]. In particular, the studies presented by NOAA Climate.gov show that environmental changes in recent decades are leading to a progressive increase in the CO_2 concentration in the atmosphere, along with human emissions. The Intergovernmental Panel on Climate Change (https://www.ipcc.ch/, accessed on 8 November 2022) takes into account possible scenarios of CO_2 emissions, showing that, in the near future (up until

2050), the CO_2 concentration will grow with different rates of increase and, in only a few cases, might be similar to that in 2015.

The environmental changes may affect CH buildings and artifacts due to the synergistic action of atmospheric agents and pollutants; see the recent studies in [6,7]. Such phenomena may determine an irreversible weakening of the mechanical strength and an increased vulnerability to the chemical aggression of porous materials through several mechanisms, such as freeze–thaw cycles, a change in precipitation, corrosion, salt crystallization cycles and an increased frequency of extreme events, just to mention a few; see [8–13]. Recently, predictive maintenance, consisting of anticipating future deterioration through appropriate diagnostic techniques, has emerged as a new tool for monitoring and protecting CH sites [14]. Such a diagnosis is carried out by collecting and analyzing, with statistical tools, data on the constitutive materials of the work of art, as well as the action of environmental factors at the CH site. Our approach is within the model-driven framework on predictive maintenance, based on mathematical models for CH (equations describing deterioration processes and the effects of conservation practices)—see, for instance, [15–18]—coupled with data derived from laboratory experiments and/or gathered by sensors suitably placed at the CH site.

Our study is focused on concrete since it is the most widely used construction material worldwide [19], as it is employed in the construction of buildings, stadiums, stairs, sidewalks and foundations; see Figure 1.

Figure 1. Two examples of concrete buildings. **Left** Panel: Termini Station in Rome, Italy. **Right** Panel: Bridge on Basento river in A2 motorway (Italy). Credit: MIBAC, 2018.

Such a material has a porous structure and its durability is mainly due to its resistance to chemical aggression. In particular, one of the most important degradation phenomena for concrete is carbonation, caused by carbon dioxide in the atmosphere. As shown in Figure 2, environmental changes in recent decades have determined a constant increase in the carbon dioxide concentration in the atmosphere. This fact may accelerate the damage process of building structures.

Concrete carbonation is a complex process whose study requires the cooperation of chemists, engineers and mathematicians in order to capture its features. It is caused by a sequence of chemical reactions that consume calcium hydroxide ($Ca(OH)_2$) and form calcium carbonate ($CaCO_3$). The above reactions are induced by carbon dioxide (CO_2), which is transported by water through the porous medium.

Although concrete is a long-lasting material, its degradation and the consequent weakening may be due to the sulfation of the cementitious matrix, freeze–thaw cycles and the corrosion of steel armor caused by carbonation. Indeed, a basic environment inside the non-carbonated concrete forms a thin film of oxide that protects the steel bars reinforcing the structures, and such a layer is maintained as long as the environment has a pH value of at least 13. In carbonated areas, the presence of carbon dioxide instead neutralizes the alkalinity of concrete with the solution within the pores, assuming a pH value inferior to 9. This causes the destruction of the protective layer and starts corroding the steel bars [20].

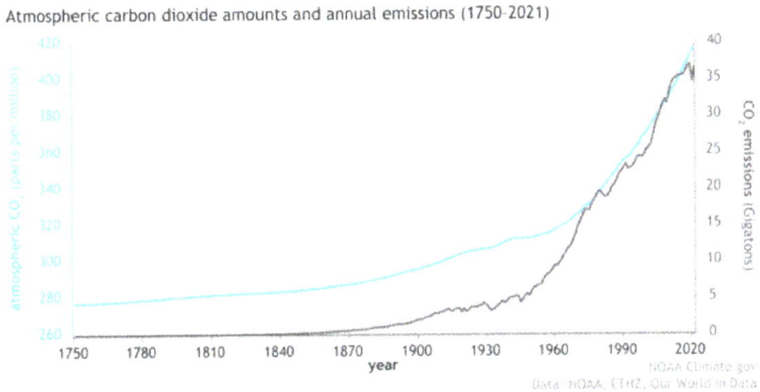

Figure 2. Increasing behavior of carbon dioxide in the atmosphere (blue line) associated to human emissions (gray line) since 1750 (start of the Industrial Revolution) until 2021. Credit: NOAA Climate.gov.

In the literature, many experiments for quantitatively evaluating the effects of the carbonation process on cement materials can be found. Since, in natural conditions, the carbonation process is very slow, most of the experiments are carried out in an accelerated regime. For fixed conditions of the CO_2 concentration, temperature, humidity and time, the samples are subjected to carbonation in a sealed chamber [21]. At the end of the test, the specimens are split, cleaned and sprayed with a phenolphthalein pH indicator, which is an organic compound that is colorless in an acid environment (carbonated part) and turns pink in a basic environment (non-carbonated part); see, for instance, [22,23]. SEM microscope observations and X-rays allow for the quantification of the effect of carbonation by identifying the presence of $CaCO_3$ formed and residual $Ca(OH)_2$ not yet reacted with CO_2. From the comparison of the sample before and after carbonation, the material, initially rich in calcium hydroxide, transforms into being mainly composed of $CaCO_3$, clearly indicating that the reaction with CO_2 has taken place [24,25]. Porosity represents a fundamental parameter for process control that allows us to evaluate the effect of carbonation on the structure of cement. To this aim, gammadensimetry and mercury intrusion analysis can be carried out.

Carbonation has attracted research interests also within the mathematical community and there is a huge amount of literature addressing the mechanism of carbonation with different mathematical models; see, for instance, [26–34].

In [35], we introduced a mathematical model describing porosity variation as the result of several intermediate chemical reactions triggered by the penetration of carbon dioxide that diffuses and is transported into the pores by water that is present.

The present paper describes the application of a mathematical model toward the simulation of CH degradation under climate change scenarios. Such scenarios are introduced in Section 2. The mathematical-based simulation algorithm is presented in Section 3. Section 4 is devoted to describe the fitting procedure of the mathematical-based simulation algorithm against laboratory data obtained in natural conditions of the carbon dioxide concentration. Then, in Section 5, we present different pollution scenarios produced by the simulation algorithm with the forecast of porosity variation occurring during the carbonation process. Moreover, a theoretical study on the effects of climate changes in terms of pollution levels on the carbonation constant K was carried out for the analyzed scenarios.

2. Review on Climate Change Scenarios

Given the scope of the paper, a brief discussion of some elements of the global climate crisis is necessary. Since a wide and deep overview on the matter is outside our purposes, we focused on the elements that are relevant for the present research: the carbon dioxide concentration in the atmosphere and global warming. For each of the above elements, a description of the literature sources of climate data that we used in our simulations is provided (see Section 5).

2.1. Atmospheric Pollution: CO_2 Emission Levels

The Intergovernmental Panel on Climate Change (IPCC)[1] is the UN body that advances knowledge of every aspect related to climate change, with a particular emphasis on human-induced climate change. IPCC "prepares comprehensive Assessment Reports about the state of scientific, technical and socio-economic knowledge on climate change, its impacts and future risks, and options for reducing the rate at which climate change is taking place".

One of the main forces of climate change is the emission of gases due to human activities (such as industry production, transportation, etc.). Among these gases, carbon dioxide has a central role and has become a reference measure of climate change. In its reports, IPCC takes into account several future scenarios of emissions based on assumptions about social, economic and technical development. These shared socio-economic pathways (SSPs) were developed in the year to describe, in a self-consistent logic, major trends of the economy, human lifestyle, technology, demography, policy, etc. There are five SSPs and each provides distinct routes that describe how societies might act in the future and how these actions will impact our environment [36].

SSP1 Sustainability—Taking the Green Road (Low challenges to mitigation and adaptation);
SSP2 Middle of the Road (Medium challenges to mitigation and adaptation);
SSP3 Regional Rivalry–A Rocky Road (High challenges to mitigation and adaptation);
SSP4 Inequality—A Road Divided (Low challenges to mitigation, high challenges to adaptation);
SSP5 Fossil-fueled Development—Taking the Highway (High challenges to mitigation, low challenges to adaptation).

Concerning atmospheric pollution, Figure 3 shows the evolution of the concentration of carbon dioxide in the atmosphere from 2015 to 2150 for several scenarios; see [37]. We can observe that, in the near future (up until 2050), the carbon dioxide concentration keeps increasing under all scenarios; the rate of increase differs for each scenario. However, after at least 2050, some scenarios foresee a slow decrease in concentration; in a few cases, the concentration of carbon dioxide in 2150 will be equal to the concentration in 2015.

In this work, we show some simulations of the carbonation of concrete using the values of carbon dioxide concentration following the scenarios developed in [37]. The actual values of carbon dioxide concentration that we used in our simulations are listed in Table 1.

Table 1. The values (percentage) of carbon dioxide concentration foreseen for 2075 for the indicated scenarios. Values taken from [37].

SSP1-1.9	**SSP4-6.0**	**SSP5-8.5**
0.041973	0.060695	0.080169

Remark 1. *The codes of the scenarios in Table 1 are identified by the following rule: SSPx-y.z, where x is the number of the related shared socio-economic pathway and y.z is the target radiative forcing expected by the year 2100. For example, the code SSP4-6.0 is related to SSP4 and foresees a value of radiative forcing of* $6.0 W/m^2$ *at the end of the century. Radiative forcing is a measure of the imbalance between the incoming solar radiation and the outgoing IR thermal emission due to the change in a variable (i.e., the increase/decrease in carbon dioxide concentration) while all other variables remain the same.*

2.2. Global Warming: Changes in Temperatures and Relative Humidity

Climate change also acts on the global temperature, since the amount of water vapor in the atmosphere is changing, which has potentially significant effects. Over the last few decades, global warming has been observed and studied and, correspondingly, climate models have been developed. Climate scientists foresee that the Earth has a high probability to continue to warm over this century and beyond, with a corresponding increase in temperatures [3,4]. Indeed, global warming is caused by the growth in carbon dioxide emissions associated to the presence of other harmful gases produced by human activities. According to a wide range of climate model simulations [38,39], the global average temperature could be between 1.1 and 5.4 °C warmer in 2100; see Figure 4. Such scenarios are the results of projections, but it is also possible that greenhouse gas concentrations may increase at higher rates with respect to those indicated in the graph. In fact, carbon dioxide emissions are increasing at a rate of more than 3% per year; if the rate would maintain the same level, in the future, the carbon dioxide concentration in the atmosphere would exceed the scenario depicted in red by the end of this century or even before.

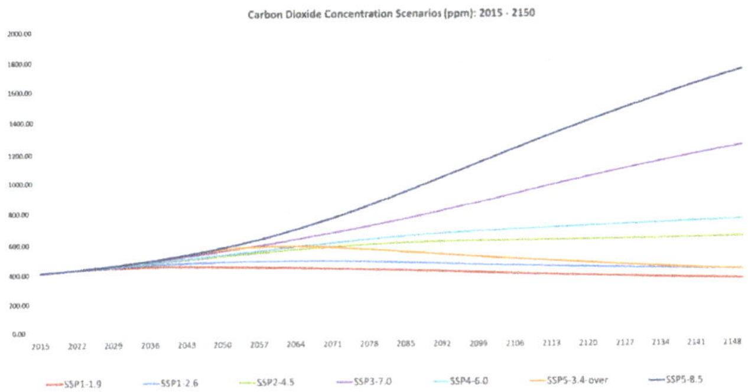

Figure 3. CO_2 concentration scenarios (data taken from [37]).

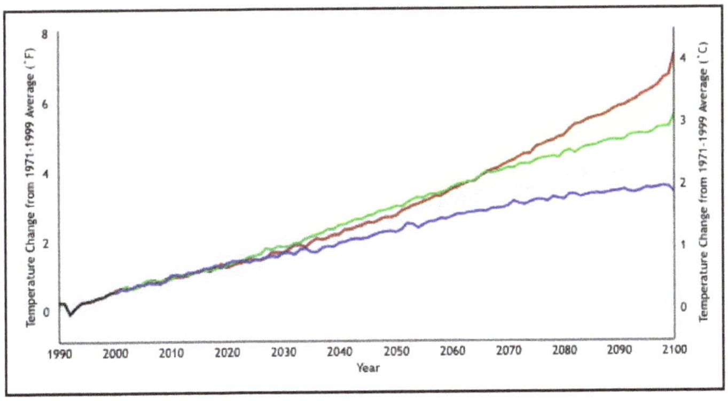

Figure 4. Future climate projections based on different human emission pathways. Credit [40]. The picture shows the average of a set of temperature simulations for the 20th century (black line), followed by projected temperatures for the 21st century based on a range of emissions scenarios (colored lines). The shaded areas around each line indicate the statistical spread (one standard deviation) provided by individual model runs.

Regarding the relative humidity, further investigations and studies are still needed to be able to predict its trend in land through the next decades. Indeed, the RH values may be related to many factors, such as changes in atmospheric circulations and land surface properties; see [3].

3. The Mathematical-Based Simulation Algorithm

Here, we refer to the mathematical model recently introduced in [35]. Such a model describes the movement of water (indicated by the letter w) within the porous stones;

carbon dioxide (a) dissolved in water; the carbonate ion (b) produced by the dissolution of carbon dioxide; the evolution of calcium hydroxide (i), which reacts with carbonate to produce calcium carbonate; the evolution of calcium carbonate (c), produced by the above reaction and later dissolving; finally, the evolution of calcium ion (e) produced by the dissolution of calcium carbonate. The model will describe the subsequent change in porosity, i.e., the fraction of the volume of voids over the total volume of the porous sample, (ε). The complete model reads as follows:

$$\begin{cases} \frac{\partial w}{\partial t} = \frac{\partial}{\partial z}\left(\left(\frac{\varepsilon - w_{min}}{\varepsilon_0 - w_{min}}\right)^{\frac{19}{6}} D(w,\varepsilon)\frac{\partial w}{\partial z}\right) + r_w \\ \frac{\partial (wa)}{\partial t} = \frac{\partial}{\partial z}\left(a\left(\frac{\varepsilon - w_{min}}{\varepsilon_0 - w_{min}}\right)^{\frac{19}{6}} D(w,\varepsilon)\frac{\partial w}{\partial z}\right) + \frac{\partial}{\partial z}\left(D_a w \frac{\partial a}{\partial z}\right) + r_a \\ \frac{\partial (wb)}{\partial t} = \frac{\partial}{\partial z}\left(b\left(\frac{\varepsilon - w_{min}}{\varepsilon_0 - w_{min}}\right)^{\frac{19}{6}} D(w,\varepsilon)\frac{\partial w}{\partial z}\right) + \frac{\partial}{\partial z}\left(D_b w \frac{\partial b}{\partial z}\right) + r_b \\ \frac{\partial i}{\partial t} = r_i \\ \frac{\partial e}{\partial t} = r_e \\ \frac{c}{m_c} = \frac{i_0}{m_i} + \frac{c_0}{m_c} + \frac{e_0}{m_e} - \frac{i}{m_i} - \frac{e}{m_e} \\ \varepsilon = \varepsilon_1 + (\varepsilon_0 - \varepsilon_1)\frac{i}{i_0} + \varepsilon_2\left(\frac{e}{e_0} - 1\right). \end{cases} \tag{1}$$

The terms r_k, $k = w, a, b, i, c, e$ represent the chemical reactions, defined according to the following equations:

$$\begin{aligned} r_w &= m_w(\omega_{ib} - \omega_a) \\ r_a &= -m_a\omega_a \\ r_b &= -m_b(\omega_{ib} - \omega_a - \omega_c) \\ r_i &= -m_i\omega_{ib} \\ r_c &= m_c(\omega_{ib} - \omega_c) \\ r_e &= m_e\omega_c \end{aligned}$$

where the terms m_k, $k = w, a, b, i, c, e$ indicate the molar masses of the respective substances, and the ωs describe the single chemical reactions and are defined as follows:

$$\begin{aligned} \omega_a &= \nu\frac{a}{m_a}, \text{ dissolution of } CO_2 \text{ in water} \\ \omega_{ib} &= \mu\left(\frac{i}{m_i}\cdot\frac{b}{m_b}\right), \text{ reaction between } Ca(OH)_2 \text{ and } CaCO_3 \\ \omega_c &= \delta\frac{c}{m_c}, \text{ dissolution of } CaCO_3 \end{aligned}$$

where ν, μ and δ are the reaction rates that represent the key parameters of our model, since their values take into account the different time scales of the chemical reactions involved in the carbonation process. In particular, ν represents the reaction coefficient between carbon dioxide and water, μ is the reaction coefficient between calcium hydroxide and carbonate ions and δ is the dissolution rate of calcium carbonate.

Remark 2. *Since porosity varies as a consequence of carbonation, a couple of words are in order to explain how we model it. Porosity changes following two processes:*

1. *Calcium hydroxide reacts with carbonate ions to form calcium carbonate; this process reduces porosity.*
2. *Calcium carbonate dissolves; this process increases porosity.*

Following the same arguments as in [35], we can represent porosity with the following expression:

$$\varepsilon = \varepsilon_1 + (\varepsilon_0 - \varepsilon_1)\frac{i}{i_0} + \varepsilon_2\left(\frac{e}{e_0} - 1\right), \tag{2}$$

where:

ε_0 *is the porosity of the non-carbonated concrete;*

ε_1 *is the porosity of (totally) carbonated concrete;*

ε_2 *represents how porosity varies when calcium carbonate dissolves.*

3.1. Initial and Boundary Conditions

The mathematical model is endowed with initial and boundary conditions for the unknown variables. Here, we consider a concrete specimen in the domain $[0, L]$, with $L > 0$. The left boundary $z = 0$ identifies the surface in contact with the environment, with which the exchange of humidity, carbon dioxide and carbonate occurs. For the unknowns, we impose initial conditions of the form

$$f(z, 0) \equiv f_0 \ \text{ for } f = w,\ a,\ b,\ c,\ i,\ e.$$

Notice that no boundary conditions are needed for the variables i, c and e since these are solutions of ODEs and their evolution in space only depends on the evolution of the other variables. For the unknowns w, a and b, we impose a zero flux condition at the right boundary $z = L$, i.e.,

$$\frac{\partial f}{\partial z} = 0 \ \text{ for } f = w,\ a,\ b.$$

At $z = 0$, we assign a Dirichlet condition for water, i.e.,

$$w(0, t) \equiv \bar{w}(t),$$

where the constant $\bar{w}(t)$ is the value of the (time-dependent) environmental moisture content computed as

$$\bar{w} = SVD(T(t))RH(t), \tag{3}$$

where SVD is the saturated vapor density in $[\text{g/cm}^3]$ computed as in [27]:

$$SVD(T(t)) = 10^{-6} \times (5.018 + 0.32321T(t) + 8.1847 \times 10^{-3}T(t)^2 + 3.1243 \times 10^{-4}T(t)^3), \tag{4}$$

where temperature T and relative humidity RH can be time-dependent if we consider real environmental settings. If instead, we refer to the experimental settings such as those reported in [22], we assume a relative humidity of $RH = 70\%$ and temperature $T = 20\ ^\circ\text{C}$.

Then, for a, we assume that the flux at $z = 0$ depends on the difference between the internal and external carbon dioxide concentration:

$$\frac{\partial a}{\partial z}(0, t) = -K_a(a(0, t) - \bar{a}(t)),$$

where K_a is an unknown constant describing the penetration rate of carbon dioxide estimated from the calibration against data and $\bar{a}(t)$ is the value of the external carbon dioxide concentration.

Remark 3. *Most often, in experimental works, the concentration of carbon dioxide is given as non-dimensional units as a percentage of the substance within the mixture; however, in our settings, we need the value $\bar{a}$ expressed in dimensional units. Given a concentration of carbon dioxide at $y\%$, $\bar{a}$ is computed in g/cm^3 as:*

$$\bar{a}(t) = \frac{\frac{m_a y}{RT_K(t)}}{10^3}, \tag{5}$$

where $T_K(t)$ is the temperature expressed in Kelvin and $R = 0.082\ L\ atm\ K^{-1}mol^{-1}$ is the gas constant.

Note that, if we consider laboratory conditions reported in [22], we have the $\bar{w}$ and $\bar{a}$ constant.

We assume the null-flux condition at the left boundary for carbonate:

$$\frac{\partial b}{\partial z}(0,t) = 0.$$

The model described above is able to describe the evolution of the carbonation process in the time interval $[0,T]$, with $T > 0$.

3.2. Numerical Algorithm

For the simulations, we assumed the model parameters listed in Table 2.

The interval $[0,L]$ is discretized with a step $\Delta z = \frac{L}{N+2}$, with $\lambda = \frac{\Delta t}{\Delta z}, z_j = j\Delta z$, $j = 0, ..., N+1$. We also set $w_j^n = w(z_j, t_n)$ as the approximation of the function w at the height z_j and at the time t_n.

We can assume that:

$$V_j^n = D_w \left(\frac{\varepsilon_j^n - w_{min}}{\varepsilon_0 - w_{min}} \right)^{\frac{19}{6}} \frac{w_{j+1}^n - w_{j-1}^n}{2\Delta z}, \quad \text{for } j = 1, \ldots, N,$$

with the boundary values set as follows:

$$\begin{aligned} V_0^n &= 0; \\ V_{N+1}^n &= 0. \end{aligned}$$

As shown in [35], if we define

$$\Delta_j(u,v) := \frac{(u_j + u_{j+1})(v_{j+1} - v_j) - (u_{j-1} + u_j)(v_j - v_{j-1})}{2\Delta z^2}.$$

the numerical algorithm is the following:

$$\begin{aligned}
w_j^{n+1} &= w_j^n + \Delta t D_w \Delta_j\left(\left(\tfrac{\varepsilon - w_{min}}{\varepsilon_0 - w_{min}} \right)^{\frac{19}{6}}, w^n \right) \\
&\quad + \Delta t m_w \left(\tfrac{\mu}{m_i m_b} \tfrac{i_j^{n+1} b_j^n + i_j^n b_j^{n+1}}{2} - \tfrac{\nu a_j^n}{m_a} \right), \ j = 1, \ldots, N, \\
a_j^{n+1} &= \tfrac{1}{w_j^n + \nu \Delta t} \Big\{ (wa)_j^n + \Delta t \tfrac{|V_{j+1}^n| a_{j+1}^n - 2|V_j^n| a_j^n + |V_{j-1}^n| a_{j-1}^n}{2\Delta z} \\
&\quad + \Delta t \tfrac{V_{j+1}^n a_{j+1}^n - V_{j-1}^n a_{j-1}^n}{2\Delta z} + \Delta t D_a \Delta_j(w^n, a^n) \Big\}, \quad j = 1, \ldots, N, \\
b_j^{n+1} &= \tfrac{1}{w_j^n + \Delta t \frac{\mu}{2 m_i} i_j^n} \Big\{ (wb)_j^n + \Delta t \tfrac{|V_{j+1}^n| b_{j+1}^n - 2|V_j^n| b_j^n + |V_{j-1}^n| b_{j-1}^n}{2\Delta z} \\
&\quad + \Delta t \tfrac{V_{j+1}^n b_{j+1}^n - V_{j-1}^n b_{j-1}^n}{2\Delta z} + \Delta t D_b \Delta_j(w^n, b^n) \\
&\quad + \Delta t \tfrac{\nu m_b}{m_a} a_j^n + \Delta t \tfrac{\delta m_b}{m_c} c_j^n - \Delta t \tfrac{\mu}{2 m_i} i_j^{n+1} b_j^n \Big\}, \\
&\quad j = 1, \ldots, N, \\
i_j^{n+1} &= \tfrac{1}{1 + \frac{\Delta t}{2} \frac{\mu b_j^n}{m_b}} \left\{ i_j^n - \tfrac{\Delta t}{2} \tfrac{\mu}{m_b} i_j^n b_j^{n+1} \right\}, \\
e_j^{n+1} &= e_j^n + \Delta t \delta \tfrac{m_e}{m_c} c_j^n, \\
c_j^{n+1} &= c_0 + m_c \left(\tfrac{1}{m_i}(i_0 - i_j^n) + \tfrac{1}{m_e}(e_0 - e_j^n) \right), \qquad (6) \\
\varepsilon_j^{n+1} &= \varepsilon_1 + (\varepsilon_0 - \varepsilon_1) \tfrac{i_j^n}{i_0} + \varepsilon_2 \left(\tfrac{e_j^n}{e_0} - 1 \right), \qquad (7)
\end{aligned}$$

with a suitable discretization of the boundary conditions for w, a and b described above. Note that the scheme above is convergent under the CFL condition $\Delta t \leq \Delta^2 z / D_a$. More details are reported in [35].

4. Model Validation and Calibration

In the present Section, we show the numerical validation of the mathematical model described above. A fine tuning of key model parameters, i.e., the reaction coefficients ν, μ and δ, was carried out in a time window of one year in order to represent realistic situations. A common approach usually adopted in order to study the carbonation of concrete is to consider specimens exposed to carbon dioxide concentrations of up to 20%, the so-called accelerated conditions. On the contrary, our aim here was to validate and calibrate the mathematical algorithm over one year for a natural concentration of carbon dioxide in the air, i.e., 0.03%.

In order to validate and calibrate the mathematical model, we referred to data from a carbonation experiment described in [22] performed on a type Portland cement specimen. As described in the work by Pan et al., concrete specimens characterized by a w/c ratio of 0.53 are first polymerized at a temperature of 20 °C and RH of 70% for 24 h; then, they are placed in a seasoning environment at 20 ± 3 °C and 90% RH for 28 days and, successively, samples are placed in a drying oven at 50°C for 48 h. The carbonation test is executed at T = 20 °C and 70% humidity for CO_2 concentrations 0.03%, 3% and 20%. In what follows, we considered only the data corresponding to the carbon dioxide concentration of 0.03%. For more details on the experimental setting, the reader may refer to the original paper [22].

In the following numerical tests, we assumed the values of the parameters, taken from the literature or calibrated against data, reported in Table 2.

Since the kinetic coefficients refer to an aqueous solution under ideal conditions, such coefficients were estimated by the simulation algorithm against data, together with the diffusivity coefficient D_a. Indeed, in the presence of a significantly lower concentration of CO_2, we expect that the diffusivity within the material is higher.

In Figures 5, we report the plots of the profiles of the quantities obtained by the mathematical-based algorithm described in Section 3 assuming parameters reported in Table 2. In particular, the left picture shows the system behavior at $t = 0$, when the diffusion of gaseous CO_2 has already taken place in the cement matrix and the carbonic acid formation reaction has just begun, whereas, on the right, the situation at t =365 days is depicted.

Looking at the curve profiles of the main quantities involved in the carbonation process and depicted in Figure 5, we obtained a qualitative validation of the model, since it perfectly describes the real phenomenon according to the following aspects:

- Carbon dioxide a (magenta line) enters within the pores and is rapidly consumed;
- Water w (blue line) is consumed by the reaction, as expected;
- The carbonate ion b described by the yellow curve is the sum of two reactions, i.e., the dissolution of carbon dioxide in water and the reaction with calcium hydroxide to form $CaCO_3$;
- Calcium hydroxide i shows an "S" shape (green line): near the face in contact with carbon dioxide, calcium hydroxide dissolves by the chemical reaction with carbonate ion; far from there, it is still close to the initial datum since the calcium ion has not yet penetrated sufficiently within the stone;
- On the other hand, the calcium carbonate c (red line) reaches its maximum at the left end of the specimen (in contact with CO_2) and decreases towards the other end, where the concentration of carbonate ions is very low; moreover, the dissolution of calcium carbonate has already started;
- The calcium ion e (black line), due to the rapid dissolution of calcium hydroxide, only participates in the dissociation reaction of calcium carbonate and it is consequently consumed.

Table 2. Parameters of the model (1).

	Description	Units	Value	Ref.
h	Specimen's height	cm	2	datum
Δz	Space step	cm	0.1	-
Δt	Time step	s	4	-
A	Shape coefficient of water diffusivity	cm^2/s	3.24×10^{-8}	[28]
B	Shape coefficient of water diffusivity	-	100	[28]
ε_0	Porosity of the unperturbed material	-	0.2	[41]
ε_1	Porosity after complete consumption of $Ca(OH)_2$	-	0.08	hypothesis/ [42]
ε_2	Porosity change due to $CaCO_3$ dissolution	-	0.025	hypothesis/ [42]
D_a	Diffusivity of CO_2 in water at 20 °C	cm^2/s	4.8×10^{-3}	calibrated against data
D_b	Diffusivity of CO_3^{2-} in water at 20 °C	cm^2/s	0.81×10^{-5}	[43]
ρ_w	Density of water	g/cm^3	1	[44]
ν	Coefficient of reaction between CO_2 and water at 20 °C and 1 atm	s^{-1}	1	calibrated against data
μ	Coefficient of reaction between $Ca(OH)_2$ and CO_3^{2-} at 20 °C and 1 atm	$cm^3/(mol\ s)$	7×10^{-2}	calibrated against data
δ	Dissolution rate of $CaCO_3$ at 20 °C and 1 atm	s^{-1}	6×10^{-9}	calibrated against data
m_a	Molecular mass of CO_2	g/mol	44.01	[45]
m_b	Molecular mass of CO_3^{2-}	g/mol	60.01	[45]
m_c	Molecular mass of $CaCO_3$	g/mol	100.09	[45]
m_e	Molecular mass of Ca^{2+}	g/mol	40.08	[45]
m_i	Molecular mass of $Ca(OH)_2$	g/mol	74.10	[45]
m_w	Molecular mass of water	g/mol	18.01	[45]
w_0	Initial water content	g/cm^3	0.622	[22]
a_0	Initial concentration of CO_2	-	0.03%	[22]
b_0	Initial concentration of CO_3^{2-}	g/cm^3	1.3×10^{-11}	hypothesis
i_0	Initial concentration of $Ca(OH)_2$	g/cm^3	5.2×10^{-2}	[22]
e_0	Initial concentration of Ca^{2+}	g/cm^3	8.9×10^{-3}	hypothesis
c_0	Initial concentration of $CaCO_3$	g/cm^3	9.8×10^{-4}	[22]
$\bar{w}$	Moisture content of the ambient air for $UR = 70\%, T = 20°C$	g/cm^3	0.25	datum
$\bar{a}$	Concentration of CO_2 at the boundary	g/cm^3	derived with formula (5) for {0.03%, 0.04%, 0.06%, 0.08%}	scenarios
K_a	Penetration rate of CO_2 in the medium	cm^{-1}	10^4	calibrated against data

Finally, it is worth noting that, due to the carbonation process, the porosity profile ε (cyan line) depicted in the bottom picture of Figure 5 shows an increasing behavior.

Now, we present the numerical procedure for a fine tuning of model parameters against laboratory data available from the literature. In particular, we used data in ([22], Table 5).

reporting the concentration of calcium hydroxide and calcium carbonate at the natural exposure of carbon dioxide in the air corresponding to 0.03% for one year. As can be noticed from experimental data, the content of calcium hydroxide always increases with depth, and calcium carbonate shows an opposite trend. Then, a qualitative validation of model outcomes was obtained looking at the relationship between the carbonation depth and the content of calcium hydroxide and calcium carbonate.

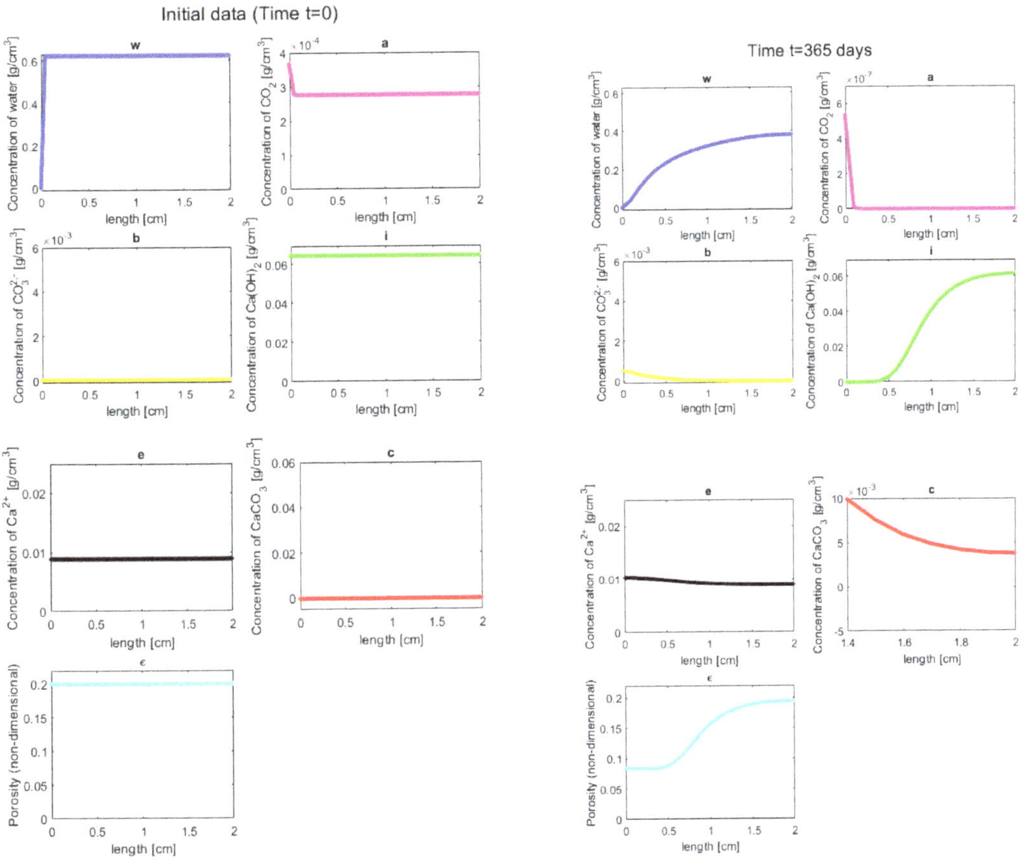

Figure 5. Profiles of numerical solutions to the system (1) at time $t = 0$ (on the left) and after $t = 365$ days (on the right) obtained with model parameters in Table 2 and with concentration of carbon dioxide of 0.03%. We depict the concentration of water, carbon dioxide, carbonate, calcium hydroxide, calcium carbonate and calcium ion (in g/cm^3) and the non-dimensional porosity profile (cyan line). The figures depict the profile on the space dimension (length of the specimen in cm).

In Figure 6, plots of the profiles derived from experimental data (line-circles) taken from [22] for a carbon dioxide concentration of 0.03% and the related numerical results obtained by the model (line-points) using parameters in Table 2 at time $t = 365$ days are shown. As can be observed from the left picture in Figure 6, with a fine tuning of model parameters describing reaction rates, not only the qualitative behavior but also the quantitative values of the calcium hydroxide profile are very close to the experimental ones after one year, and the same occurs for the calcium carbonate depicted in the right picture of Figure 6.

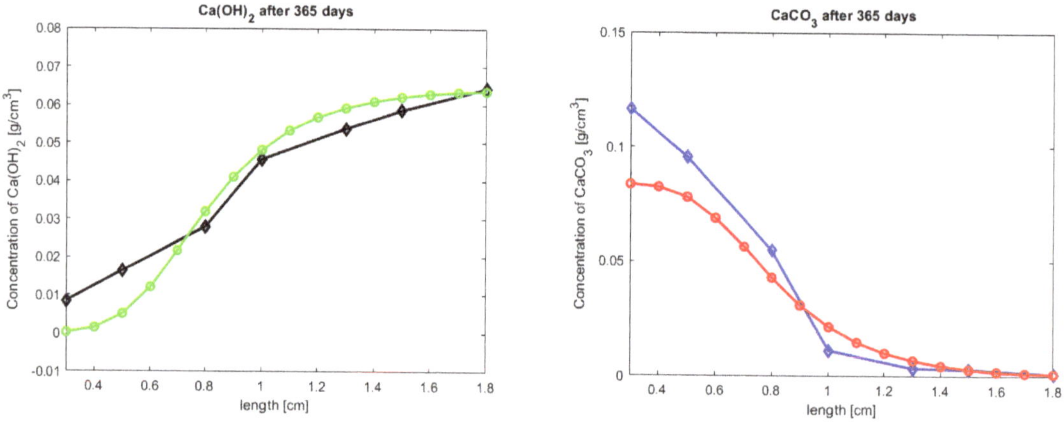

Figure 6. Plots of the profiles derived from experimental data (line-circles) taken from [22] and numerical results obtained by the model (line-points) at time t = 365 days. On the left: profile of measured VS-computed $Ca(OH)_2$ and on the right: profile of measured VS-computed $CaCO_3$. The y-axis depicts the concentration of the substances (in g/cm^3), whereas the x-axis shows the length of the specimen (in cm).

In conclusion, a qualitative and quantitative validation of the model is obtained. Indeed, the main feature of our model is its capability to reproduce the mechanism of the creation and consumption of $CaCO_3$ on one hand and, on the other hand, the ability to predict the time evolution of the system, even including the long-time behavior.

The forecasting algorithm was implemented in Matlab © and the computational time taken for a simulation on the complete model with fixed parameters until time $t = 365$ days was approximately 2000 seconds on an Intel(R) Core(TM) i7-3630 QM CPU 2.4 GHz.

5. Mathematical-Based Forecasting Algorithm: Damage Scenarios

5.1. Scenario 1. Laboratory Setting VS. Real Environmental Conditions

In Figure 7, the profiles of the temperature (T) and relative humidity (RH) detected in the laboratory setting (constant values reported in paper [22]) and variable values detected in the environment over a year (2017) by sensors of Arpa Lazio [46] and linearly interpolated on the computational grid are reported. The corresponding boundary conditions for water and carbonic acid obtained by using, respectively, Formulas (3) and (5) are reported in Figure 8.

In Figure 9, the porosity profile in the first point of the material across the time window of $[0, 365]$ days is depicted. In particular, we depict the porosity at point x_0 computed by the mathematical model (1) at time $T = 356$ days for a concentration of carbon dioxide of 0.03% in laboratory (red dotted line) vs. real (blue line) environmental conditions. As can be observed, in both cases, the porosity initially decreases but, after approximately 150 days, it changes its behavior and starts increasing. Moreover, as can be noticed from the superposition between the blue and red curves, the behavior of the porosity obtained by the model (1) using inflow boundary conditions coming from laboratory conditions is reproduces the curves obtained using boundary conditions coming from real environmental conditions quite well. This is probably due to the fact that the inflow of carbon dioxide $\bar{a}$ in the case of constant laboratory conditions is more or less an averaged value of the real values, as can be seen in the right panel of Figure 8.

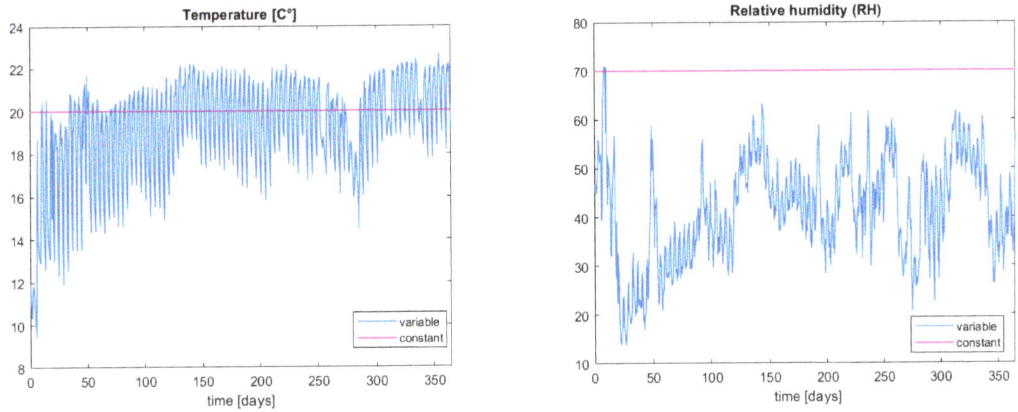

Figure 7. Temperature and relative humidity in fixed (constant) laboratory conditions vs. interpolated real conditions detected over a year in 2017 in Rome (Arpa Lazio [46]).

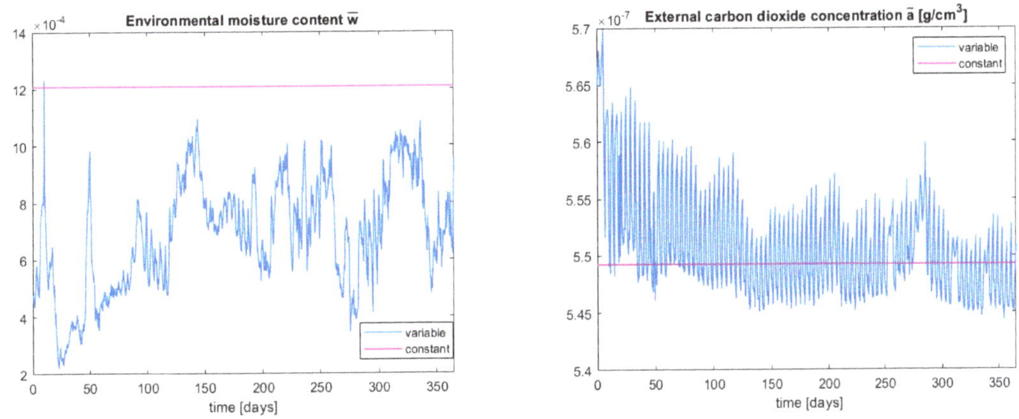

Figure 8. Fixed in laboratory vs. numerical boundary conditions computed with real data for $\bar{w}$ and $\bar{a}$ using, respectively, Formulas (3) and (5).

If we run the simulation algorithm until a time of 5 years, we can see that the porosity still increases, as can be observed in Figure 10, thus possibly determining a weakening of the structure of the building material.

5.2. Scenario 2. Effects of CO_2 Pollution Growth on Porosity

Here, using the forecasting tool calibrated against laboratory data as described in Section 4, we consider the carbon dioxide concentration scenarios depicted in Figure 3. In particular, we focused on the concentrations of carbon dioxide reported in Table 1 and we simulated the possible scenarios occurring for the porosity profile in these situations; see Figure 11.

As can be seen, in the case of a higher concentration represented by the worst case scenario SSP5-8.5 (approximately 0.08%), the porosity assumes its minimum after approximately 1 month, i.e., approximately three months earlier with respect to the present case, with a 0.03% carbon dioxide concentration. For the intermediate scenarios (SSP4-6.0 and SSP1-1.9), we can observe that the minimum is reached, respectively, after 50 and 80 days.

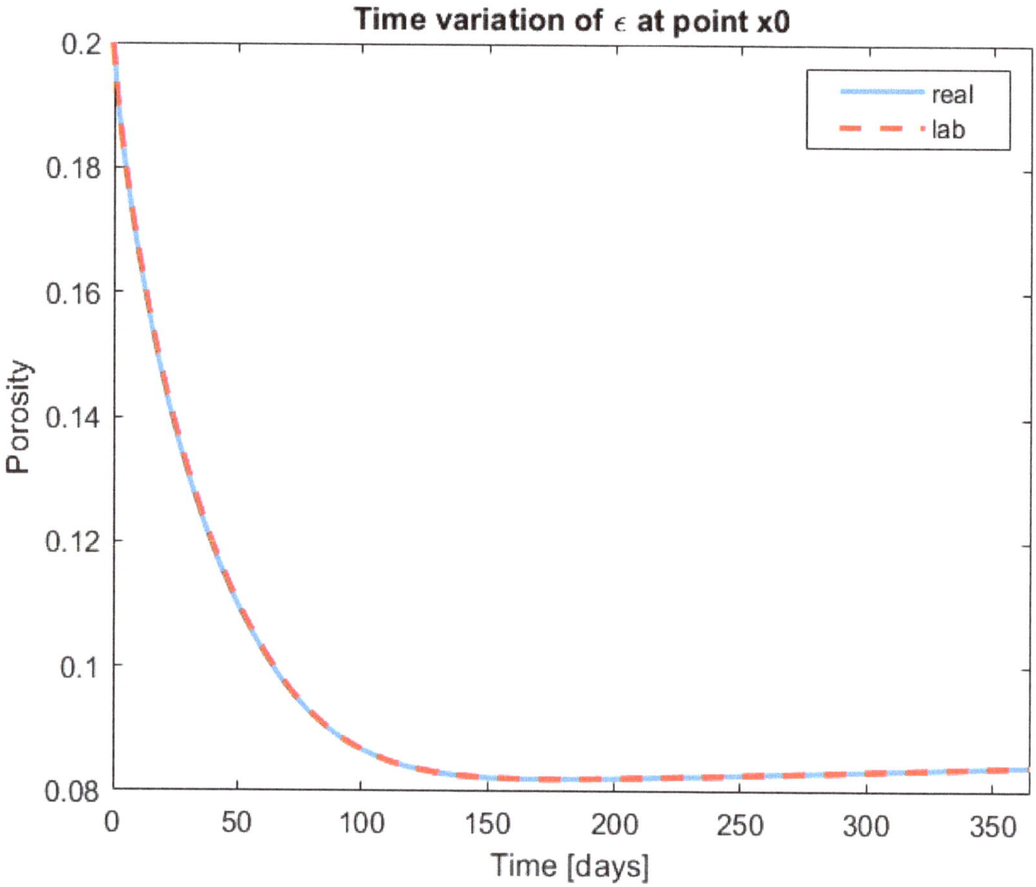

Figure 9. Profile of the porosity at point x_0 computed by the mathematical model (1) at time $T = 356$ days for a concentration of carbon dioxide of 0.03% in laboratory vs. real environmental conditions.

For longer times, when the porosity starts increasing, all of the curve profiles have the same qualitative and quantitative behavior. This phenomenon should depend on the fact that the porosity growth does not depend on CO_2 dissolved in water but on the dissolution of calcium carbonate.

A Theoretical Estimate on the Effect of Changes in CO_2 Levels on the Carbonation Front

When studying carbonation processes, one often evaluates the advancement in the carbonation front (i.e., the boundary separating the carbonated from the non-carbonated zone) in the interior of concrete samples. From an experimental point of view, the carbonated zone can be identified through several techniques, as explained in the introduction. Here, we want to briefly discuss how the carbon dioxide concentration might affect the advancement in the carbonation front.

If we indicate with $\sigma = \sigma(t)$ the position of the carbonation front, we have that our specimen is divided in two regions:

Carbonated region $\mathcal{C}(t)$ defined by $x \in [0, \sigma(t)]$;

Uncarbonated region $\mathcal{U}(t)$, $x \in [\sigma(t), L]$.

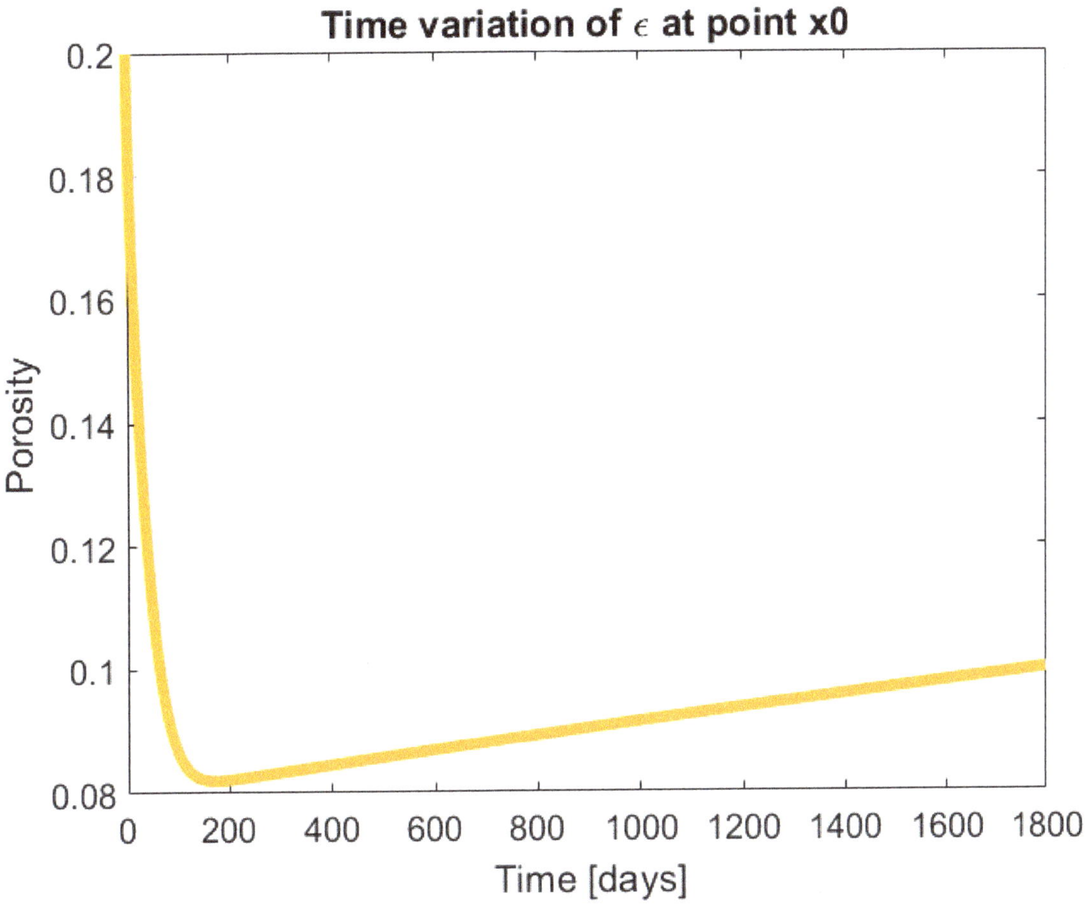

Figure 10. Profile of the porosity at point x_0 computed by the mathematical model (1) at time $T = 1825$ days (5 years) obtained for a concentration of carbon dioxide of 0.03% in laboratory conditions.

The front moves into the specimen following a square root in time law [47], i.e.,

$$\sigma(t) = K\sqrt{t}$$

where K $(\mathrm{mm} \cdot \mathrm{s}^{-1/2})$ is a constant that depends on the material properties, as well as the concentration of carbon dioxide.

In our previous work [27], we determined the value of K as:

$$K = \sqrt{2\varepsilon_1 \frac{D_a}{1-\omega} \frac{\bar{a}}{\mu_h m_a}} \tag{8}$$

where μ_h is the molar density of calcium hydroxide, i.e.,

$$\mu_h = 0.02984 \ \mathrm{mol} \cdot \mathrm{cm}^{-3}$$

and $\omega = 0.0723$ is a dimensionless parameter taking into account the shrinkage of concrete as an effect of carbonation.

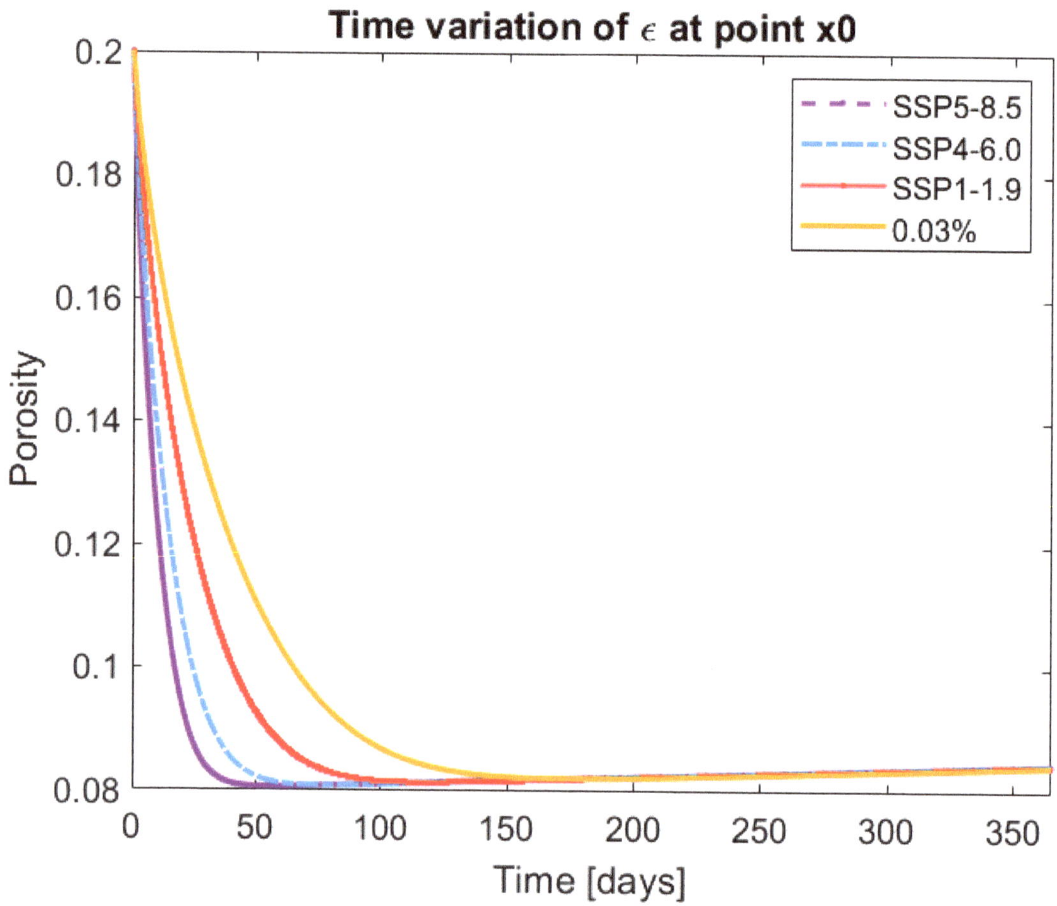

Figure 11. Profile of the porosity at point x_0 computed by the mathematical model (1) at time $T = 356$ days for a concentration of carbon dioxide of 0.03% (orange line), 0.04% (red dotted line), 0.06% (blue dotted line) and 0.08% (purple dotted line).

If we insert the values of carbon dioxide for the base case 0.03% in (8), we obtain, after one year, that the front reached the position:

$$\sigma_{0.03\%}(1 \text{ year}) = 0.3282 \text{ cm}.$$

Such a value is smaller than that reported in [22] (i.e., 7.8 mm). However, they reported only one value for the experiments in natural conditions; on the other hand, the value that we obtained here is of the same order of magnitude. Thus, we might assume $\sigma_{0.03\%}(1 \text{ year})$ as a base case to compare the carbonation effect under different pollution scenarios. In other words, here, we evaluated how much the front modifies its position in the three scenarios SSP1-1.9, SSP4-6.0 and SSP5-8.5. Table 3 lists the values of $\sigma(t)$ after one year using the concentration of carbon dioxide from Table 1.

Table 3. The position of the carbonation front (in cm) after one year under the considered climate change scenarios.

	SSP1-1.9	SSP4-6.0	SSP5-8.5
σ (1 year)	0.3882	0.4668	0.5365

As one can imagine, since the concentration of carbon dioxide is higher in 2075 than in the present days (according to all three scenarios), the carbonation front penetrates more deeply in concrete stone. It is interesting to see the extent of the penetration with respect to the base case. If we calculate the percentage of increase, i.e.,

$$d\sigma \ (1 \text{ year}) = \frac{\sigma \ (1 \text{ year}) - \sigma_{0.03\%} \ (1 \text{ year})}{\sigma_{0.03\%} \ (1 \text{ year})} \cdot 100$$

we can find the values in Table 4.

Table 4. Percentage of change in front position due to the increased CO_2 concentration with respect to the base case.

	SSP1-1.9	SSP4-6.0	SSP5-8.5
$d\sigma$ (1 year)	18.2836	42.2381	63.4717

For example, in the scenario SSP1-1.9, which foresees a concentration of carbon dioxide that is approximately 40% higher than the base case, the carbonation front increases its depth by "only" 18%.

5.3. Scenario 3. Global Warming Effects on Porosity

The present paragraph is devoted to the numerical study on the influence of the temperature change on the carbonation progress. If we suppose that the RH stays more or less constant and consider a scenario taking into account a temperature increase of 5 °C as predicted in [40], we do not expect to have a significant impact on porosity only depending on this factor. Indeed, as can also be seen in [32], where the authors take into account the effect of the temperature increase in the formulation of the diffusion coefficient, the results show minimal effects (of the order of 1%).

In the next Figure 12, where we depict the porosity profile after 50 days for CO_2 levels corresponding to the SSP5-8.5 pollution scenario, we can observe that the profiles obtained at a temperature of $T = 20$ °C (cyan line) and at $T = 25$ °C (blue dotted line) are nearly the same.

A Theoretical Study of the Effect of Temperature on Carbonation Front

The mathematical model that we used in this paper does not explicitly describe the action of temperature in carbonation processes. However, we can give some estimate of how temperature affects carbonation in a changing environment. First of all, as we can find in the literature (see for example [32]), we might reformulate the diffusion coefficient through the well-known Arrhenius equation to include temperature, i.e.,

$$D_a(T) = D_a \exp\left\{ \frac{E}{R} \left(\frac{1}{T_0} - \frac{1}{T} \right) \right\}, \tag{9}$$

where E is the activation energy (40 kJ/mol), R is the gas constant and T_0 is a reference temperature (expressed in Kelvin) that we chose here to equal 20 °C . Using the above formula for $D_a(T)$ and assuming in our settings that the temperature remains constant, the coefficient K in the equation is modified as follows:

$$K(T) = \sqrt{2\varepsilon_1 \frac{D_a(T)}{1-\omega} \frac{\bar{a}}{\mu_h m_a}}. \tag{10}$$

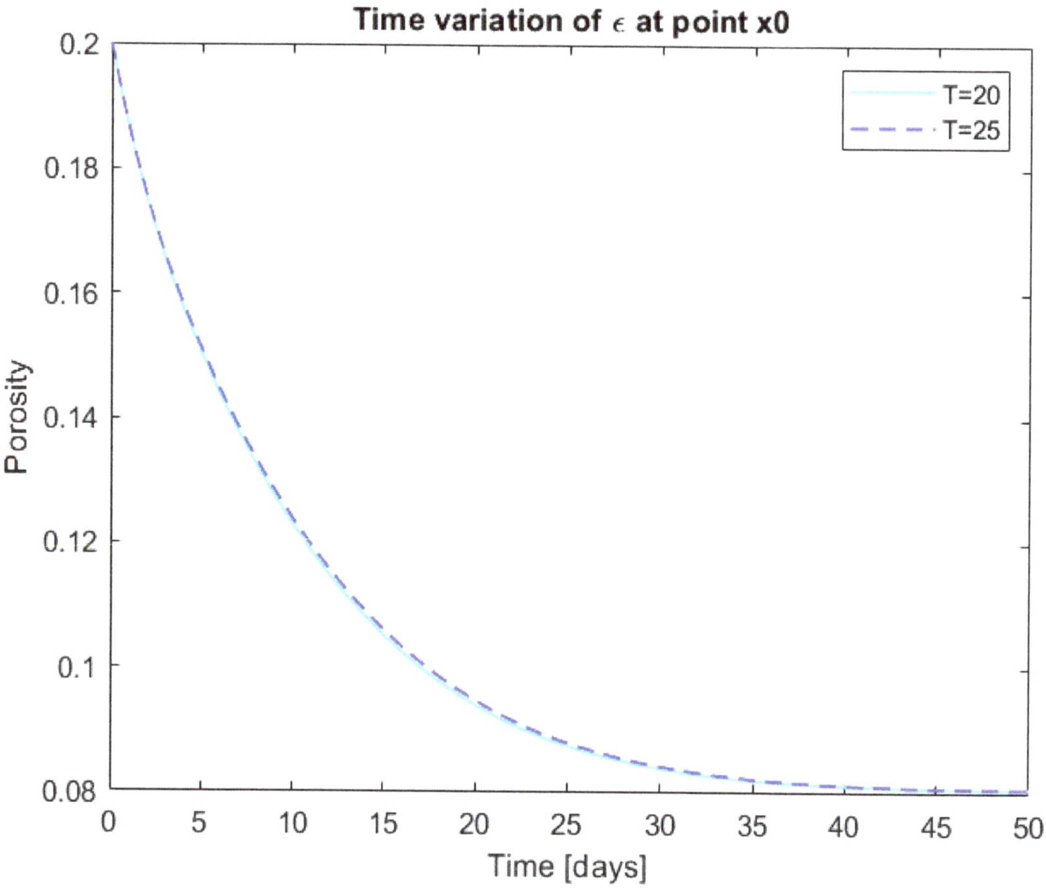

Figure 12. Profile of the porosity at point x_0 computed by the mathematical model (1) at time $T = 50$ days for a concentration of carbon dioxide of 0.08% at a temperature of $T = 20$ °C (cyan line) and at $T = 25$ °C (blue dotted line).

If we calculate the front σ using this new coefficient for a temperature of 25 °C, we find the values in Table 5:

Table 5. The position of the carbonation front (in cm) after one year under the considered climate change scenarios for a temperature of 25 °C.

	SSP1-1.9	SSP4-6.0	SSP5-8.5
σ (1 year)	0.4417	0.5311	0.6104

We can see that, after one year, increasing the temperature by 5 °C induces an advancement in the carbonation front by a few tenths of a millimeter.

Thus, temperature does not seems to play a substantial role in carbonation with the considered carbon dioxide concentration. This might seems to contradict other research results, such as in [32]. However, this contradiction is more apparent than real since the authors there considered accelerated conditions (a 20% concentration of CO_2). Indeed, if we repeat the same calculations above for a concentration of 20% of CO_2, we obtain the values in Table 6. We can see that, although our model overestimates the position of the front with respect to the values in [32] [Table 13], we can find that the difference is between the two values (approximately 0.34 cm) and is consistent with the experimental findings.

Table 6. The position of the carbonation front (in cm) after 7 days for a carbon dioxide concentration of 20%.

	20 °C	30 °C
$\sigma(t)$	1.1734	1.5126

Thus, we can conclude that temperature does not play a fundamental role in the carbonation processes for "small" concentrations of carbon dioxide.

6. Conclusions

The present paper focused on the forecasting capability of the mathematical model of carbonation - and the related simulation algorithm presented in Section 3. The principal aim of our work was to obtain a reliable simulation algorithm to be used as a numerical forecasting tool for predicting the effects of climate changes (i.e., CO_2 emission levels, global warming) on the conservation of Portland cement in terms of the porosity variation and penetration depth of the carbonation front.

In this framework, a preliminary qualitative validation of the model followed by a quantitative calibration of its key parameters, i.e., the reaction rates of the chemical reactions involved in the carbonation process, was successfully carried out in Section 4. Indeed, the pictorial representation of the curve profiles of the quantities mainly involved in the carbonation process allows for a first qualitative validation of the model. Then, a quantitative calibration of the model in the case of exposure to a natural carbon dioxide concentration (0.03%) over a year was carried out by comparing laboratory data available from the literature and model outcomes, for both calcium hydroxide and calcium carbonate concentrations, thus resulting in a fine tuning of model parameters. Moreover, a comparison between real vs. controlled laboratory settings was proposed, suggesting no effects in terms of porosity variation. Once calibrated, the simulation algorithm was used as a forecasting tool for predicting damage scenarios in order to evaluate the effects of the possible future increase in the carbon dioxide concentration, as well as in the temperature, as predicted in climate change scenarios reported in Section 2. Finally, since our model does not explicitly describe the advancement in the carbonation front, some theoretical estimates on the effects of climate changes on the penetration of carbon dioxide within cement is proposed in Section 5. In more detail, a study on the effects of climate changes in terms of pollution levels and temperature rise on the carbonation constant K was carried out for the analyzed scenarios, and a substantial agreement with both the experimental and numerical outcomes was obtained.

Then, we can summarize the procedure by the steps reported below:

- A qualitative validation and fine tuning of model parameters against laboratory data;
- A numerical simulation of different damage scenarios quantifying the modification of the front position;
- A theoretical verification of experimental and numerical findings for the analyzed scenarios.

This represents, to our knowledge, the first study on the effects of climate changes on the porosity of Portland cement.

Our future work will follow two interconnected lines of research.

First, we will further enhance the forecasting tool by applying it to different kinds of concrete and with a combination of multiple damaging factors. To this aim, not only data available from literature but ad hoc laboratory experiments and sensor measurements are needed to acquire data (chemical/physical properties such as porosity analyses, gas penetration rates and the presence of other harmful substances). The second line of research will focus on the impact of climate change on CH sites. Indeed, climate change implies modifications in several environmental aspects, such as temperature, precipitation, moisture content, wind intensity, sea level rise and the occurrence of extreme events. All of these changes will affect the degradation mechanisms of monuments or other artifacts in several ways. Thus, we will use our forecasting tool to study how the variation in the above environmental processes impacts cultural assets. This will be interesting from a scientific point of view, as well as for heritage management practice. The final goal of our research in the future is indeed to build a digital-twin prototype of the monitored object in order to develop a mathematical tool for predictive maintenance. Using mathematical algorithms to predict damages from chemical actions and from other mechanisms for building heritage can bring economic savings and allow for the optimal planning of conservative intervention strategies.

Author Contributions: Conceptualization, G.B. and M.C.; Methodology, G.B. and M.C.; Software, G.B.; Validation, G.B. and M.C.; Investigation, G.B. and M.C.; Data curation, G.B. and M.C.; Writing—original draft, G.B. and M.C.; Writing—review & editing, G.B. and M.C.; Visualization, G.B. All authors have read and agreed to the published version of the manuscript.

Funding: This research received no external funding.

Institutional Review Board Statement: Not applicable.

Informed Consent Statement: Not applicable.

Data Availability Statement: All data are contained within the article.

Conflicts of Interest: The authors declare no conflict of interest.

Notes

[1] https://www.ipcc.ch/ (accessed on 1 October 2022).

References

1. Fatorić, S.; Seekamp, E. Are cultural heritage and resources threatened by climate change? A systematic literature review. *Clim. Chang.* **2017**, *142*, 227–254. [CrossRef]
2. Orr, S.A.; Richards, J.; Fatorić, S. Climate change and cultural heritage: A systematic literature review (2016–2020). *Hist. Environ. Policy Pract.* **2021**, *12*, 434–477. [CrossRef]
3. Byrne, M.P.; O'Gorman, P.A. Trends in continental temperature and humidity directly linked to ocean warming. *Proc. Natl. Acad. Sci. USA* **2018**, *115*, 4863–4868. [CrossRef]
4. Douville, H.; Qasmi, S.; Ribes, A.; Bock, O. Global warming at near-constant tropospheric relative humidity is supported by observations. *Commun. Earth Environ.* **2022**, *3*, 1–7. [CrossRef]
5. Pasqui, M.; Di Giuseppe, E. Climate change, future warming, and adaptation in Europe. *Anim. Front.* **2019**, *9*, 6–11. [CrossRef]
6. Ballard, C.; Baron, N.; Bourgès, A.; Bucher, B.; Cassar, M.; Daire, M.Y.; Daly, C.; Egusquiza, A.; Fatoric, S.; Holtorf, C.; et al. Cultural Heritage and Climate Change: New Challenges and Perspectives for Research. Available online: https://digital.csic.es/bitstream/10261/279867/1/White-Paper-March-2022.pdf (accessed on 8 November 2022) .
7. Cassar, J.; Galdies, C.; Muscat Azzopardi, E. A New Approach to Studying Traditional Roof Behaviour in a Changing Climate—A Case Study from the Mediterranean Island of Malta. *Heritage* **2021**, *4*, 3543–3571. [CrossRef]
8. Doehne, E. Salt weathering: a selective review. *Geol. Soc. London, Spec. Publ.* **2002**, *205*, 51–64. [CrossRef]
9. Scherer, G.W.; Valenza, J. Mechanisms of frost damage. *Mater. Sci. Concr.* **2005**, *7*, 209–246.
10. Franzoni, E.; Sassoni, E. Correlation between microstructural characteristics and weight loss of natural stones exposed to simulated acid rain. *Sci. Total Environ.* **2011**, *412*, 278–285. [CrossRef]
11. Sesana, E.; Gagnon, A.S.; Ciantelli, C.; Cassar, J.; Hughes, J.J. Climate change impacts on cultural heritage: A literature review. *Wiley Interdiscip. Rev. Clim. Chang.* **2021**, *12*, e710. [CrossRef]
12. Sardella, A.; Palazzi, E.; von Hardenberg, J.; Del Grande, C.; De Nuntiis, P.; Sabbioni, C.; Bonazza, A. Risk mapping for the sustainable protection of cultural heritage in extreme changing environments. *Atmosphere* **2020**, *11*, 700. [CrossRef]
13. Bonazza, A.; Sardella, A.; Kaiser, A.; Cacciotti, R.; De Nuntiis, P.; Hanus, C.; Maxwell, I.; Drdácký, T.; Drdácký, M. Safeguarding cultural heritage from climate change related hydrometeorological hazards in Central Europe. *Int. J. Disaster Risk Reduct.* **2021**, *63*, 102455. [CrossRef]
14. Rodrigues, F.; Cotella, V.; Rodrigues, H.; Rocha, E.; Freitas, F.; Matos, R. Application of Deep Learning Approach for the Classification of Buildings' Degradation State in a BIM Methodology. *Appl. Sci.* **2022**, *12*, 7403. [CrossRef]

15. Bracciale, M.; Bretti, G.; Broggi, A.; Ceseri, M.; Marrocchi, A.; Natalini, R.; Russo, C. Mathematical modelling of experimental data for crystallization inhibitors. *Appl. Math. Model.* **2017**, *48*, 21–38. [CrossRef]
16. Giavarini, C.; Santarelli, M.; Natalini, R.; Freddi, F. A non-linear model of sulphation of porous stones: Numerical simulations and preliminary laboratory assessments. *J. Cult. Herit.* **2008**, *9*, 14–22. [CrossRef]
17. Bretti, G.; Filippo, B.D.; Natalini, R.; Goidanich, S.; Roveri, M.; Toniolo, L. Modelling the effects of protective treatments in porous materials. In *Mathematical Modeling in Cultural Heritage*; Springer: Berlin/Heidelberg, Germany, 2021; pp. 73–83.
18. Saba, M.; Quiñones-Bolaños, E.; López, A.L.B. A review of the mathematical models used for simulation of calcareous stone deterioration in historical buildings. *Atmos. Environ.* **2018**, *180*, 156–166. [CrossRef]
19. Meyer, C. Concrete as a green building material. In Proceedings of the Construction Materials Mindess Symposium, Vancouver, BC, Canada, 22–24 August 2005; Volume 10.
20. Pedeferri, P. *Corrosione e Protezione Dei Materiali Metallici. Vol. 1 e 2*; Polipress: Milan, Italy , 2010.
21. Chang, C.F.; Chen, J.W. The experimental investigation of concrete carbonation depth. *Cem. Concr. Res.* **2006**, *36*, 1760–1767. [CrossRef]
22. Pan, G.; Shen, Q.; Li, J. Microstructure of cement paste at different carbon dioxide concentrations. *Mag. Concr. Res.* **2018**, *70*, 154–162. [CrossRef]
23. Villain, G.; Thiery, M. Gammadensimetry: A method to determine drying and carbonation profiles in concrete. *Ndt E Int.* **2006**, *39*, 328–337. [CrossRef]
24. Thiery, M.; Villain, G.; Dangla, P.; Platret, G. Investigation of the carbonation front shape on cementitious materials: Effects of the chemical kinetics. *Cem. Concr. Res.* **2007**, *37*, 1047–1058. [CrossRef]
25. Villain, G.; Thiery, M.; Platret, G. Measurement methods of carbonation profiles in concrete: Thermogravimetry, chemical analysis and gammadensimetry. *Cem. Concr. Res.* **2007**, *37*, 1182–1192. [CrossRef]
26. Ashraf, W. Carbonation of cement-based materials: Challenges and opportunities. *Constr. Build. Mater.* **2016**, *120*, 558–570. [CrossRef]
27. Bretti, G.; Ceseri, M.; Natalini, R. A moving boundary problem for reaction and diffusion processes in concrete: Carbonation advancement and carbonation shrinkage. *Discret. Contin. Dyn. Syst.-S* **2022**, *15*, 2033–2052.
28. Chapwanya, M.; Stockie, J.M.; Liu, W. A model for reactive porous transport during re-wetting of hardened concrete. *J. Eng. Math.* **2009**, *65*, 53–73. [CrossRef]
29. Chen, T.; Gao, X.; Qin, L. Mathematical modeling of accelerated carbonation curing of Portland cement paste at early age. *Cem. Concr. Res.* **2019**, *120*, 187–197. [CrossRef]
30. Freddi, F.; Mingazzi, L. Phase-field simulations of cover cracking in corroded RC beams. *Procedia Struct. Integr.* **2021**, *33*, 371–384. [CrossRef]
31. Kashef-Haghighi, S.; Shao, Y.; Ghoshal, S. Mathematical modeling of CO_2 uptake by concrete during accelerated carbonation curing. *Cem. Concr. Res.* **2015**, *67*, 1–10. [CrossRef]
32. Peng, J.; Tang, H.; Zhang, J.; Cai, S. Numerical simulation on carbonation depth of concrete structures considering time-and temperature-dependent carbonation process. *Adv. Mater. Sci. Eng.* **2018**, *2018*, 2326017
33. Peter, M.A.; Muntean, A.; Meier, S.A.; Böhm, M. Competition of several carbonation reactions in concrete: A parametric study. *Cem. Concr. Res.* **2008**, *38*, 1385–1393. [CrossRef]
34. Torgal, F.P.; Miraldo, S.; Labrincha, J.; De Brito, J. An overview on concrete carbonation in the context of eco-efficient construction: Evaluation, use of SCMs and/or RAC. *Constr. Build. Mater.* **2012**, *36*, 141–150. [CrossRef]
35. Bretti, G.; Ceseri, M.; Natalini, R.; Ciacchella, M.C.; Santarelli, M.L.; Tiracorrendo, G. A forecasting model for the porosity variation during the carbonation process. *Gem-Int. J. Geomath.* **2022**, *13*, 1–24. [CrossRef]
36. Riahi, K.; Van Vuuren, D.P.; Kriegler, E.; Edmonds, J.; O'neill, B.C.; Fujimori, S.; Bauer, N.; Calvin, K.; Dellink, R.; Fricko, O.; et al. The shared socioeconomic pathways and their energy, land use, and greenhouse gas emissions implications: an overview. *Glob. Environ. Chang.* **2017**, *42*, 153–168. [CrossRef]
37. Meinshausen, M.; Nicholls, Z.R.; Lewis, J.; Gidden, M.J.; Vogel, E.; Freund, M.; Beyerle, U.; Gessner, C.; Nauels, A.; Bauer, N.; et al. The shared socio-economic pathway (SSP) greenhouse gas concentrations and their extensions to 2500. *Geosci. Model Dev.* **2020**, *13*, 3571–3605. [CrossRef]
38. Friedlingstein, P.; Houghton, R.A.; Marland, G.; Hackler, J.; Boden, T.A.; Conway, T.J.; Canadell, J.; Raupach, M.; Ciais, P.; Le Quéré, C. Update on CO_2 emissions. *Nat. Geosci.* **2010**, *3*, 811–812. [CrossRef]
39. Solomon, S.; Qin, D.; Manning, M.; Averyt, K.; Marquis, M. *Climate Change 2007-the Physical Science Basis: Working Group I Contribution to the Fourth Assessment Report of The IPCC*; Cambridge University Press: Cambridge, UK, 2007; Volume 4.
40. Climate Change: Global Temperature Projections. Available online: https://www.climate.gov/news-features/understanding-climate/climate-change-global-temperature-projections (accessed on 12 December 2022).
41. Papadakis, V.G.; Vayenas, C.G.; Fardis, M. A reaction engineering approach to the problem of concrete carbonation. *AIChE J.* **1989**, *35*, 1639–1650. [CrossRef]
42. Alì, G.; Furuholt, V.; Natalini, R.; Torcicollo, I. A mathematical model of sulphite chemical aggression of limestones with high permeability. Part I. Modeling and qualitative analysis. *Transp. Porous Media* **2007**, *69*, 109–122. [CrossRef]

43. Zeebe, R.E. On the molecular diffusion coefficients of dissolved CO_2, HCO_3^-, and CO_3^{2-} and their dependence on isotopic mass. *Geochim. Cosmochim. Acta* **2011**, *75*, 2483–2498. [CrossRef]
44. Ball, P. Water—an enduring mystery. *Nature* **2008**, *452*, 291–292.
45. Shriver, D.; Atkins, P.; Cooper, H. *Inorganic Chemistry*; Freeman: New York, NY, USA, 1990.
46. Arpa Lazio. Available online: https://www.arpalazio.it/web/guest/base-dati. (accessed on 12 December 2022).
47. Leemann, A.; Moro, F. Carbonation of concrete: the role of CO_2 concentration, relative humidity and CO_2 buffer capacity. *Mater. Struct.* **2017**, *50*, 1–14. [CrossRef]

heritage

MDPI

Article

Moisture as a Driver of Long-Term Threats to Timber Heritage—Part I: Changing Heritage Climatology

Jenny Richards [1,2] and Peter Brimblecombe [3,*]

1 St John's College, Oxford University, Oxford OX1 3JP, UK
2 School of Geography and the Environment, Oxford University, Oxford OX1 3QY, UK
3 Department of Marine Environment and Engineering, National Sun Yat-Sen University, Kaohsiung 80424, Taiwan
* Correspondence: p.brimblecombe@uea.ac.uk

Abstract: Timber is widely used in the construction of buildings on a global scale, but it is sensitive to degradation. Moisture notably poses a risk to timber decay, and this is likely to change significantly during the 21st century if a high emission scenario occurs. Global HadGEM3 model output was used to map projected changes in relative humidity range, seasonality of relative humidity, time of wetness, wind-driven rain, salt transitions and potential for fungal attack (Scheffer Index). In the Congo Basin, Great Plains (USA) and Scandinavia, humidity ranges are likely to increase along with seasonal change. In many parts of the tropics, time of wetness is likely to decrease by the end of the century. Increases in days of wind-driven rain are projected for western Russia, eastern Europe, Alaska, western Canada and Southern Brazil and Paraguay. Drylands have historically had a low salt risk, but this is projected to increase. In the future, a broad extension of fungal risk along the Himalayas and into central China seems likely, driven as much by temperature as rainfall. The picture presented suggests a slightly less humid heritage climate, which will redistribute the risks to heritage. Mapping global pressures of timber decay could help policymakers and practitioners identify geographically disparate regions that face similar pressures.

Keywords: deterioration; global climate change; cultural heritage; built heritage; wood; HadGEM3; heritage management

Citation: Richards, J.; Brimblecombe, P. Moisture as a Driver of Long-Term Threats to Timber Heritage—Part I: Changing Heritage Climatology. *Heritage* **2022**, *5*, 1929–1946. https://doi.org/10.3390/heritage5030100

Academic Editor: Francesco Soldovieri

Received: 9 July 2022
Accepted: 26 July 2022
Published: 29 July 2022

1. Introduction

Timber has been widely used since prehistoric times in both major buildings and more humble dwellings. It has also been present in other forms of heritage from structures, such as bridges, through to movable objects, such as furniture, sculpture and musical instruments. Given its wide application, the process of timber deterioration has long been of concern. The pressures driving deterioration may increase under a changing climate as expressed in the ICOMOS *Principles for the Conservation of Wooden Built Heritage* as a need to "recognize the vulnerability of structures made wholly or partially of wood in varying environmental and climatic conditions, caused by (among other things) temperature and humidity fluctuations, light, fungal and insect attacks, wear and tear, fire, earthquakes or other natural disasters, and destructive actions by humans".

Water and moisture relations are especially important for wood. These can drive physical changes and mediate biological and chemical processes that cause deterioration in timber [1,2]. To assess the impact of climate on historic timber, traditional meteorological and climatological parameters need to be tuned and refined to reflect potential threats as a *heritage climate*. Heritage climatology is an expression of meteorological characteristics that affect tangible heritage [3–5]. The projected change to global climate over the 21st century poses a complex challenge to the management of heritage, with the nature of change uncertain in many regions. Previous work on the impact of climate on heritage has been predominantly studied at site [6], country [7,8] and regional [3,9] scales and less so on the global level. Nevertheless, a global scale gives a context that can inform management of a specific site or policy for a region.

1.1. Distribution

Timber tends to dominate as a construction material in areas where (i) suitable timber is available, (ii) it is an economical resource, and (iii) it is socially and architecturally desirable. As examples, in Northern Europe, abundant forests have led to a long tradition of wooden construction [10], and timber is an important material in the vernacular architecture of Africa [11]. Additionally, wood has found a prominent place in major buildings of China and Japan [12].

The wide geographical distribution of timber heritage means that it interfaces with a wide range of hygrometric conditions from drylands to rainforests. Given the sensitivity of timber heritage to moisture, long-term climate change could exacerbate deterioration processes in timber. This threat has been particularly noted in areas where the frequency of extreme rainfall and winds, along with the potential for warmer and possibly damper climates, occurs [13–15]. Similarly, the effect of climate change on insects that burrow into timber has been studied, with insects being more abundant and able to complete more reproduction cycles under future climates [16,17].

This paper focuses on timber heritage and the impact of rainfall, temperature and relative humidity to explore the importance of changing heritage climates.

1.2. Threats

Moisture is an important control on the physical dimensions of wood, with fluctuations in moisture content causing swelling and shrinking of the material. Persistent variations in relative humidity and moisture interacting on the carved wood surfaces (e.g., in sculpture) can cause such objects to weaken and crack.

Rot is another common process threatening timber heritage. Fungal attack is mediated by climate factors: temperature, water or exposure to high humidity [1,18]. The changes are often the result of fungi, such as *Monilinia fructicola, M. laxa, Serpula lacrymans, Gloeophyllum trabeum* and *Coniophora puteana*, and are commonly described as brown rot (dry rot) [19], white rot or soft rot [20]. Brown rot fungi are often from tropical climates and southern temperate zones [21]. White rot fungi are active over the temperature range 18–32 °C, but may be active up to 45 °C. [22].

Timber is also threatened by insects [23,24], such as carpenter ants (*Camponotus* spp.), termites (Epifamily: Termitoidae), bark beetles (subfamily Scolytinae), longhorn beetles (Family: Cerambycidae), weevils (Superfamily: Curculionoidea), and powderpost beetles (superfamily Bostrichoidea). The deathwatch beetle (*Xestobium rufovillosum*) is especially well-known in historic structures as it prefers aged oak timber rather than softwood. Wood is generally susceptible to insect attack at high humidity. Under humid, damp conditions, timber can be softened through fungal decay, meaning that insect larvae can more easily tunnel into the wood using the cellulose and hemicellulose as a food [25].

As with many other heritage building materials (e.g., stone, brick and earth), the deterioration of timber can also be caused by salts [26,27]. For example, salt deterioration in timber has been described from polar regions in explorer huts [27]. Salts can also arise from groundwater or due to the activities occurring within the buildings (e.g., fish curing). One of the most common salts is sodium chloride, which dissolves or crystallises at 75.5% relative humidity and is derived from sea spray and road salts [26], but in buildings located inland, nitrates and urea contribute to the salts present. This state change can cause physical stresses within the timber, and in cases of high salt concentrations, defibration can also occur [27]. The severity of salt deterioration processes can be estimated from environmental conditions, typically using changes in relative humidity.

Other mechanisms of damage, such as abrasion, wear and tear, fire, earthquakes or other natural disasters, are not discussed here due to our focus on moisture, although they can be readily explored as for example at the temples and shrines of Nikkō and expedition huts on Antarctica [28,29]. In addition, we focus on outdoor timber, so we do not discuss indoor heritage [9,30] or waterlogged archaeological wood [31,32].

1.3. Approach

This study investigates the moisture-related pressures imposed by climate on timber heritage, here taken primarily as outdoor buildings. These pressures can represent threats to conservation by driving processes of deterioration. We used global climate models to identify regions exposed to high moisture-related threats and to examine the changes over time and space to reveal regions of particular sensitivity.

We used maps as the primary tool for assessing the results from the climate model. Maps are widely used in the heritage sciences as a method to convey spatial data [33–36]. Although the transfer of the data presented in maps into practice is nontrivial [14], they are useful for gaining a broad-scale understanding of the effect of change on heritage across the world.

Global maps also provide a useful tool to represent projections of future conditions and give an understanding of processes and impacts that are not constrained to a single country or region. In a world where so much heritage research has focussed on mid-latitude, high-income countries, these global scale outputs are important [37,38]. This scale of approach facilitates assessment and comparison of threats from across multiple regions, informing decision making at a strategic level. A high-level understanding of threats is important for agenda setting within regional- and national-level institutions. The importance of addressing these risks at a local scale is also recognised and is discussed in a later publication [39]. In particular, in this study, we aim to globally assess the risk posed by past and future moisture-related processes, which drive deterioration of historic timber.

2. Methods

This study used global climate data available from historic, contemporary and future projections derived from the HadGEM3-GC31-MM model within the CMIP6 ensemble [40]. We used four 30-year time periods: 1850–1879 (historic; to set a baseline); 1984–2013 (recent past; many observational datasets available); 2025–2054 (near future; current planning period) and 2070–2099 (far future; sense of overall direction), with future projections determined using the Shared Socio-economic Pathways 585, a scenario based on a high emission future (i.e., a worst-case scenario). We used mean daily relative humidity, temperature, surface wind speed and precipitation (all at 2 m above ground level) from the climate model. These outputs were converted to heritage climate variables, relevant to external pressures on timber. The use of daily timesteps, rather than using higher temporal resolution, captures the response time of historic timber to moisture changes studied in this paper.

Six heritage climate parameters were calculated as moisture-related drivers of timber deterioration: (i) annual relative humidity range (ΔRH_a); (ii) seasonality of relative humidity, (iii) time of wetness, ToW; (iv) wind-driven rain, WDR; (v) salt transitions; and the (vi) Scheffer Index for fungal risk (*Sch*). No distinction was made between untreated timber and that with pesticides or surface coatings, although these would respond more slowly to climate and biological threats; they are looked at in more detail in Brimblecombe and Richards [39].

- **Relative humidity range** was calculated as the annual range in mean monthly relative humidity (%), where the annual range (ΔRH_a) is $RH_{max} - RH_{min}$; RH_{max} is the RH of the month with the highest mean RH in a given year, and RH_{min} is the minimum mean monthly RH in the same year. These differences were summed and divided by 30 to give the average range over the 30-year period. It should be noted that this value is not the same as calculating the difference in the highest and lowest RH values from an averaged 30-year dataset, which are commonly presented in climate summaries. The range expressed in such climatological norms does not reflect the humidity stress experienced by wood each year. Assessing the difference between the highest and lowest month per year (rather than over a 30-year timeframe) captures greater variations in humidity conditions, so the ranges are notably larger than represented by 30-year climatologies.

- **Seasonality of relative humidity** identified the month in a given year that had the highest and lowest mean RH. The modal month for each 30-year time period characterised the seasonality.
- **Time of wetness** is often determined as the number of days per year when relative humidity is >80%, and temperature is >0 °C [41]. This was adopted here, and we also note that a relative humidity of 80% is part of standard methods for laboratory evaluation of insect (e.g., termite) damage to and consumption of wood [42].
- **Wind-driven rain** is defined as the number of days per year when rain is >4 mm; mean wind speed is >2 m s^{-1}, and temperature is >0 °C; adapted from Rydock et al. [2].
- **Salt transitions** were expressed as the number of cycles per year where the mean daily RH crossed 75.5%, to account for sodium chloride crystallisation. A daily transition was adopted to account for change in the phase of salts at the surface as in Grossi et al. [43].
- **Scheffer Index** estimates the risk of fungal attack expressed in the equation: $Sch = \Sigma (T_m - 2)(D - 3)/16.7$, which represents the sum over twelve months for the monthly mean temperature (T_m) and number of days in the month with $\geq$0.3 mm of rain [44,45]. This index has been frequently used at specific locations (e.g., Japan [29], Korea [7,15], Norway [8], Switzerland [6], the UK [46] and the USA [47]).

The heritage parameters were calculated for each 30-year period and averaged to provide a mean value. In the case of WDR and the Scheffer Index, which use multiple climate variables, these parameters were also run with only one climate variable active to assess the contribution of individual parameters. Box and whisker plots were used to present the RH range data. The box is bounded by the 25th and 75th percentiles, with the median denoted by the central line in the box. The whiskers represent the range of all other points, except those that are deemed as outliers. An outlier is considered to be any value that lies over 1.5 times the interquartile range below and above the 25th and 75th percentiles. We used the Mann–Whitney test, rather than the *t*-test, to assess significant differences in results between time periods, because of the nonparametric nature of our data.

3. Results

Our results show how the threat posed by heritage climate to timber heritage differs over space and time. These contribute to a strategic understanding of the past, present and future risks posed by moisture parameters to timber heritage across the world. The results should be interpreted in the context of broad groups of timber heritage within given regions, as microclimatic variations will influence the extent of risk posed by the heritage parameter at scale of sites, buildings or objects [6].

3.1. Relative Humidity Range

Figure 1 shows the global change in the RH range across the four time periods: the 19th century (1850–1879); recent past (1984–2013); the near future (2025–2054); and the far future (2070–2099). In general, the lowest RH ranges are found in polar and equatorial regions and areas with subtropical high pressure (e.g., the Sahara). In the 19th century and recent past, a low annual variability in the Amazon and Congo Basins and in Southeast Asia is particularly notable. Changes can be subtle, especially in the past, so we provide animated versions of the global maps in the supplement. An increase in the RH range is projected (Figure 1d) for the Great Plains of North America (extending into Northern Canada) and Southeast China, and at higher latitudes, this is also apparent across Western Europe and Scandinavia. This is likely to affect the timber heritage in these regions, which includes vernacular wooden structures in North America, ancient wooden temples and pagodas of China and Europe's timber-framed buildings, stave churches and carved wooden statues.

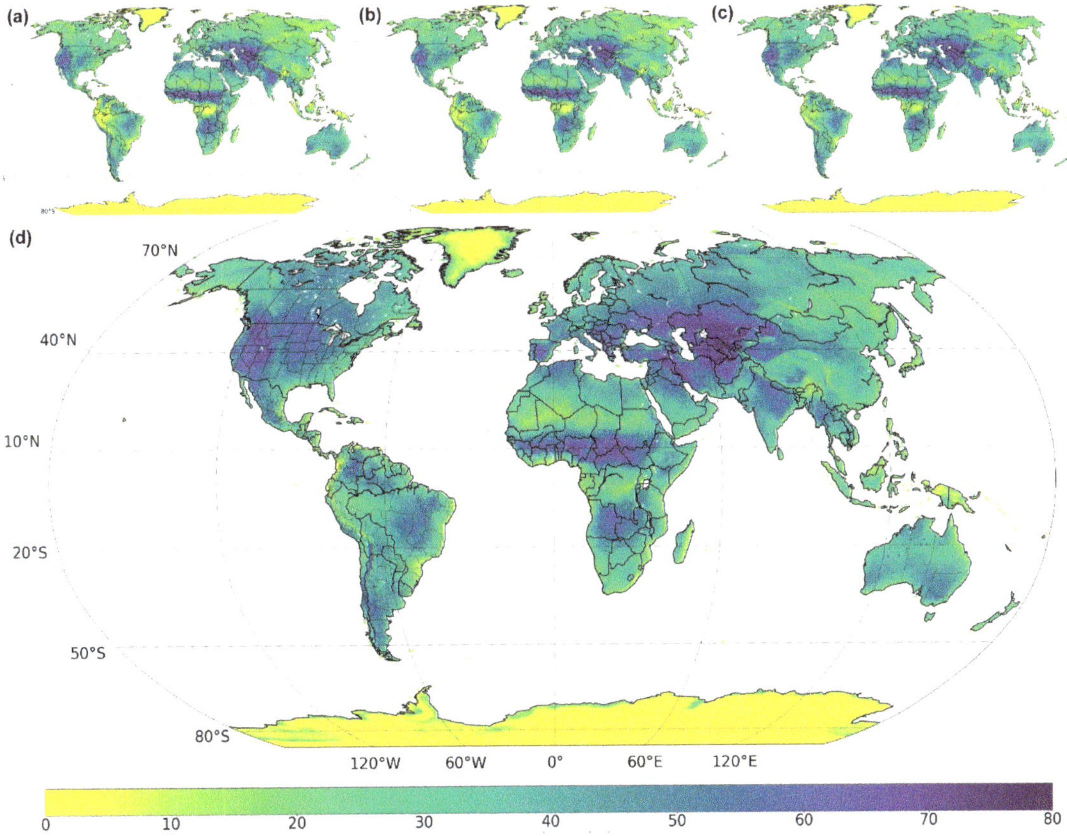

Figure 1. Mean annual relative humidity range (ΔRH_a) for the periods (**a**) 1850–1879, (**b**) 1984–2013, (**c**) 2025–2054 and (**d**) 2070–2099. See also Video S1 (as an animated GIF) in Supplementary Materials.

The RH of the Congo is high throughout the year, but by the end of the 21st century, the RH range in the tropics is projected to increase substantially, a result of less humid conditions becoming more common in a warmer climate (Figure 2a). In the Congo, all months are projected to be less humid by the end of the 21st century, which is particularly apparent between January and March (Figure 2a).

Other mid- and high-latitude locations show similar trends in the RH range to that seen in the Congo, with increases throughout the 21st century (Figure 1). The Great Plains region of the USA, here as Wisconsin, Iowa and Minnesota, shows a notable decrease in summer RH for the far future (Figure 2b), and Scandinavia shows an increasing range, but winters remain damp (Figure 2c). The increase in the RH range in these mid- to high-latitude regions is predominantly caused by lower humidity in summer months, with some winter months also showing slightly higher humidity. In the Congo, the month with the maximum RH has shifted to later in the year; from 1850 to 2054, it predominantly occurred in April or May, but in the last 30 years of the 21st century, it is projected to be in July. The Mann–Whitney test suggests these periods are significantly different ($n_a = 90$; $n_b = 30$; $p_2 = 0.0271$). Typically, it is temperature that is seen as an important control on insect growth [16]. However, under a drier future in tropical areas, it is likely that a humidity decrease could be a more important factor than temperature in the declining diversity of termites in tropical forests [48]. In these regions, the prevalence of drier conditions means less moisture is likely to be available for deterioration processes, particularly those biologically mediated ones.

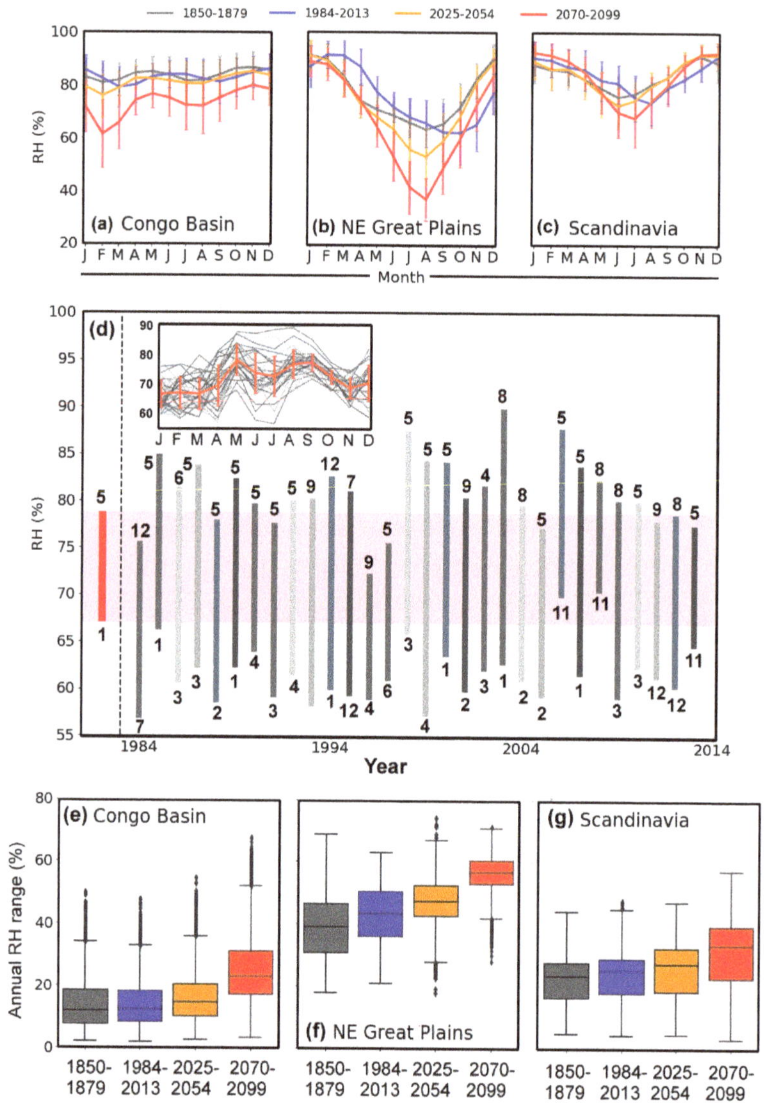

Figure 2. (**a**–**c**) The 30-year mean monthly relative humidity (RH%) for the time periods 1850–1879 (grey), 1984–2013 (blue), 2025–2054 (yellow) and 2070–2099 (red) for (**a**) the Congo Basin, (**b**) the Northeast Great Plains region of the USA, here as Wisconsin, Iowa and Minnesota and (**c**) Scandinavia (Norway, Sweden and Finland). Error bars show one standard deviation from the mean. (**d**) The range of RH between 1984 and 2013 for a single grid cell from the Congo Basin. The range of the 30-year mean monthly RH (red) is compared with each of the annual RH ranges (grey) for each of the 30 individual years. The months with the highest and lowest RH are denoted by the two numbers at the bottom and top of the bars. The inset displays the annual variation for each of the 30 years, and the red line displays the mean monthly RH and standard deviation for the period. (**e**–**g**) Box and whisker plots of the 30-year mean relative humidity range for the four time periods in (**e**) the Congo Basin, (**f**) the Northeast Great Plains and (**g**) Scandinavia.

It is important to consider that when the range is calculated from the average monthly humidity across a 30-year period, the range can seem quite small compared to when the range is calculated for each individual year in a 30-year period and then averaged. In this cell, the shorter red bar shows that the range in averaged monthly RH over a 30-year period is 11.7% (i.e., May (5) maximum 78.7% and January (1) minimum of 67%). The averaged range in RH of each of the 30 years is larger (19.4 ± 3.5%), shown in Figure 2d by the longer grey bars.

3.2. Seasonality of RH

The month most frequently associated with maximum RH (modal month) in the near past and the far future is shown in Figure 3. Globally, there is often a shift to the most humid months occurring earlier in the year. Large changes in seasonality are present in North and South America, with the most humid month in the USA moving from February to March to earlier in the winter, i.e., December–January, while in Central Brazil, there is a notable shift from the maximum RH occurring in April–May to February–March. In Western Australia, the season moves from August to September in the recent past back to June–July by the end of the 21st century. Smaller shifts in the most humid month are found in many other locations. For example, from February–March to December–January across much of Europe and Northern and Southern Africa and from September–October to August–September in Central and Western Africa. For India, Nepal and the Tibetan region, the most humid month in the recent past has been September, but by the end of the century, this is projected to occur in August. While shifts of a month or so seem small, these could result in noticeable changes to the timing of growing cycles. It may also be important to the cycle of human and cultural activities [29,49] or to heritage managers in determining when would be the most effective point in the year to implement conservation and management strategies.

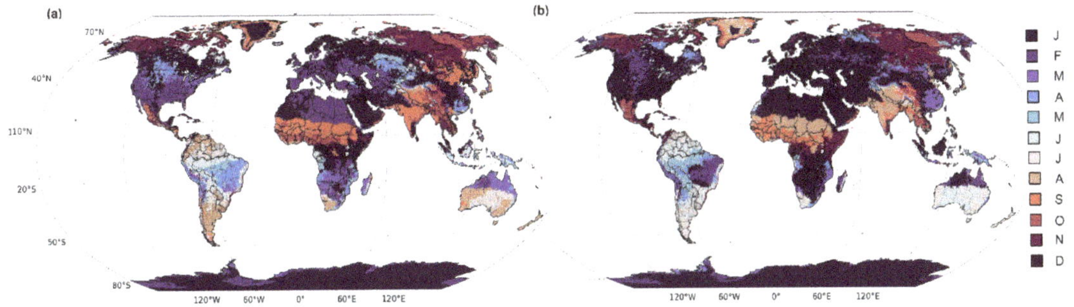

Figure 3. Modal month with the maximum monthly RH (**a**) 1984–2013 and (**b**) 2070–2099. See Supplementary Materials, Figure S1 and Video S1 for all four time periods. See also Video S2 (as an animated GIF) in Supplementary Materials.

Changes in seasonality in the modal month with minimum monthly RH were also found, but the changes were less notable. The results can be found in Supplementary Materials, Figure S2 and Video S3.

3.3. Time of Wetness

Figure 4 shows the ToW across the globe from the 19th century through to the end of the 21st century. The changes and global geographical distribution of pressures in the future for ToW is similar to that of the RH variation, with notable increases in the upper reaches of the Amazon, the Congo Basin and Southeast Asia. These regions have had lengthy periods of wetness in the past (>300 days per year), but the areas where this is persistent are projected to shrink in the near future (2025–2054), with substantial declines seen by the end of the century (Figure 4d). In cities and smaller settlements in these regions, e.g., Tefé, a municipality in Amazonas in northern Brazil or in the Congo, Kisangani (formerly Stanleyville), the capital of Tshopo province, where timber has been used in both vernacular and old colonial buildings, such decreases in time of wetness could reduce the risk of insect damage as timber will have shorter periods of high moisture content.

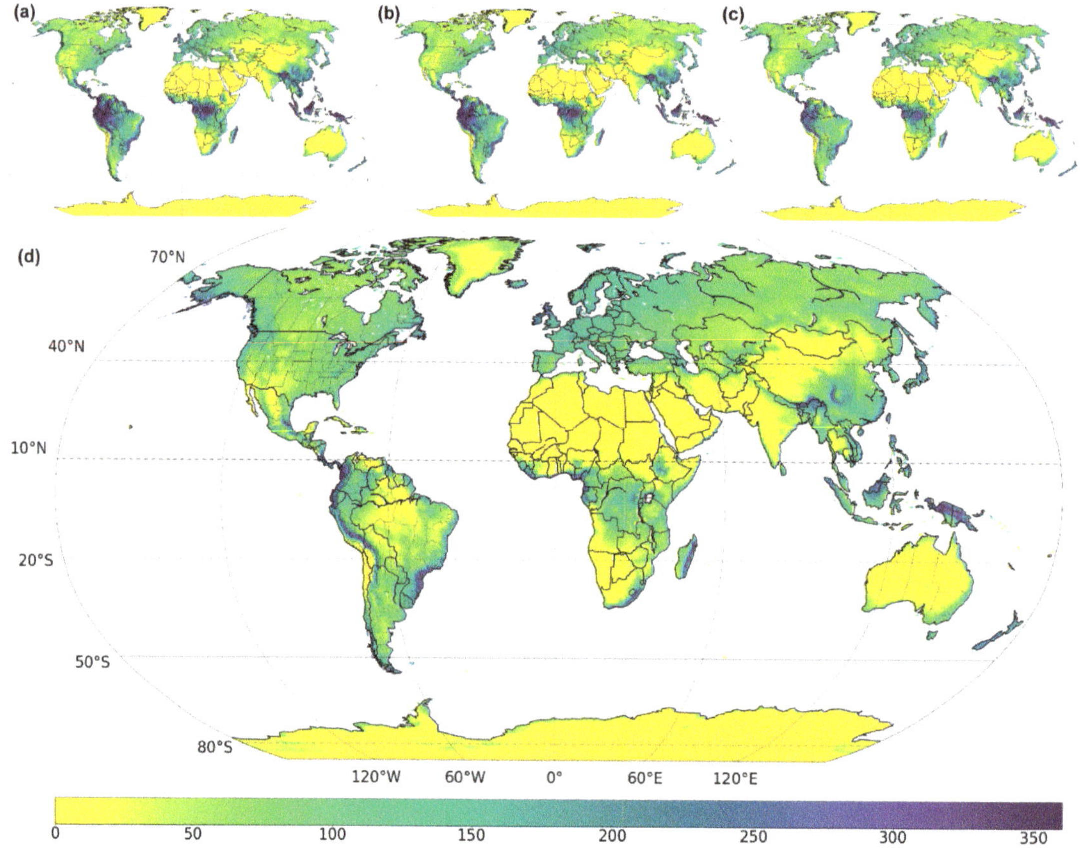

Figure 4. Time of wetness (days per year) for the periods (**a**) 1850–1879, (**b**) 1984–2013, (**c**) 2025–2054 and (**d**) 2070–2099. See also Video S4 (as an animated GIF) in Supplementary Materials.

The length of the wetness period in the high-altitude areas of the New Guinea Highlands and the Tibetan Plateau region are projected to remain high over the 21st century (Figure 4). This continued pressure from lengthy periods of wetness means that the risk of insect attack on historic timber in these regions remains high.

Perhaps of greater concern are increases in future ToW in regions where warming means more days above freezing, e.g., Russia, Canada and Antarctica (Figure 4). For example, in coastal areas of Antarctica, projections suggest that rising temperatures could lead to a more than three-fold increase in days of wetness by the end of the 21st century. Antarctic wooden heritage, such as the buildings associated with polar exploration [27] or industry [50], would be exposed to new drivers of deterioration over the next century. As timber heritage in these regions is already exposed to deteriorative conditions [28], the addition of further pressures from ToW could have a substantial impact on future deterioration.

3.4. Wind-Driven Rain

Figure 5 shows the pressure on timber heritage caused by WDR across the four time periods used in this study. Globally, changes in future pressures from WDR are low in dryland regions, but most frequent in the tropics, mid-latitudes and coastal regions with strong prevailing winds (e.g., eastern North America, south-eastern South America, north-western Europe, West Africa and coastal regions of Southeast Asia). Many regions are projected to face increasing days of WDR by the end of the 21st century (e.g., Western Russia, Eastern Europe, Alaska and Western Canada). There are also increases in southern Brazil and Paraguay, where the style of timber architecture heritage has been influenced by German and Italian immigration. Additionally, increasing WDR will be expected in an area stretching from Sierra Leone and Liberia, across Central Africa to Sudan. In West Africa, in particular Freetown, Sierra Leone, it may affect North American-style timber houses found as a result of immigration after the abolition of slavery in the 19th century. As many heritage sites are found near the coast, this overlap of geographic distribution of pressures and location of heritage increases challenges for the conservation of timber heritage.

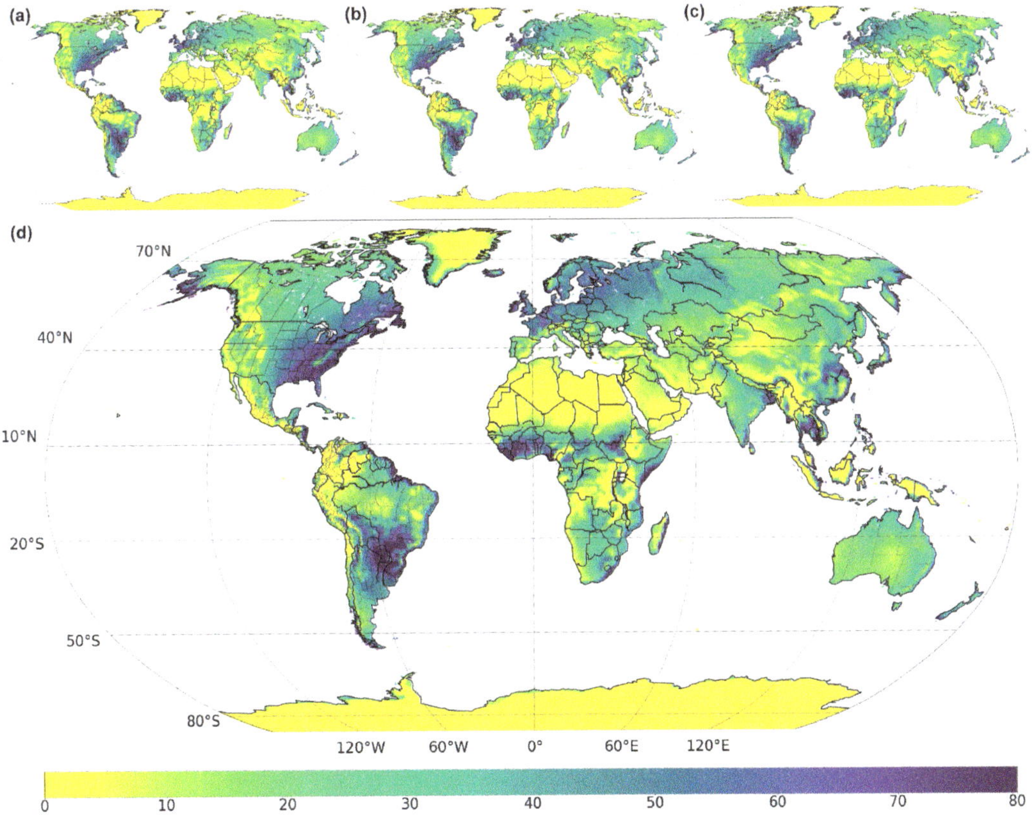

Figure 5. Average number of days of wind-driven rain each year for the periods (**a**) 1850–1879, (**b**) 1984–2013, (**c**) 2025–2054 and (**d**) 2070–2099. See also Video S5 (as an animated GIF) in Supplementary Materials.

Future change in pressures imposed by WDR are predominantly driven by changes in rainfall rather than wind speed (Figure 6). As climate models, such as HadGEM3, can underestimate rainfall in regions, such as parts of Southern Africa [51,52], the risk posed by WDR in these areas may be uncertain.

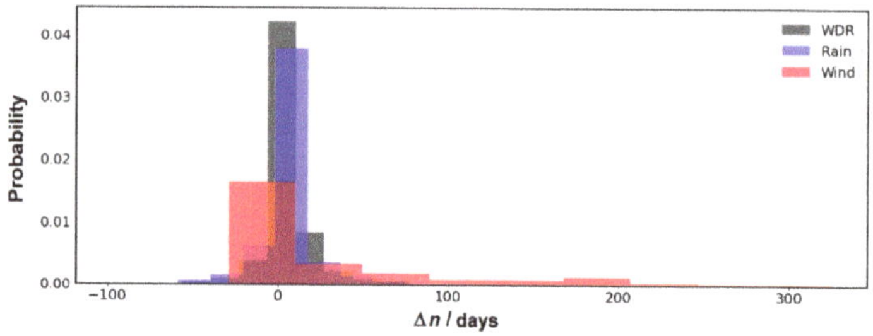

Figure 6. The probability of the difference in the number of days (Δn) of (i) wind-driven rain (WDR), (ii) days with rain >4 mm d^{-1} and (iii) days with wind >2 m s^{-1} between the far future (2070–2099) and the near past (1984–2013) across all land masses.

3.5. Salt Transitions

The modelled number of salt transitions for sodium chloride, occurring in timber heritage between the 19th and 21st centuries is shown in Figure 7. Regions experiencing frequent salt transitions of more than 30 per year are found across the globe (e.g., North America, Europe, China, Japan, New Zealand and southern South America and coastal Antarctica, Greenland, Australia and South Africa).

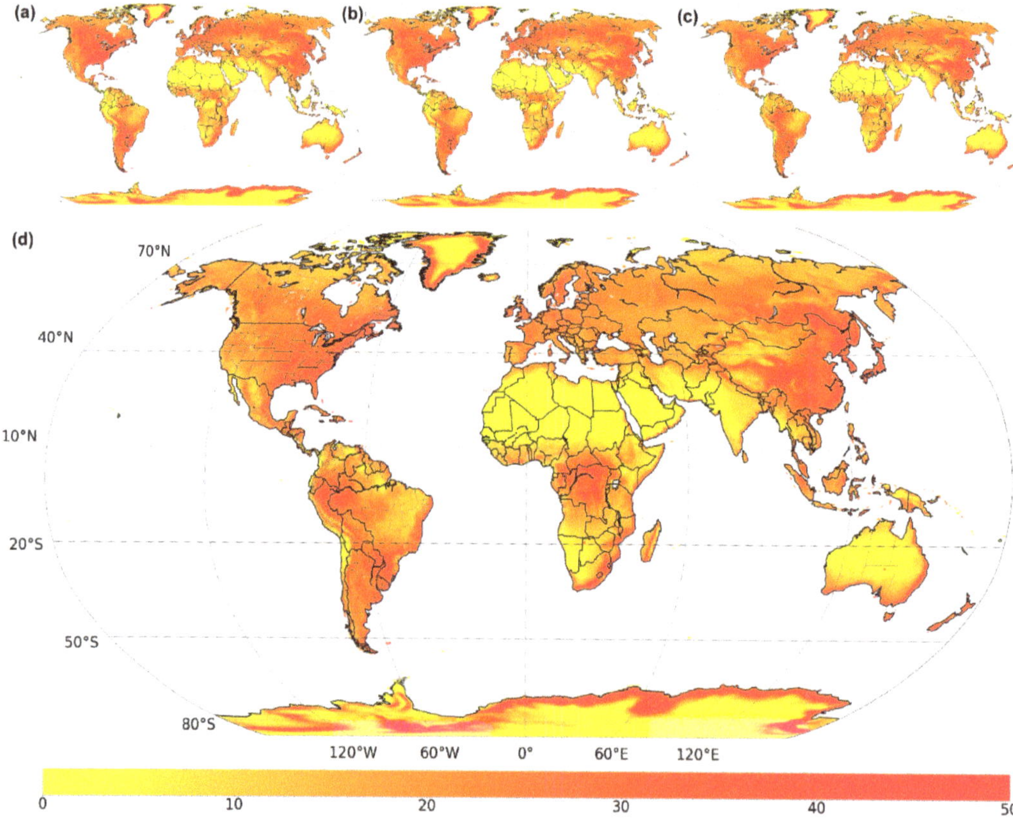

Figure 7. Salt transitions per year for the periods (**a**) 1850–1879, (**b**) 1984–2013, (**c**) 2025–2054 and (**d**) 2070–2099. See also Video S6 (as an animated GIF) in Supplementary Materials.

In terms of change, Figure 7 shows that some dryland regions that historically have had few salt transitions (e.g., Central Australia, the Middle East and the Sahara) will in future show increases in the projected pressure as they become more humid. By the end of the century, large increases are projected for western South America, Central Africa and Southeast Asia. These regions also experience an increased RH range (Figure 2), largely because the drier months will have lower RH in the future, which drives an increase in the number of transitions. Some small decreases are also projected in the eastern USA, Mediterranean and Europe as noted by Grossi and Brimblecombe [43].

3.6. *Scheffer Index*

The change in Scheffer Index for the four periods under investigation was small in comparison to the range of index values calculated. Even changes in index values of 50 appeared as indistinguishable using our map projections (see Supplementary Materials, Figure S3), whereas changes of this magnitude are highly important for deterioration [45,53]. Therefore, Figure 8 shows the Scheffer Index for the period 1850–1879 (Figure 8a) and how it is projected to change from this historic baseline by the end of the 21st century (Figure 8b).

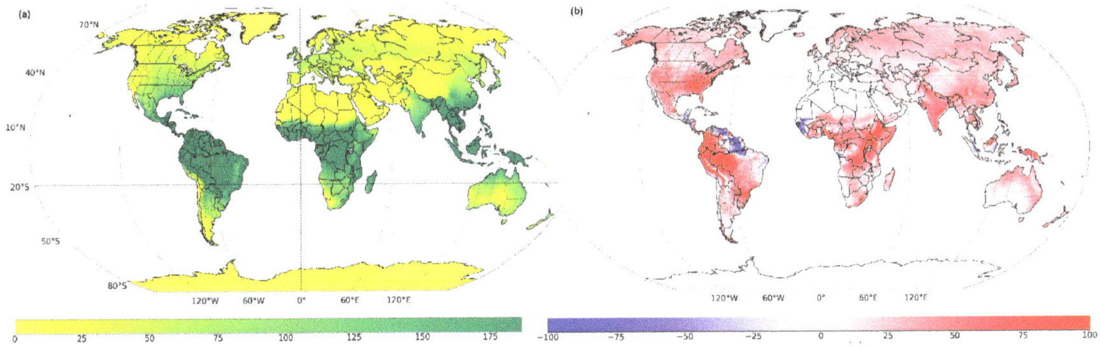

Figure 8. Scheffer Index (*Sch*) (**a**) in 1850–1879 and (**b**) the projected change in Scheffer Index between the historic past (1859–1879) and the far future (2070–2099). See Supplementary Materials, Figure S3 for maps of the four time periods. See also Video S7 (as an animated GIF) in Supplementary Materials.

Figure 8a shows that the greatest risk associated with the Scheffer Index is located in the region 20 degrees north and south of the Equator, with areas having an index score of more than 175. Minimal risk is found in polar regions and hyper arid drylands. Large areas of Europe, Central and Eastern Asia, Australia and northern North America have a Scheffer Index above 50. In previous studies, the value for high decay risk has varied with suggested critical index values including 48 [53] and 65 [45]. As such, even areas within our map that have comparably moderate index values still face notable risks.

By the end of the 21st century, projections show that the spatial area with an index value of more than 175 will continue to increase on all continents (e.g., along the Himalayas, northwards into Central China, south-eastern USA and the northwest coast of Australia). These changes affect a large range of timber heritage including large temples and pagodas in China to plantation houses in southern US states. Smaller but still notable increases in the Scheffer Index are projected across Europe where the index has increased from 50 (1850–1879) to 75 (2070–2099), which means according to the thresholds previous outlined by Lisø et al. [45] and Tajet and Hygen [53] this region is projected to fall within the high-risk threshold by the end of the century.

Changes in temperature have previously been identified as the primary driver of the Scheffer Index in Japan [29]. In areas between 0 and 60° N/S with large increases in the Scheffer Index (>+30), we similarly find temperature as the main driver of change over the 21st century (Figure 9a,b). However, for areas where there have been smaller increases, or reductions in the Scheffer Index, changes in rainfall will likely be the primary driver. Polewards of 60°, large increases in the Scheffer Index (≤100) are projected to be caused by increases in rainfall (Figure 9c).

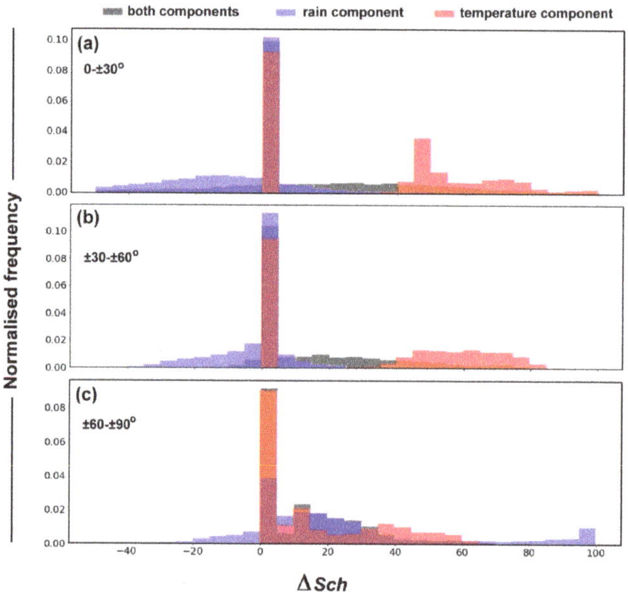

Figure 9. The normalised frequency of Scheffer Index values (Δ*Sch*) between the far future (2070–2099) and the near past (1984–2013) when assessing (i) all components of the Scheffer Index (grey), (ii) the rain component (blue) and (iii) the temperature component (red) for areas at latitudes (**a**) 0–±30°, (**b**) ±30–±60° and (**c**) ±60–±90°.

4. Discussion

Previous studies of sites and regions broadly agree with our results. For example, the UK is likely to experience periods that are less humid than at present, in alignment with earlier findings, which showed that the UK would experience a ≤ 10% decrease in RH and a reduction in ToW [54,55]. If the drying of timber is prolonged, it could reduce the risk posed by rot that requires damp material, and insect infestation may be limited if wood is very dry [12]. An increase in the RH range was generally caused by less humid conditions in the drier months, increasing physical stress. Additionally, global shifts in the Scheffer Index over the next century align with previous country-wide assessments [7,29,46,55].

Our study presents a picture of a less humid heritage climate that contrasts with a frequent view of a future wetter world [32]. There is a need to distinguish between specific humidity and relative humidity. The amount of water in the atmosphere (specific humidity) may increase under a warmer climate and hence drive heavier precipitation. However, higher temperatures mean the atmosphere can hold more moisture in a given volume of air, so that relative humidity could nevertheless be lower. Importantly, it is relative humidity that is a critical parameter for the water content of wood, fungal growth and insect activity [1]. We need to consider both temperature and moisture when assessing RH pressures, as changes to these can operate in the same or opposite directions, amplifying or suppressing the impact on timber deterioration.

4.1. Spatial and Temporal Scales

Understanding the global threat of moisture to historic timber enables a strategic approach for implementing effective conservation and management. Previous work assessing the environmental and climatic pressures on timber heritage has been undertaken predominantly for specific sites [6,28] or at regional scales [45,46]. These smaller scales are useful for capturing local variations in climate that will affect specific sites and objects and particularly local issues pertaining to their use, such as the need for heating [56]. Recent findings of Brischke and Selter [6] highlight the importance of small-scale mapping at high resolution to account for decay processes in wood, due to variations imposed by local topographical and hydrological conditions. Finer spatial scales can limit the comparison between geographically, politically or economically disparate locations. Indeed, by using a global scale assessment, our findings show that many disparate areas can face similar threats to timber heritage. The identification of shared threats provides an exciting opportunity for new collaborations in strategic research and management, even where these do not arise from contiguous regions. This approach therefore speaks to the need raised by Richards and Brimblecombe [57] that if models are to be useful within heritage science, the model or model results need to be transferable into practice or aid policy decision making. In a subsequent paper, we explore changes in climate at specific sites, with particular types of coatings, wood and buildings [39], enabling us to further assess the tensions between the fine details provided by site scale approaches with the need for a strategic understanding of processes.

In addition, a global approach can show the pressures facing regions that have been less intensively studied to date. Our approach speaks to the needs highlighted by Orr et al. [37] and Simpson et al. [38] that advocate for additional research to address the imbalance in the geographical distribution of heritage research. Our results both broaden our understanding of risks facing timber heritage and highlight regions that may require additional research focus and identify narrow regions that are sensitive to change, which could easily be overlooked if a global scale approach was not taken.

4.2. Dominant Pressure

Figures 1, 3–5 and 7 illustrate the multiplicity of drivers of timber deterioration. It is difficult to assess combined or synergistic pressures without considering the specific nature of the interactions between climate and the wooden elements at sites, many of which are unknown. Instead, we identify the dominant pressure present in a given area. We define this dominance in a specific area, relative to the global distribution (as defined in Figure 10). However, the dominant pressure does not directly transfer into a quantifiable deterioration risk, as it does not account for thresholds in damage mechanisms. Nevertheless, it gives the geographical distribution of key processes that are likely to be important for conservation.

Figure 10 shows changes in the dominant pressure on heritage across the globe. In the future, there will be increases in the area dominated by fungal risk, most notably the Amazon and Congo Basins. However, for many regions, the RH range remains the dominant pressure across the period 1850–2099. Wind-driven rain is constrained to localised regions (e.g., Eastern Canada). Such enhanced ranges may place well-ventilated wooden interiors at risk, but these also represent locations that will become drier in the future. This figure also highlights the need for management strategies to recognise that pressures on historic timber may shift from physical to biological threats. As an example, the dominant pressure in Central Africa and Southeast Asia shifts from a historic threat from ToW to fungal risk by the end of the 21st century. This increasing importance of the Sheffer Index could suggest that issues relating to mould and insects may become more prevalent by the end of the 21st century, thus requiring new management strategies.

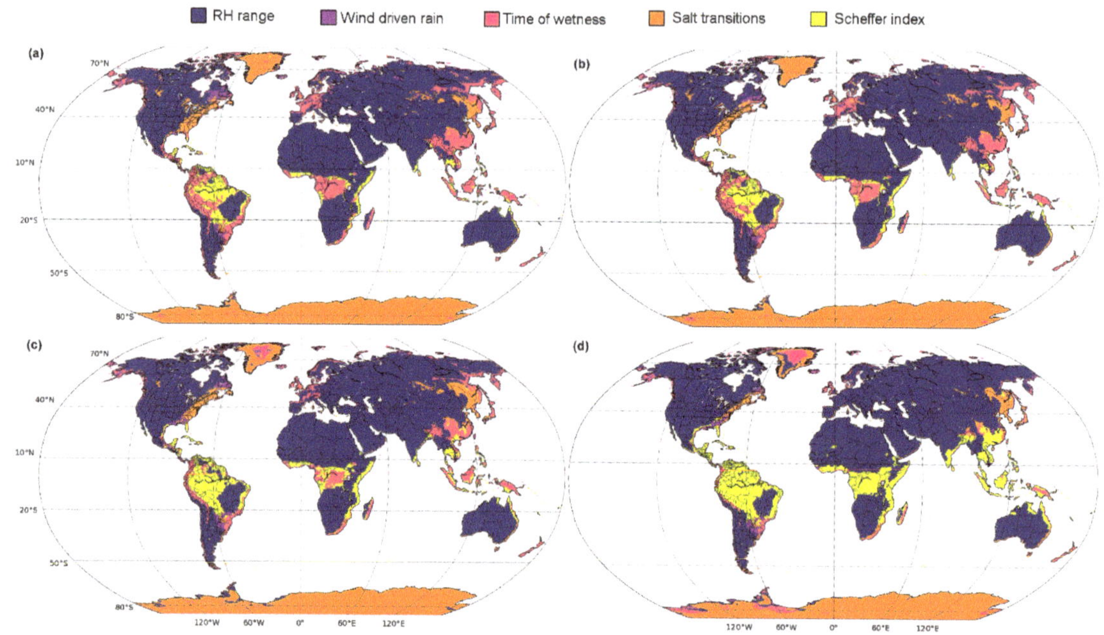

Figure 10. Dominant heritage climate pressure on timber (**a**) 1850–1879, (**b**) 1984–2013, (**c**) 2025–2054 and (**d**) 2070–2099. The dominant pressure was determined by comparing the percentile value for each of five pressures (excludes RH seasonality) for a given location compared to the global terrestrial datasets. The pressure with the greatest percentile value was assigned as the dominant pressure. For example, if in a given location the pressure percentiles were: RH range = 91%, WDR = 72%, ToW = 10%, Salt = 25% and Scheffer = 85%, the RH range would be assigned as the dominant pressure. See also Video S8 (as an animated GIF) in Supplementary Materials.

4.3. Propagation of Errors and Implication for Management

Our study engages with models to provide both the input data (HadGEM3-GC31-MM) and quantify climate pressures on timber heritage. The outcomes of any heritage model are not a perfect representation of reality [57], and this inevitably means that there will be differences between the modelled pressures and that experienced at specific heritage sites [39].

Climate models are less effective at capturing the climate in some regions; e.g., there is a lack of consensus over the sign of directional change in modelled precipitation and biases in surface wind speed over Central and West Africa [58,59]. While developments in regional climate modelling will improve the resolution of localised processes and the accuracy of model outputs, it is likely that present projections differ depending on the model selected [58]. Some climate parameters used in our assessment, such as temperature and humidity, are less affected by specific weather tracks, so they may be more resistant to error [60]. However, we need to be mindful that the results from regions associated with model uncertainty may be sensitive to model choice, so future use of an ensemble of models is likely to reduce the uncertainty in projected heritage climates.

Our study, which focussed on a future worst-case scenario, aimed to provide an understanding of the nature of the threat to timber. Even if the worst-case is never realised, this approach gives a sense of what threats likely need to be managed. Additionally, it also highlights regions that might be sensitive to tipping points or thresholds. Identifying regions that are projected to undergo large changes, allows resources to be appropriately distributed in addressing threats in a timely fashion. Change in climate pressures across the seasons may be important in heritage management, as cultural events, visitor activities and the landscape are strongly dependent on time of year [49].

4.4. Strategic Policy for Timber Heritage

Our approach has highlighted the challenge of developing meaningful heritage climate parameters at an appropriate scale for policy or management. We assessed the pressures at a global scale, enabling inter-regional comparisons to be made. This approach fills the need for a strategic approach to heritage climate pressures. However, it is hard to interpret global projections in terms of managing the threat posed to individual sites, buildings or objects, so these are explored in Brimblecombe and Richards [39].

As heritage is constantly interacting with and being influenced by environmental surroundings [61], heritage policymakers, managers and practitioners must constantly plan and react to the inevitability of change at heritage sites. Being able to identify regions that are likely to experience future change or cross thresholds for decay processes can enable conservation strategies to be preventative, rather than reactive. For timber heritage, regions likely to experience notable increases in moisture-related pressures, this global approach could allow international collaboration between regions experiencing similar climatic decay processes.

Our study shows the benefits of including heritage climate parameters within policy. Traditional parameters used in climate science, such as average change in temperature or specific humidity over a 30-year period, can miss key interactions between heritage and climate that drive deterioration. If further developed, the notion of heritage climatology could extend to the concepts of risk that are more directly transferable to heritage practice. By adopting climate parameters attuned to heritage, institutions, such as the UNESCO WHC, could be in a better position to embrace the notion of climate change impacts on heritage.

5. Conclusions

We conclude that the future heritage climate will lead to a redistribution of risks to timber heritage under a high emission scenario. The changes in the magnitude and spatial distribution of the pressures in this heritage reflect shifts in the relative humidity range, seasonality, time of wetness, wind-driven rain, salt transitions and potential for fungal attack. It is noticeable that in the future, tropical areas, such as the Congo and Amazon Basins, along with parts of Southeast Asia show less humid conditions. Some temperate regions will show increasing humidity ranges, with the most humid months shifting earlier in the year. We find projected increases in days of wind-driven rain for some temperate regions. Salt transitions may become more evident in drylands. By 2070, a broad extension of fungal attack on timber will spread from the Himalayas into Central China, driven as much by temperature as rainfall. Our picture presents a less humid global heritage climate, and it contrasts with common views that the future will be wetter. However, it is important to distinguish specific and relative humidity as the former is likely to increase with temperature, while the latter is likely to decrease. Relative humidity is of particular importance to timber heritage because this parameter affects fungal growth, insect activity and water content of timber.

Our study used a single climate model under a worst-case scenario, giving us a sense of the direction of change. However, future work might address uncertainties associated with climate models by using ensemble outputs. Additional work could also focus on how multiple pressures synergistically interact to increase timber deterioration, particularly in regions where the dominant process is projected to shift over the coming century. More collaboration between heritage researchers and practitioners might help clarify the incorporation of uncertainties associated with projections into management plans.

Our model outputs used global maps, which can be a useful tool for dissemination and discussion both in terms of (i) policy making, particularly at a strategic level and (ii) raising public awareness by providing visual representations. Shifts in threats over time can provide an engaging narrative for a public interested in climate change. However, heritage is frequently managed at the level of a specific site, so such broad pictures will hopefully provoke thoughts about choices to be made at the site level.

Supplementary Materials: The following supporting information can be downloaded at: https://www.mdpi.com/article/10.3390/heritage5030100/s1: Figure S1: Modal month with the maximum monthly RH for the periods (a) 1850–1879, (b) 1984–2013, (c) 2025–2054 and (d) 2070–2099; Figure S2: Modal month with the minimum monthly RH for the periods (a) 1850–1879, (b) 1984–2013, (c) 2025–2054 and (d) 2070–2099; Figure S3: Scheffer Index for the periods (a) 1850–1879, (b) 1984–2013, (c) 2025–2054 and (d) 2070–2099; Video S1: Animated GIF of Figure 1 showing RH range (a) 1850–1879, (b) 1984–2013, (c) 2025–2054 and (d) 2070–2099; Video S2: Animated GIF of Figure 3 showing modal month with the maximum monthly RH for the periods (a) 1850–1879, (b) 1984–2013, (c) 2025–2054 and (d) 2070–2099. Video S3: Animated GIF showing modal month with the minimum monthly RH for the periods (a) 1850–1879, (b) 1984–2013, (c) 2025–2054 and (d) 2070–2099. Video S4: Animated GIF of Figure 4 showing ToW for the periods (a) 1850–1879, (b) 1984–2013, (c) 2025–2054 and (d) 2070–2099. Video S5: Animated GIF of Figure 5 showing WDR for the periods (a) 1850–1879, (b) 1984–2013, (c) 2025–2054 and (d) 2070–2099. Video S6: Animated GIF of Figure 7 showing salt transitions for the periods (a) 1850–1879, (b) 1984–2013, (c) 2025–2054 and (d) 2070–2099. Video S7: Animated GIF of Figure 8 showing the Scheffer Index for the periods (a) 1850–1879, (b) 1984–2013, (c) 2025–2054 and (d) 2070–2099. Video S8: Animated GIF of Figure 10 showing combined pressures for the periods (a) 1850–1879, (b) 1984–2013, (c) 2025–2054 and (d) 2070–2099.

Author Contributions: Conceptualization, formal analysis, investigation and writing—review and editing, P.B. and J.R.; software and data curation, J.R.; original draft preparation P.B. All authors have read and agreed to the published version of the manuscript.

Funding: This research received no external funding.

Institutional Review Board Statement: Not applicable.

Informed Consent Statement: Not applicable.

Data Availability Statement: Data are available from the links noted in the text.

Acknowledgments: We would like to thank Sebastian Engelstaedter for his technical support with the climate model data.

Conflicts of Interest: The authors declare no conflict of interest.

References

1. Skaar, C. *Wood-Water Relations*; Springer Science & Business Media Location: Berlin/Heidelberg, Germany, 2012.
2. Rydock, J.P.; Lisø, K.R.; Førland, E.J.; Oswaldová, K.; Thue, J.V. A driving rain exposure index for Norway. *Build. Environ.* **2015**, *40*, 1450–1458. [CrossRef]
3. Sabbioni, C.; Brimblecombe, P.; Cassar, M. (Eds.) *The Atlas of Climate Change Impact on European Cultural Heritage: Scientific Analysis and Management Strategies*; Anthem Press: London, UK, 2010.
4. Brimblecombe, P. Heritage Climatology. In *Climate Change and Cultural Heritage*; Lefevre, R.-A., Sabbioni, C., Eds.; Edipuglia: Bari, Italy, 2010; pp. 57–64.
5. Brimblecombe, P. Policy Relevance of Small Changes in Climate with Large Impacts on Heritage. In *Cultural Heritage Facing Climate Change*; Lefevre, R.-A., Sabbioni, C., Eds.; Edipuglia: Bari, Italy, 2018; pp. 25–32.
6. Brischke, C.; Selter, V. Mapping the decay hazard of wooden structures in topographically divergent regions. *Forests* **2020**, *11*, 510. [CrossRef]
7. Kim, T.G.; Ra, J.B.; Kang, S.M.; Wang, J. Determination of decay hazard index (Scheffer index) in Korea for exterior above-ground wood. *J. Korean Wood Sci. Technol.* **2011**, *39*, 531–537. [CrossRef]
8. Hygen, H.O.; Øyen, C.F.; Almås, A.J. Assessment of climate vulnerability in the Norwegian built environment. *Adv. Sci. Res.* **2011**, *6*, 151–153. [CrossRef]
9. Leissner, J.; Kilian, R.; Kotova, L.; Jacob, D.; Mikolajewicz, U.; Broström, T.; Ashley-Smith, J.; Schellen, H.L.; Martens, M.; van Schijndel, J.; et al. Climate for Culture: Assessing the impact of climate change on the future indoor climate in historic buildings using simulations. *Herit. Sci.* **2015**, *3*, 1–5. [CrossRef]
10. Bakken, K. (Ed.) *Preserving the Stave Churches. Craftsmanship and Research*; Pax Forlag: Oslo, Norway, 2016.
11. Prussin, L. An introduction to indigenous African architecture. *J. Soc. Archit. Hist.* **1974**, *33*, 183–205. [CrossRef]

12. Brimblecombe, P.; Hayashi, M. Sustaining Wooden Architectural Heritage. In *Sustainable Development: Asia-Pacific Perspectives*; Low, P.S., Ed.; Cambridge University Press: Cambridge, UK, 2022; pp. 52–67.
13. Feng, Z.; Leung, L.; Hagos, S.; Houze, R.A.; Burleyson, C.D.; Balaguru, K. More frequent intense and long-lived storms dominate the springtime trend in central US rainfall. *Nat. Commun.* **2016**, *7*, 13429. [CrossRef] [PubMed]
14. Brimblecombe, P.; Hayashi, M.; Futagami, Y. Mapping climate change, natural hazards and Tokyo's built heritage. *Atmosphere* **2020**, *11*, 680. [CrossRef]
15. Oh, J.J.; Choi, Y.S.; Kim, G.S.; Kim, G.H. Assessment of the effects of projected climate change on the potential risk of wood decay in Korea. *J. Cult. Herit.* **2022**, *55*, 43–47. [CrossRef]
16. Brimblecombe, P.; Lankester, P. Long-term changes in climate and insect damage in historic houses. *Stud. Conserv.* **2013**, *58*, 13–22. [CrossRef]
17. Querner, P.; Sterflinger, K.; Derksen, K.; Leissner, J.; Landsberger, B.; Hammer, A.; Brimblecombe, P. Climate Change and Its Effects on Indoor Pests (Insect and Fungi) in Museums. *Climate* **2022**, *10*, 103. [CrossRef]
18. Choidis, P.; Kraniotis, D.; Lehtonen, I.; Hellum, B. A modelling approach for the assessment of climate change impact on the fungal colonization of historic timber structures. *Forests* **2021**, *12*, 819. [CrossRef]
19. Ringman, R.; Pilgård, A.; Brischke, C.; Richter, K. Mode of action of brown rot decay resistance in modified wood: A review. *Holzforschung* **2014**, *68*, 239–246. [CrossRef]
20. Blanchette, R.A.; Held, B.W.; Jurgens, J.A.; McNew, D.L.; Harrington, T.C.; Duncan, S.M.; Farrell, R.L. Wood-destroying soft rot fungi in the historic expedition huts of Antarctica. *Appl. Environ. Microbiol.* **2004**, *70*, 1328–1335. [CrossRef]
21. Ryvarden, L. Tropical Polypores. In *British Mycological Society Symposium Series*; Cambridge University Press: Cambridge, UK, 2013; Volume 19, pp. 149–170.
22. Daniel, G.; Nilsson, T. Developments in the Study of Soft Rot and Bacterial Decay. In *Forest Products Biotechnology*; CRC Press: Boca Raton, FL, USA, 1997; pp. 47–72.
23. Pournou, A. Wood Deterioration by Insects. In *Biodeterioration of Wooden Cultural Heritage*; Springer: Cham, Switzerland, 2020; pp. 425–526.
24. Querner, P. Insect Pests and Integrated Pest Management in Museums, Libraries and Historic Buildings. *Insects* **2015**, *16*, 595–607. [CrossRef] [PubMed]
25. Findlay, G.W. (Ed.) *Preservation of Timber in the Tropics*; Springer Science & Business Media: Berlin/Heidelberg, Germany, 2013.
26. Kirker, G.; Glaeser, J. Salt damage to wood "fuzzy wood" often confused with fungal decay. *Pile Driv.* **2011**, *Q3*, 85–86.
27. Blanchette, R.; Held, B.; Farrell, R. Defibration of wood in the expedition huts of Antarctica: An unusual deterioration process occurring in the polar environment. *Polar Rec.* **2002**, *38*, 313–322. [CrossRef]
28. Harrowfield, D. Analysis of timber weathering and wind velocity at Cape Adare, with comments on other historic sites in Antarctica. *Polar Rec.* **2006**, *42*, 291–307. [CrossRef]
29. Brimblecombe, P.; Hayashi, M. Pressures from long term environmental change at the shrines and temples of Nikkō. *Her. Sci.* **2018**, *6*, 1–12. [CrossRef]
30. Lankester, P.; Brimblecombe, P. The impact of future climate on historic interiors. *Sci. Total Environ.* **2012**, *41*, 248–254. [CrossRef] [PubMed]
31. Sauer, E. Alchester—In search of Vespasian. *Curr. Archaeol.* **2005**, *196*, 168–176.
32. Brimblecombe, P. Refining climate change threats to heritage. *J. Inst. Conserv.* **2014**, *37*, 85–93. [CrossRef]
33. Accardo, G.; Giani, E.; Giovagnoli, A. The risk map of Italian cultural heritage. *J. Arch. Conserv.* **2003**, *9*, 41–57. [CrossRef]
34. De la Fuente, D.; Vega, J.M.; Viejo, F.; Díaz, I.; Morcillo, M. Mapping air pollution effects on atmospheric degradation of cultural heritage. *J. Cult. Herit.* **2013**, *14*, 138–145. [CrossRef]
35. Wu, P.S.; Hsieh, C.M.; Hsu, M.F. Using heritage risk maps as an approach to estimating the threat to materials of traditional buildings in Tainan (Taiwan). *J. Cult. Herit.* **2014**, *15*, 441–447. [CrossRef]
36. Wang, J.J. Flood risk maps to cultural heritage: Measures and process. *J. Cult. Herit.* **2014**, *15*, 210–220. [CrossRef]
37. Orr, S.A.; Richards, J.; Fatorić, S. Climate Change and Cultural Heritage: A Systematic Literature Review (2016–2020). *Hist. Environ. Policy Pract.* **2021**, *12*, 434–477. [CrossRef]
38. Simpson, N.P.; Clarke, J.; Orr, S.A.; Cundill, G.; Orlove, B.; Fatorić, S.; Sabour, S.; Khalaf, N.; Rockman, M.; Pinho, P.; et al. Decolonizing climate change–heritage research. *Nat. Clim. Chang.* **2022**, *12*, 210–213. [CrossRef]
39. Brimblecombe, P.; Richards, J. Moisture as a driver of long-term threats to timber heritage—Part II risks imposed at local sites. *Heritage* **2022**, *in press*.
40. Met Office Hadley Centre. WCRP CMIP6: Met Office Hadley Centre (MOHC) HadGEM3-GC31-MM Model Output Collection. Centre for Environmental Data Analysis. 2020. Available online: http://catalogue.ceda.ac.uk/uuid/7fbf7e27ef554981bfff17d6f14f9d8a (accessed on 10 May 2022).
41. Corvo, F.; Pérez, T.; Martin, Y.; Reyes, J.; Dzib, L.R.; González-Sánchez, J.; Castañeda, A. Time of wetness in tropical climate: Considerations on the estimation of TOW according to ISO 9223 standard. *Corr. Sci.* **2008**, *50*, 206–219. [CrossRef]
42. Unsal, O.; Kartal, S.N.; Candan, Z.; Arango, R.A.; Clausen, C.A.; Green III, F. Decay and termite resistance, water absorption and swelling of thermally compressed wood panels. *Int. Biodeterior. Biodegrad.* **2009**, *63*, 548–552. [CrossRef]
43. Grossi, C.M.; Brimblecombe, P.; Menéndez, B.; Benavente, D.; Harris, I.; Déqué, M. Climatology of salt transitions and implications for stone weathering. *Sci. Total Environ.* **2011**, *409*, 2577–2585. [CrossRef]

44. Scheffer, T.C. A climate index for estimating potential for decay in wood structures above ground. *Prod. J.* **1971**, *21*, 25–31.
45. Lisø, K.R.; Hygen, H.O.; Kvande, T.; Thue, J.V. Decay potential in wood structures using climate data. *Build Res. Inf.* **2006**, *34*, 546–551. [CrossRef]
46. Curling, S.F.; Ormondroyd, G.A. Observed and projected changes in the climate based decay hazard of timber in the United Kingdom. *Sci. Rep.* **2020**, *101*, 1–9. [CrossRef] [PubMed]
47. Lebow, P.K.; Carll, C.G. Investigation of shift in decay hazard (Scheffer) index values over the period 1969–2008. *Proc. Am. Wood Prot. Assoc.* **2010**, *106*, 118–125.
48. Woon, J.S.; Boyle, M.J.; Ewers, R.M.; Chung, A.; Eggleton, P. Termite environmental tolerances are more linked to desiccation than temperature in modified tropical forests. *Insectes Sociaux* **2019**, *66*, 57–64. [CrossRef]
49. Brimblecombe, P.; Hayashi, M. Perception of the Relationship Between Climate Change and Traditional Wooden Heritage in Japan. In *Public Archaeology and Climate Change*; Oxbow Books: Barnsley, UK, 2017; pp. 288–302.
50. Hacquebord, L.; Avango, D. Industrial heritage sites in Spitsbergen (Svalbard), South Georgia and the Antarctic Peninsula: Sources of historical information. *Pol. Sci.* **2016**, *10*, 433–440. [CrossRef]
51. Andrews, M.B.; Ridley, J.K.; Wood, R.A.; Andrews, T.; Blockley, E.W.; Booth, B.; Burke, E.; Dittus, A.J.; Florek, P.; Gray, L.J.; et al. Historical simulations with HadGEM3-GC3.1 for CMIP6. *J. Adv. Model Earth Syst.* **2020**, *12*, e2019MS001995. [CrossRef]
52. Lim Kam Sian, K.T.; Wang, J.; Ayugi, B.O.; Nooni, I.K.; Ongoma, V. Multi-decadal variability and future changes in precipitation over Southern Africa. *Atmosphere* **2021**, *12*, 742. [CrossRef]
53. Tajet, H.T.T.; Hygen, H.O. Potential Risk of Wood Decay. 2017. Available online: https://www.met.no/publikasjoner/met-report/met-report-2017/_/attachment/download/c7df823f-5c98-4968-81fc-694e6fb6c49b:f495ddf4c9d7358398f610f0ed735c8382dad535/MET-report-08-2017.pdf (accessed on 3 May 2022).
54. Brimblecombe, P. Temporal humidity variations in the heritage climate of South East England. *Herit. Sci.* **2013**, *1*, 1–11. [CrossRef]
55. Brimblecombe, P.; Grossi, C.M.; Harris, I. The effect of long-term trends in dampness on historic buildings. *Weather* **2006**, *61*, 278–281. [CrossRef]
56. Soboń, M.; Bratasz, Ł. A method for risk of fracture analysis in massive wooden cultural heritage objects due to dynamic environmental variations. *Eur. J. Wood Wood Prod.* **2022**, *19*, 1–3. [CrossRef]
57. Richards, J.; Brimblecombe, P. The transfer of heritage modelling from research to practice. *Her Sci.* **2022**, *10*, 1–10. [CrossRef]
58. Dosio, A.; Jones, R.G.; Jack, C.; Lennard, C.; Nikulin, G.; Hewitson, B. What can we know about future precipitation in Africa? Robustness, significance and added value of projections from a large ensemble of regional climate models. *Clim. Dyn.* **2019**, *53*, 5833–5858. [CrossRef]
59. Akinsanola, A.A.; Ogunjobi, K.O.; Abolude, A.T.; Salack, S. Projected changes in wind speed and wind energy potential over West Africa in CMIP6 models. *Environ. Res. Lett.* **2021**, *16*, 044033. [CrossRef]
60. Fan, X.; Duan, Q.; Shen, C.; Wu, Y.; Xing, C. Global surface air temperatures in CMIP6: Historical performance and future changes. *Environ. Res. Lett.* **2020**, *15*, 104056. [CrossRef]
61. Richards, J.; Orr, S.A.; Viles, H. Reconceptualising the relationships between heritage and environment within an Earth System Science framework. *J. Cult. Herit. Manag. Sustain. Dev.* **2019**, *10*, 122–129. [CrossRef]

heritage

MDPI

Article

Moisture as a Driver of Long-Term Threats to Timber Heritage—Part II: Risks Imposed on Structures at Local Sites

Peter Brimblecombe [1] and Jenny Richards [2,3,*]

[1] Department of Marine Environment and Engineering, National Sun Yat-Sen University, Kaohsiung 80424, Taiwan
[2] St. John's College, Oxford University, Oxford OX1 3JP, UK
[3] School of Geography and the Environment, Oxford University, Oxford OX1 3QY, UK
* Correspondence: jennifer.richards@sjc.ox.ac.uk

Abstract: Timber heritage sites are vulnerable to damage from moisture. Simple meteorological descriptions of climate need to be tuned to capture drivers that threaten heritage, including dimensional change, insect attack and mould growth. Global climate models often provide projections through to the end of the 21st century but need to be translated to a local level to reveal processes of deterioration at specific sites. Translation to a local level can be challenging and requires the use of local information from a range of sources. This translation is explored over a range of sites facing different climate pressures, including fungal and insect risk at Harmondsworth Great Barn, England; changes in humidity range, salt risk and algal growth in rural timber buildings in the Midwestern states, USA; wind-driven rain impacts on board houses in Freetown, Sierra Leone; and rainfall and humidity range on timber buildings among the tropical rainforests of the Amazon, Congo Basin and Southeast Asia. Evidence-based narratives provide a tool to incorporate a multiplicity of local information to enrich projections and the interpretation of the model output. These could build trust and aid decision-making based on future projections, which are inherently uncertain.

Keywords: climate change; heritage climates; cultural heritage; built heritage; HadGEM3; narrative; deterioration

Citation: Brimblecombe, P.; Richards, J. Moisture as a Driver of Long-Term Threats to Timber Heritage—Part II: Risks Imposed on Structures at Local Sites. *Heritage* **2022**, *5*, 2966–2986. https://doi.org/10.3390/heritage5040154

Academic Editor: Francesco Soldovieri

Received: 22 August 2022
Accepted: 1 October 2022
Published: 5 October 2022

1. Introduction

Managing water relations is fundamental to the long-term protection of heritage [1]. Water affects heritage materials not only in terms of precipitation and humidity but also in its expression as flooding [2], water table changes and soil moisture content. Constructional timber, while widely used, has long been recognised as vulnerable to damage from fire and wear, and is sensitive to the ambient environment and climate, especially water-related aspects. For heritage sites constructed of timber, changes in moisture can drive fungal growth [3], affect biological processes in terms of insect attack [4] and exert mechanical stress on porous materials through wetting and drying and salt crystallisation [5,6].

Classical meteorological descriptions of climate often fail to capture the drivers that threaten heritage [7,8]. Climate parameters need to be tuned to the context of heritage threat, enabling these pressures to be expressed as particular heritage climatologies [1], which are taken to be particular climate forms that express a potential threat to heritage [9]. Richards and Brimblecombe [7] mapped six key moisture-related heritage climate parameters for timber heritage: Range in annual relative humidity (ΔRH), relative humidity (RH) seasonality, time of wetness (ToW), number of wind-driven rain (WDR) days, salt transitions and the Scheffer index, a function of timber decay.

Heritage climate is particularly important in the 21st century [3], especially as climate change is likely to impose shifts in weather patterns that threaten heritage [10]. The IPCC's *Sixth Assessment Report* suggests that by the end of the current century, temperatures are very likely to be 1.0–1.8 °C and 3.3–5.7 °C higher than last century, under low and high greenhouse gas emission scenarios, respectively [11]. Projections of future precipitation are more uncertain than temperature, as precipitation is highly variable on small spatial scales. Nevertheless, projections suggest an intensification of wet and dry periods of weather [11]. These changes are likely to alter potential moisture-based threats to timber heritage [7], e.g., prolonged desiccation of wood could reduce biological attacks by fungi or insects, which require a high water content.

An assessment of heritage deterioration often needs to incorporate dose or damage functions [12] and may be achieved by assessing occurrences and accumulation of environmental parameters. However, there is a need for a local, and even microscale [13], understanding of climate to improve the relevance of environmental observations to heritage management [14]. At the local scale, people responsible for heritage protection have worried about climate change impacts and have seen changes in precipitation and atmospheric moisture as a particular problem [15]. It is important to be able to transfer global-scale heritage climate assessments to a local level. Therefore, this paper aims to investigate the transferring of global pressures to site-specific risk. Our study focuses on meteorologically derived parameters (temperature, precipitation, relative humidity and wind speed) and does not specifically consider soil moisture change [8], surface flooding [16] and more extreme events, e.g., violent storms and landslides [17,18].

Approach

Previous research has commonly assessed global pressures using modelling [7,10]. However, as models are a simplification of reality, they require processes to be abstracted from their wider context. Therefore, when they are applied to unique heritage sites and objects, this simplification and abstraction can cause models to be viewed with varying levels of scepticism [14]. As summarised by Currie and Sterelny [19] (p. 14), "Where models often achieve isolation and precision at the cost of simplification and abstraction, narratives can track complex changes in a trajectory over time at the cost of simplicity and precision". Such narrative approaches can therefore be seen as bridging a communication gap between research, modelling and action [20]. The role of language in effective communication of climate change is critical to dialogues between researchers, policy makers and the general public, especially for issues of uncertainty [21,22].

We combine modelling and narrative, following Mike Hulme's thoughts found in *Why We Disagree about Climate Change* [23], where he advocates using case studies that are almost stories of a future, when describing the impact of climate change. Scientists, often positivist, can be resistant to stories [19,24,25], but this does not mean that imagined and imaginary geographies of heritage and climate change are detached from reality [26]. Furthermore, narratives can help identify diverse forms of data [19], enabling quantitative and qualitative data from multiple sources (e.g., observational records, cultural references such as books and films, policy documents and heritage reports) to be linked to model outputs.

2. Methods

2.1. Heritage Sites

In this study, we focus on heritage sites located in regions highlighted by Richards and Brimblecombe [7] as having notable changes in climate pressures over the coming century. The sites span a range of climates, heritage typologies and conservation challenges. Additionally, the amount and type of available data for each site vary, with some having rich historical records while others are more limited.

We focus on the following sites: (i) Harmondsworth Barn, a medieval barn in England (Figure 1d,e); (ii) rural timber buildings in Iowa, the USA, such as Dibble House (Figure 1a–c); (iii) timber board-houses in Freetown, Sierra Leone, built after the abolition of slavery (Figure 1f,g); and (iv) timber heritage located in rainforests across the Amazon (Figure 1i), Congo Basin (Figure 1h) and Southeast Asia, including Dayak houses and prisoner of war huts in Sarawak, Malaysia (Figure 1j,k). It is not possible to represent the enormous global range of timber heritage, given the small number of sites that can be discussed in a paper. Therefore, we selected sites that could speak more broadly, either in the type or the magnitude of the threat, the amount of data available or political and management contexts. It is worth noting that the sites selected were constructed to fulfil a range of purposes with some built to last many decades (e.g., Harmondsworth Barn), while others were intended to have much shorter intended lifespans (e.g., prisoner of war huts), influencing the construction materials and techniques used.

Figure 1. Map of key areas and photographs of sites discussed in this work. (**a**) Map of Midwestern USA; (**b**) painting *American Gothic* by Grant Wood, 1930; (**c**) Dibble House in Eldon, Iowa; (**d**) Harmondsworth Great Barn in Hillingdon Borough, London; (**e**) map showing barn location; (**f**) map of Sierra Leone; (**g**) old board house in Freetown; (**h**) map of the Congo; (**i**) map of the Upper Amazon; (**j**) map of Borneo; (**k**) Punjabi Barracks at Batu Lintang Camp, Kuching. Notes: Copyright details are in Supplementary Information Table S1.

2.2. *Datasets and Data Processing*

2.2.1. Modelled Heritage Climate Pressures

Moisture-related drivers of timber deterioration were modelled in Richards and Brimblecombe [7] using HadGEM3-GC31-MM over the time period 1850–2099. Future projections used a high-emission scenario (SSP585) to illustrate the extent of change under a worst-case scenario. Here, we also use the modelled heritage climate parameters:

- **Relative humidity range:** Annual range in mean monthly relative humidity (%), where the annual range (ΔRH_a) is $RH_{max} - RH_{min}$, and RH_{max} is the RH of the month with the highest mean RH in a given year and RH_{min} is the minimum mean monthly RH in the same year.
- **Relative humidity seasonality:** Month with the highest and lowest mean RH.
- **Time of wetness:** Number of days per year RH > 80% and temperature > 0 °C.
- **Wind-driven rain:** Number of days per year when rain is > 4 mm, mean wind speed is >2 m s^{-1} and temperature > 0 °C.
- **Salt transitions:** Number of cycles per year where the mean daily RH crossed below 75.5%, to account for sodium chloride crystallisation.
- **Scheffer index:** Risk of fungal attack expressed as $Sch = \Sigma(T_m - 2)(D - 3)/16.7$, which represents the sum over twelve months for the monthly mean temperature (T_m) and the number of days (D) in the month with ≥0.3 mm of rain [27,28].

Additionally, we used climate projections made as part of national studies (e.g., EPA report *Seasonality and Climate Change* [29]) where projections were tuned to local needs (e.g., the US reports *Impacts, Risks, and Adaptation in the United States: Fourth National Climate Assessment* [30]), and in the UK, reports on the impact of climate change, such as that from the National Trust [31] and academic research, as cited at the appropriate points in the text.

2.2.2. Observational Data

We used observations as these directly represent weather and climate at a specific location, although they may be made some distance from the site of interest. Extensive observations from national weather agencies or well-established meteorological institutions capture many meteorological parameters at a high resolution. Records can be lengthy, e.g., monthly rainfall is available from Kew in England from 1697 [32], while more regionally, the Central England Temperature Record begins with monthly data at 1659 and daily from 1772 [33]. The Harmondsworth Barn site in the UK (51.4897 N, −0.4799 E) is located close to the Heathrow weather station (51.4787 N, −0.44904 E). Weather records used from this station included hourly rainfall, daily temperature and wind speed, dating back to 1950. These are freely available from the UK Meteorological Office's MIDAS database [34–36] and were used to calculate the Scheffer index for Harmondsworth Barn over the period 1950–2020.

The discussion of humidity change in the American Midwest relied on some published climatological studies of recent and projected changes [30,37,38]. The section on the board-houses in Freetown, Sierra Leone, accessed present-day climate data from the Sierra Leone Meteorological Agency [39]. We used the current average annual rainfall data, as the daily and hourly precipitation was unavailable.

Homogenised data were sometimes used, in our study, because of the limited availability of observations at individual sites; while global, it may be at a lower spatial and temporal resolution. We used climate normals available at a global scale, and these cover the 30-year periods 1901–1930, 1931–1960, and 1961–1990. They are accessible online for more recent periods [40], with the next World Meteorological Organisation mandated normals for 1991–2020 under preparation, to be released 1 August 2023. This study also used the high-resolution gridded datasets maintained by the Climate Research Centre (CRU) at the University of East Anglia [41], which are typically at a monthly resolution. They can also be accessed via the WorldBank portal [42].

Meteorological observations can be affected by gaps, discontinuities or absent parameters. This means users have to be flexible in adopting such data and willing to accept judicious extrapolation, adjustment or proxy data. As an example, there was a discontinuity in the modelled wind speeds for Sierra Leone between the historic and future scenario model runs. To adjust for this, the difference in the decadal average on either side of the discontinuity (1995–2014; 2015–2024) was calculated and applied to the modelled wind speed data from the time period 2015–2099.

2.2.3. Heritage Documentation

Documentation about sites, materials and degradation processes was available from academic publications, policy documents, institutional reports and condition assessments. Especially useful were the *Preservation Briefs* (1978–present) [43–46] and the *Preservation Tech Notes* (1984–present) from the National Park Service, Washington, DC, and the consultation report *Review of the Monuments and Relics Act and Recommendations for New Heritage Legislation for Sierra Leone* [47]. Site guides were also useful, e.g., that of English Heritage [48] or Dibble House [49] and heritage sites in Freetown [50]. Where documentation is less readily available, broader comparisons can be made to the care and maintenance of timber heritage at other sites. We also searched for outputs from relevant fieldwork but found that such data were difficult to find for the sites we focused on.

2.2.4. Site Scale Material Damage

Climate pressures can be converted to estimates of damage through the use of the Scheffer index or other damage functions [12]. Insects and fungi represent a major threat to the deterioration of timber, so it is necessary to consider the environmental conditions conducive to their growth. As an example, our study of Harmondsworth Barn used observed temperature and relative humidity from Heathrow [36] to model the percentage of time when the optimum conditions occurred (1950–2020) for a selection of beetles and rot that commonly cause damage to timber in the UK. We modelled the optimum climate conditions for the powderpost beetle genera (*Lyctus* spp.), house longhorn beetle (*Hylotrupes bajulus*), deathwatch beetle (*Xestobium rufovillosum*), dry rot (*Serpula lacrymans*), cellar rot (*Coniophora puteana*) and oak rot (*Donkioporia expansa*). Powderpost beetles are a genus rather than specific species. Both powderpost and longhorn beetles predominantly attack sapwood because of their inability to digest lignin and cellulose. Furthermore, they pose little threat to wood over six to eight decades old as they are unable to digest degraded starch and sugars in the wood. However, we include this genus in this analysis due to their widespread occurrence [51], and they pose a risk to repairs to historic structures or new timber buildings. The optimal climate conditions were based on those outlined in [51] and are detailed in Supplementary Table S2.

2.3. Statistics

Non-parametric methods have frequently been used in this study. Trends in the observational data were determined using the Theil–Sen estimator to construct the line of best fit. The estimator uses the median slope and is a more robust method with noisy data due to its insensitivity to outliers. The Kendall τ statistic was used to test the significance of trends over time (analogous to the familiar regression statistic r^2).

2.4. Compiling Site Information

We assembled information from multiple sources to assess the deterioration of timber heritage sites. This process aimed to (i) provide a cultural context in which the heritage is located, so we could draw upon the history of the area, artistic works or local customs; (ii) bring the understanding of the effect of climate change on timber heritage to a broader audience, which could include relevant stakeholders such as the general public, local communities, conservators and heritage practitioners; (iii) suggest pathways for improved management, acknowledging that decision making has to balance financial, material and human resources, yet recognise local sensitivities.

3. Results

The results present analyses of local conditions from specific sites. These give a sense of the range of issues likely to confront timber heritage as a result of climate change threats outlined in an earlier publication [7].

3.1. Medieval Barns and Biological Risk

Richards and Brimblecombe [7] showed that timber heritage in western Europe is exposed to a range of moisture pressures. In particular, the Sheffer index is likely to increase over the coming century (Figure 2a,b). There are many iconic forms of European timber heritage including stave churches in Scandinavia and Russia, Germanic Fachwerk buildings and medieval barns, which could be affected by these changes. Old historic barns are significant as they provide insight into European medieval life and have also inspired the design of other buildings, including churches and libraries [48]. They are also interesting from a heritage climate perspective as many historic barns are uninhabited, without heating systems or insulation, meaning that their interior climates are similar to those outdoors though sheltered from direct rain.

In England, the Harmondsworth Barn (Figure 1c) is argued to rank "alongside the Houses of Parliament and Westminster Abbey for its exceptional architectural and historic interest"; it has been likened to a cathedral [48]. The barn is one of the largest (60 m long) ever built in Britain, made predominantly from oak with aisles running down the length of the structure. It was built in the early 1400s by Winchester College and provides a rich understanding of medieval farming practices and the production of wealth, and has been witness to natural changes in climate, e.g., the Little Ice Age. It was Grade I Listed in 1950 for its architectural and historic interest coupled with its rarity [52]. However, in the late 20th century, it was neglected, resulting in damage that included holes in the roof and root damage to the stone plinth [53], prior to English Heritage taking over its management and conservation. These repair materials will be sensitive to the impact of future climates.

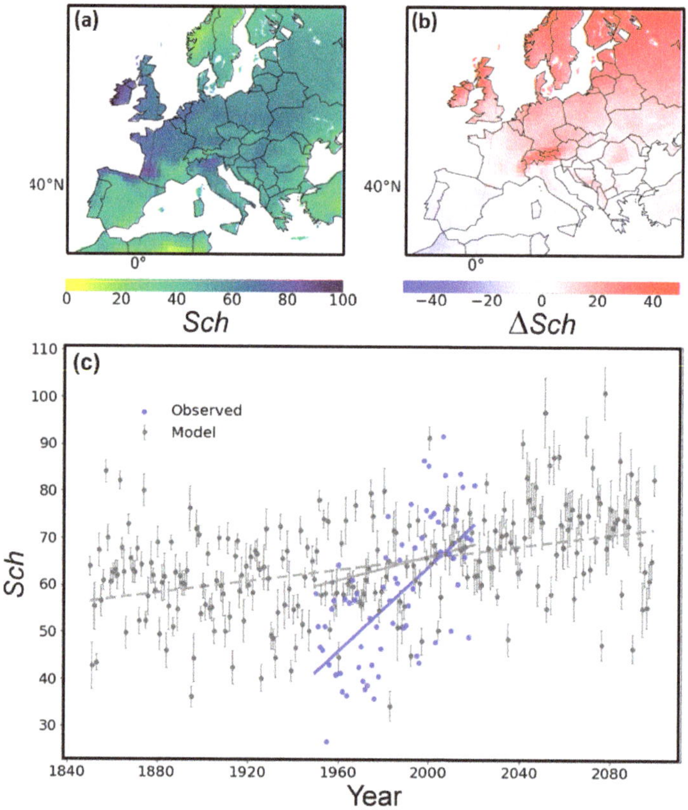

Figure 2. (**a**) Scheffer index for Western Europe, 1850–1879 and (**b**) the change in Scheffer index for this region between 2070–2099 and 1850–1879. (**c**) Annual Scheffer index for Harmondsworth calculated using (i) observed daily data from Heathrow 1950–2020 (blue) and (ii) modelled daily output from HadGEM3-GC31-MM 1850–2099 (grey). The modelled output is the mean of values from the nine grid cells centred on Heathrow, with error bars a standard deviation from the mean. Theil–Sen trends are fitted for the observed data (1950–2020) as a blue line with slope 0.45 a^{-1} ($\tau = 0.42$); the modelled output (1950–2020) as a grey line with slope 0.13 a^{-1} ($\tau = 0.17$); and the modelled output (1850–2099) as a grey dotted line with slope 0.06 a^{-1} ($\tau = 0.29$).

The historic neglect and changing ownership of Harmondsworth Barn mean it does not have such detailed conservation records, compared to some English buildings that have useful maintenance and expenditure records spanning many hundred years [54]. The academic research at the site has also been limited, and where undertaken, has tended to focus on digital documentation [55]. However, the close proximity of Harmondsworth Barn to the Heathrow weather station enables us to examine heritage climate pressures calculated using both observations and modelled data (Figure 2c).

3.1.1. Scheffer Index

The Kendall τ test for observations and modelled output for Harmondsworth Barn suggest that the Scheffer index is likely to increase significantly in the future (Figure 2c; observed data (1950–2020) $n = 70$, $\tau = 0.42$, $p_2 < 0.0001$; model (1850–2099) $n = 250$, $\tau = 0.29$, $p_2 < 0.0001$; model (1950–2020) $n = 70$, $\tau = 0.17$, $p_2 = 0.038$). This suggests fungal attack at the barn is likely to become more common through the 21st century.

The Scheffer index calculated using the observations and model output has similar ranges (Figure 2c). However, the index values calculated using the observed data were (i) generally lower, and (ii) the calculated rate of change over the last seven decades was 3.5 times greater than the model output (Figure 2c). It was possible this might be caused by a geographical misalignment, but when looking at the change in the Scheffer index for Scotland, this did not provide a full explanation for the rapid change in the observed data (Figure S1).

Despite global models not fully capturing local conditions, they can project the past and future, beyond observational records. For Harmondsworth Barn, the modelled output provides an important context beyond the period with observations, e.g., if we used the observed data to linearly project the Scheffer index to 2100, it would rise to 108. However, the modelled output suggests this is unlikely, with the Scheffer index levelling between 65 and 85. Combining observations and models provides conservation managers with the opportunity to see the local changes within a wider temporal and spatial context.

3.1.2. Optimum Climate Conditions for Individual Organisms

Figure 3 shows the percentage of days per year with optimal conditions at Harmondsworth Barn for insects and rot (1950–2020). For the three types of rot, the outdoor climate is rarely (<10 d a^{-1}) in the range required for the growth and reproduction of the species. This suggests that unheated structures are less likely to be subject to fungal attack than those warmed for human comfort. However, there has been a significant increase in the occurrence of optimum conditions for dry rot (Kendall $\tau = 0.29$, $p_2 < 0.001$) with the Theil–Sen slopes suggesting an increase in optimal conditionals by an extra day every two decades.

A significant increase in optimal conditions was also seen for powderpost beetles (0.07% per annum, Kendall $\tau = 0.39$, $p_2 < 0.0001$) with an extra optimal day added every four years. This increase will have little impact on timber older than six to eight decades but suggests an increase in the vulnerability of repaired timber or new timber structures to insect attacks in southern England over the coming century. There is no significant change for the house longhorn beetle, deathwatch beetle, cellar rot or oak rot (Kendall τ, $p_2 \gtrsim 0.05$).

We also found marked changes in the seasonality of optimal conditions (Figure 4). In July and August, the percentage of time spent in optimal conditions for dry rot increased between 1950 and 2019 (Figure 4a). Smaller increases were also noted in spring and autumn, which could extend the period in which this fungus would thrive at Harmondsworth Barn (Figure 4a). In addition, these seasonal changes were also seen in species where no significant change had occurred over an annual period. For example, the amount of time in the optimal conditions for the deathwatch beetle decreased in the summer but increased in winter and autumn (Figure 4b). This could mean that routine monitoring for the presence of deathwatch beetles in the Barn may need to be undertaken more frequently in autumn and winter.

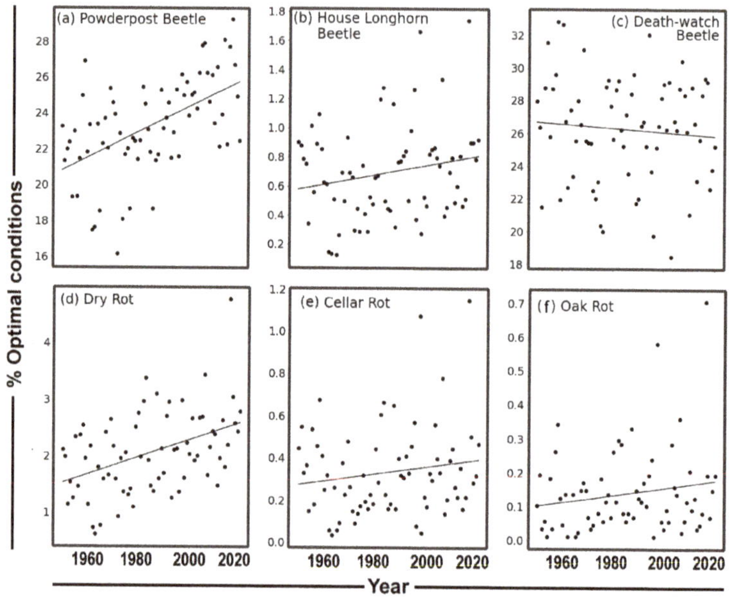

Figure 3. The percentage of time per year spent in optimal conditions between 1950 and 2020 for (**a**) the powderpost beetle; (**b**) house longhorn beetle, (**c**) deathwatch beetle, (**d**) dry rot, (**e**) cellar rot and (**f**) oak rot, with lines showing Theil–Sen slopes.

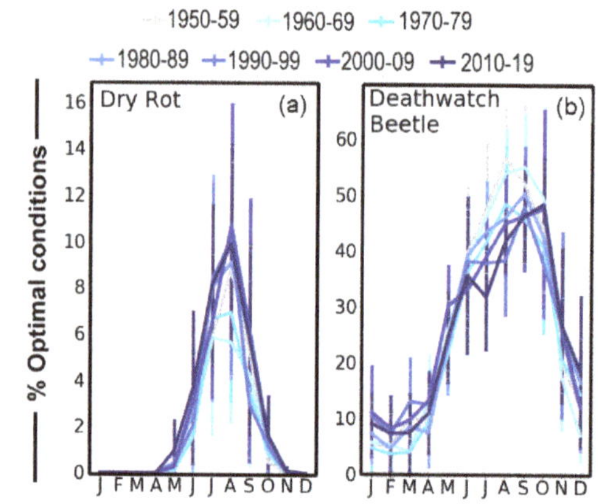

Figure 4. The mean percentage of time per month spent in optimal conditions for (**a**) dry rot, (**b**) deathwatch beetle, per decade. Error bars show one standard deviation from the mean.

3.2. Rural Timber Buildings and Humidity Threats

Richards and Brimblecombe [7] suggest that the annual RH range in many regions will increase through the 21st century, including the Great Plains of the North American Midwest (Figure 1). The Midwest has a humid continental climate with hot or warm summers and without a dry season. Angel et al. [30] describe annual temperatures in the warmer months as increasing more than in any other region of the US. In Iowa, temperatures have been increasing throughout the 20th century, but these are most notable at night with asymmetric warming that has been attributed to rising humidity [56]. The Midwest is projected to have an increased annual range in RH (Figure 5), which could be driven in part by intense heat waves that are often accompanied by high humidity [56]. In the central USA, Feng et al. [38] argued for a future with higher specific humidity (0.04 g kg^{-1} a^{-1})

and thus more storms. There may have been little change in annual RH (1947–2010), but there has been a noticeable decrease in March–May, and an increase in June–August [37]. The vapour pressure deficit is likely to increase, so plants and soils will become drier [30]. Change may be spread unevenly across the state, but in general, in the 20th century, there has been a notable anomaly in the summer humidity. This is a result of a moistening that has been characteristic of the Midwest [37], although our calculations [7] suggest the warmer months are likely to become less humid (lower RH) through the current century.

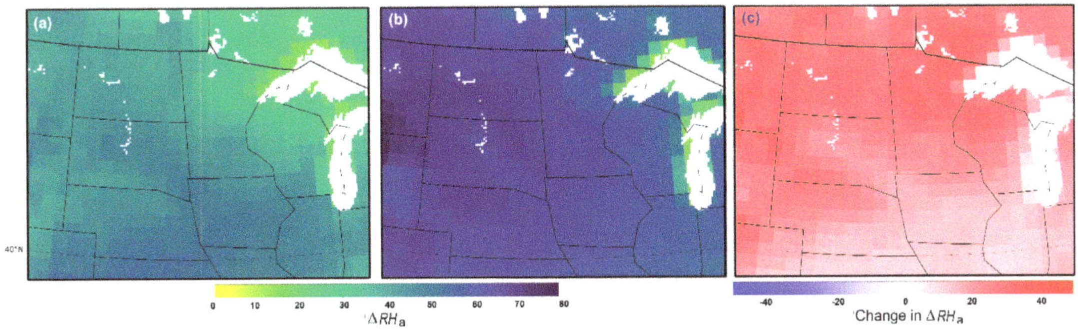

Figure 5. Mean annual RH range (ΔRH_a) for the periods (**a**) 1850–1879 and (**b**) 2070–2099 and (**c**) the difference between 2070–2099 and 1850–1879, from the HadGEM3 model data in the CMIP6 ensemble (HadGEM3-GC31-MM).

In Wisconsin, Iowa and Minnesota, there are numerous rural wooden heritage buildings associated with a 19th century expansion of agriculture, as cheap land encouraged European immigrants to settle. The farming activity is represented by the region's homes, rural workshops, old railway stations (Hatton railroad depot near West Millgrove, Ohio [57]) and mills [58]. Notable Prairie School architects, such as Frank Lloyd Wright and William Gray Purcell, constructed imaginative wooden buildings, e.g., the Town Hall (1915) at Jump River, Wisconsin [59]. Dibble House (designed by carpenters Busey and Herald), Eldon, Iowa, is an iconic gothic style house that features in Grant Wood's painting *American Gothic* (Figure 1b,c). The house is a testament to the gothic revival brought to the state by German immigrants. Wood's 1930 painting is doubtless one of the best-known images of interwar rural America; admired as much as lampooned. Dibble House is owned by the State Historical Society of Iowa. It currently operates as a museum managed by the Molalla Area Historical Society (American Gothic House Center [49]). It was refurbished in 1976 [60], but despite lengthy preservation efforts (1960–1990), there are few formal academic studies of the environmental threats to the house. There are also important wooden buildings at Herbert Hoover National Historic Site, West Branch, Iowa, including the Birthplace Cottage and a blacksmith shop [61].

3.2.1. Multiple Threats

The future climate is likely to expose timber heritage in the Midwest to multiple threats. The increasing humidity range predicted is likely to cause a greater dimensional change in wood, which may be a serious mechanism for damage in wood sheltered from the rain. In historic buildings without active mechanical ventilation, changing climate conditions will propagate indoors. The temperature of the interiors may be slightly higher due to solar gain [62], thereby RH is lowered further. Indoor wood, particularly painted wood, is sensitive to humidity change [63], so there could be increased cracking of coatings. Such changes are likely to occur in different seasons, as the month of maximum humidity (Figure S2) will move from the boreal spring (1984–2013) to earlier in the year (2070–2099), potentially requiring a change to the timing of maintenance regimes.

There is also likely to be a modest increase in the Scheffer index, and thus mould risk [7] driven by rising temperatures. The algal growth on paint work and roofing (e.g., shingles) has become an increasing concern in the USA [64]. The discolouration of painted surfaces or roofs by alga [65], while in some senses an aesthetic issue, can develop into broader maintenance problems [64]. The stains are often caused by the cyanobacterium, with *Gloeocapsa sp* a recognised problem for heritage [66]. Furthermore, the termite threat from species such as *Coptotermes formosanus* and *Reticulitermes flavipes* will expand northward through the Midwest over the coming decades [67].

Salt transitions should decrease from 50 (1850–1879) to 30 per year (2070–2099) [7]. Thus, threats to salt-laden timber may decline even though road salt continues to be used and persists in shallow aquifers [68]. However warmer winters [29] should mean that timber buildings near roads will be exposed to less road salt.

3.2.2. Maintenance Strategies

Smaller, rural buildings can find it hard to attract large flows of visitors or resources, so conservation is often an act of love provided by the local community. Often little academic literature is available to describe the conservation of these buildings. Focused and informed maintenance is stressed in National Parks Service guidance, in particular regarding protection against moisture [45]. The Service has published a range of briefs and technical notes dealing with special issues such as doors [69], porches [44] shingle roofs [46] and exterior paint on historic woodwork [43]. Additionally, moisture is a critical factor in the biological growth of paintwork and thrives under damp conditions or in the absence of sunlight [45]. Managing possible future damage at heritage sites in the Midwest will likely involve diligent maintenance, which could adopt Parks Service guidelines, in the absence of specific local practices. In many cases, such maintenance will require a continuation of current approaches but implemented more frequently or in different seasons to reduce the threat from a changing climate.

Increasing concern over algal growth and termites has led to considerable commercialisation of roof cleaning [70] and the protection of timber from insect attack [71]. In a world of changing climates, the algal darkening of roofs increases solar gain [72] at a time when decreases would be desirable. However, the aged look may be valued as a patina, encouraged by coating newly replaced parts of a roof with yoghurt [73].

3.3. *Board Houses and Wind-Driven Rain Risk*

West Africa has been noted as an area vulnerable to climate change exacerbated by rapid population growth and with high densities in urban areas, and variable community resilience [74,75]. The seasonal climate is dominated by the West African monsoon and the Harmattan winds from the Sahara. Recent studies have suggested that WDR (Figure 6a,b), extreme rainfall and consecutive dry days are likely to increase in future, but uncertainties in whether annual precipitation will increase or decrease remain [7,76,77]. The frequent occurrence of WDR along the Sierra Leonean coast suggests that it poses a threat to timber heritage (Figure 6a), and that by the end of the 21st century, this threat will have expanded inland (Figure 6b).

Within West Africa, Sierra Leone has been noted as being particularly vulnerable to climate change [75,78], affecting its timber heritage. In Sierra Leone, wood has commonly been used to create objects such as statues, ceremonial masks and instruments, as well as the construction of homes and monuments. On the Sierra Leone Heritage website (a digital resource aimed at virtually bringing together collections from Sierra Leone's rich cultural heritage), almost a quarter (24%) of the listed items are made from wood [79]. However, the colonial history of the country means that many of these objects are housed in European museums [80]. The timber board houses in Freetown (Figure 1g) remain in Sierra Leone, seen by residents as providing an important link to the city's history and the effects of slavery and colonialism. The houses' style reflects late 18th and 19th century American eastern seaboard cabins. These were built with the arrival of former American slaves to

Sierra Leone after the abolition of slavery [81]. They are now seen as an important form of vernacular architecture that provides a distinctive character to the city [47].

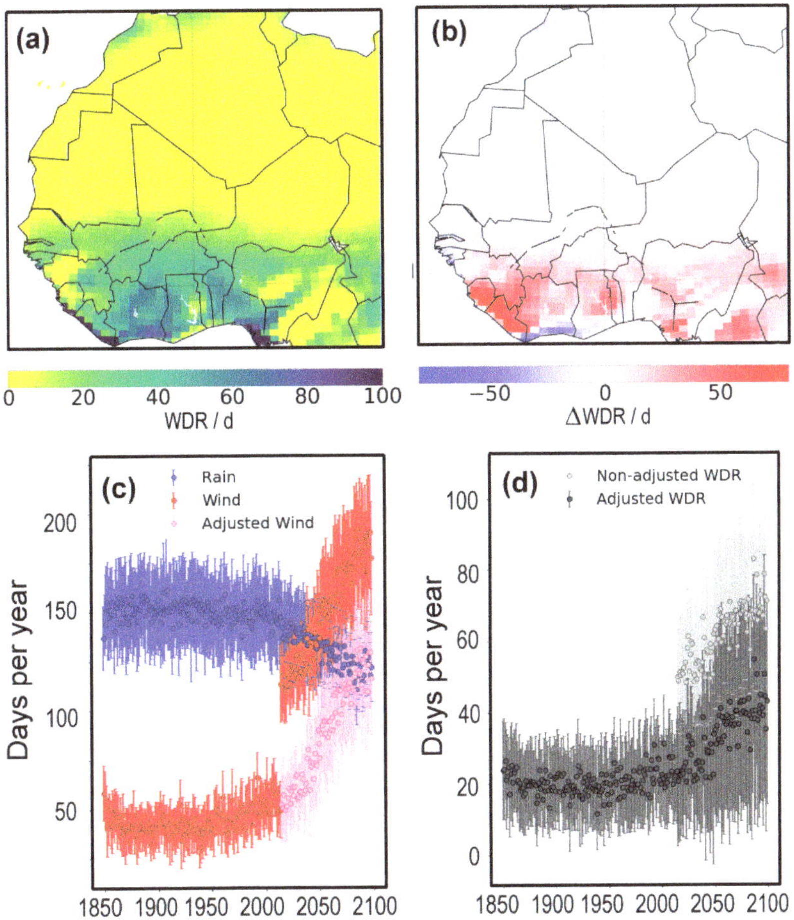

Figure 6. (**a**) Days of wind-driven rain per year for West Africa, 1850–1879 and (**b**) the change in days of wind-driven rain for this region between 2070–2099 and 1850–1879. (**c**) Number of days per year for Freetown where rain is >4 mm (blue) and mean wind speed is >2 m s^{-1} (red). The adjusted wind speed (2015–2099) is shown in pink. (**d**) Annual number of wind-driven rain days for Freetown calculated using the non-adjusted (light grey) and adjusted (dark grey) inputs. The output is the mean of values from the nine grid cells around Freetown, with error bars showing one standard deviation from the mean.

3.3.1. Climate and Climate Pressures

There are eight automatic weather stations located across Sierra Leone. Data from these stations show that average annual rainfall across the country is zonal, with coastal regions receiving >3500 mm a^{-1} (Figure S3). These meteorological records have been used by researchers such as Taylor et al. [82] to assess precipitation dating back to the 1960s. Richards and Brimblecombe [7] found that the projected changes of climate pressure to heritage in this region were driven by increasing threats from WDR (Figure 6a,b) and a change in RH seasonality. Freetown could experience a 50% increase in the projected number of wind-driven days between 2015 and 2099 (Figure 6d).

The occurrence of pressures on timber heritage in Freetown was found to be seasonal. Wind-driven rain and ToW occurred most frequently in August and September, coinciding with the peak months of the rainy season (Figure 7). In contrast, salt transitions peak between February and June, during the dry season, associated with higher temperatures. With future increases in temperature and warmer weather in the rainy season, wood might dry more quickly between rain events, increasing the number of wetting–drying cycles, and thus driving mechanical damage. However, drier conditions could also reduce biological growth, altering the present balance between the drivers of deterioration. Such an outcome may mean that future conservation will need an increasing focus on addressing mechanical rather than biological damage.

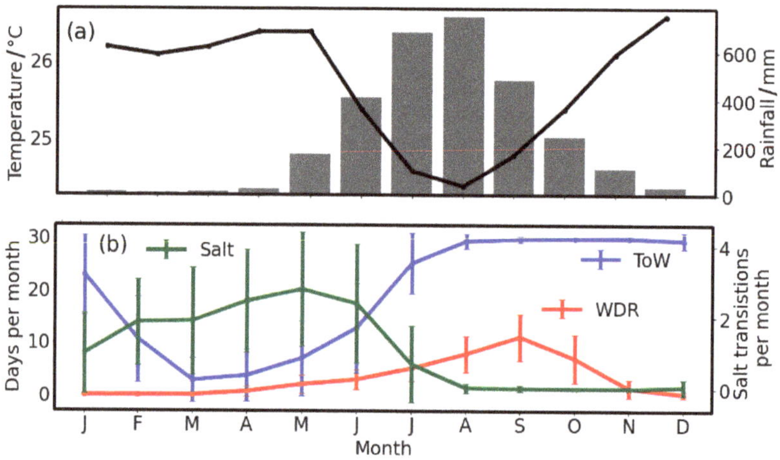

Figure 7. Freetown's (**a**) average monthly temperature and rainfall (1991–2021—raw data from [83]) and (**b**) days of wind-driven rain, red; time of wetness in days, blue; and salt transitions, green (1984–2013).

3.3.2. Conservation Challenges

In addition to pressures caused by a shifting heritage climate, the board houses in Freetown face threats from the growing popularity of stone and concrete as building materials. Thus, the 20th century has seen a decline in the condition and number of board houses [47]. Furthermore, Sierra Leone faces many other challenges, which include widespread impoverishment, high levels of youth unemployment and poor infrastructure, along with disease outbreaks including Ebola, cholera and COVID-19 [84]. In this context, the conservation of vernacular-built heritage can easily be overlooked. However, as argued in a consultation report for the Sierra Leonean Ministry of Tourism and Cultural Affairs, cultural heritage "is an integral part of the broader cultural life of a nation and can provide a vehicle for economic development, for building social cohesion and stability, for achieving environmental sustainability, and for fostering resilience in communities . . . that needs to be carefully managed and safeguarded through national legislation that is adequately supported and effectively implemented by the State" [47] (p3).

These board houses critically require effective conservation, through the implementation of monitoring systems, maintenance and planning regulations [47]. The management of these sites could provide an opportunity to foster international relations between Sierra Leone and other countries facing similar heritage climate pressures and conservation challenges. Networks could be developed virtually, with online meetings and workshops becoming increasingly common since the COVID-19 pandemic.

3.4. Built Heritage of the Wet Tropics and Time of Wetness

It is easy to overlook built heritage in tropical areas as they are so widely recognised as being rich in natural heritage and biodiversity. The rainforests that characterise so much of the tropics reflect very warm and wet climates. These conditions place much pressure on the preservation of timber, as continuous high humidity takes its toll on organic materials. Major areas of tropical rain forests include the great river basins of the Amazon and the Congo along with the forests of Southeast Asia. The seasonal cycle of precipitation from these regions is shown in Figure 8. Total annual rainfall varies between regions, and often presents a distinct seasonal cycle. There are only slight trends (1901–2021) in the annual rainfall amount, which increases by 1 mm a^{-1} (p_2 = 0.0025) in Amazonas, an insignificant 0.07 mm a^{-1} (p_2~0.5) in the Congo and 1.6 mm a^{-1} (p_2 = 0.08) in Sarawak, Malaysia. In earlier work [7], we gave a picture of a less humid tropical climate, with wider ranges of humidity, and reduced times of wetness (Figure 9), such that surfaces remain wet for shorter times [85].

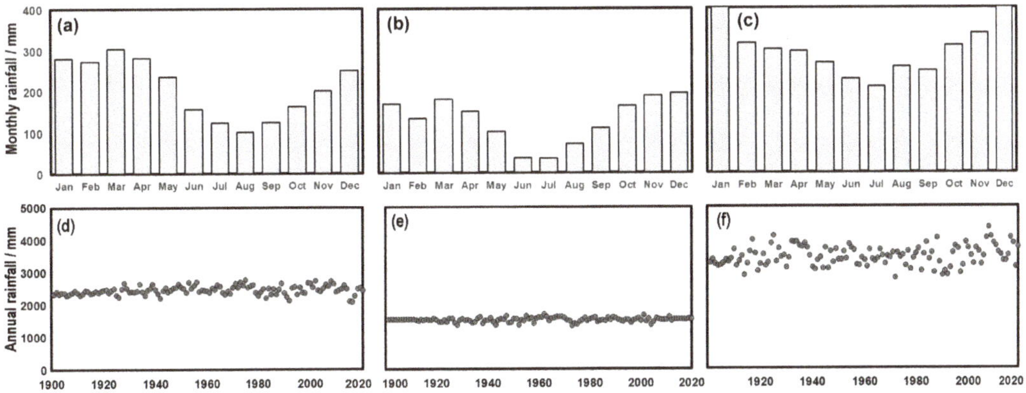

Figure 8. Seasonal cycles of precipitation averaged across 30 years (1992–2021) for (**a**) Amazonas State, Brazil, (**b**) Maniema Province, Democratic Republic of Congo and (**c**) Sarawak, a Malaysian state on the island of Borneo. Annual rainfall (1901–2021) shown for (**d**) Amazonas, (**e**) Maniema and (**f**) Sarawak. Datasets: [41,42].

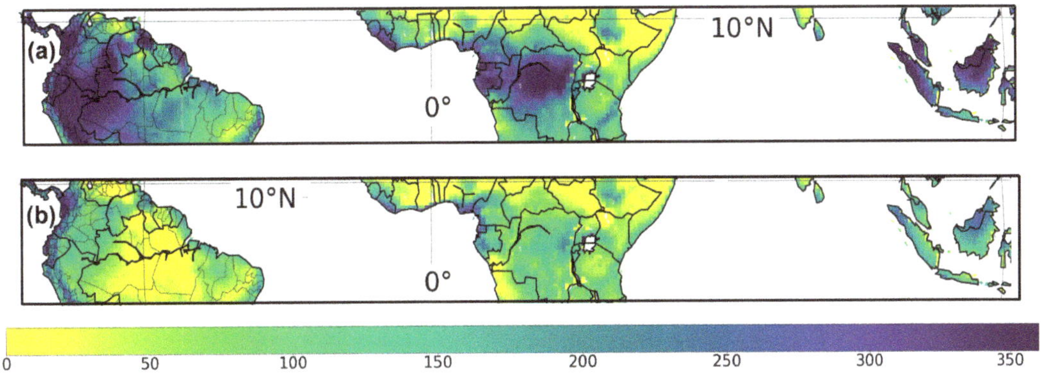

Figure 9. The time of wetness in days (**a**) 1850–1879 and (**b**) 2070–2099.

3.4.1. Amazonia

Manaus (Brazil) and Iquitos (Peru) are Amazonian cities, notable for art, architecture and culture and were a backdrop for Werner Herzog's film *Fitzcarraldo* (1982), where a misguided visionary is determined to attract funds for a new opera house in the Amazon. Additionally, other smaller municipalities have interesting heritage. In the upper Amazon, Tefé, has historic sites such as the Barreira das Missões. The town was the headquarters of the scientific commission (1781–91) that settled the Spanish–Portuguese boundaries and was a centre for subsequent exploration [86].

Tefé has a pronounced seasonal cycle in rainfall (Figure 10a), with more than 300 mm mth^{-1} March–May, through to as little as 100 mm mth^{-1} by August. This pattern follows the general climate of Amazonas State quite closely (Figure 8a). Esquivel-Muelbert et al. [87] have summarised the changing climate across the Amazonian region and suggest that, in recent decades, the dry season has been more intense. There have been repeated drought events and precipitation has declined in the south and south-east. However, precipitation has increased during the wet season, with episodes of extreme rainfall. As a result, ecological changes are observed in forests that have become increasingly dominated by large trees. The seasonal cycle of humidity in Amazonas from HadGEM3-GC31-MM shows that, in the future, RH is likely to decline, but there will be a more distinct seasonal cycle with April–June remaining humid (Figure 10b).

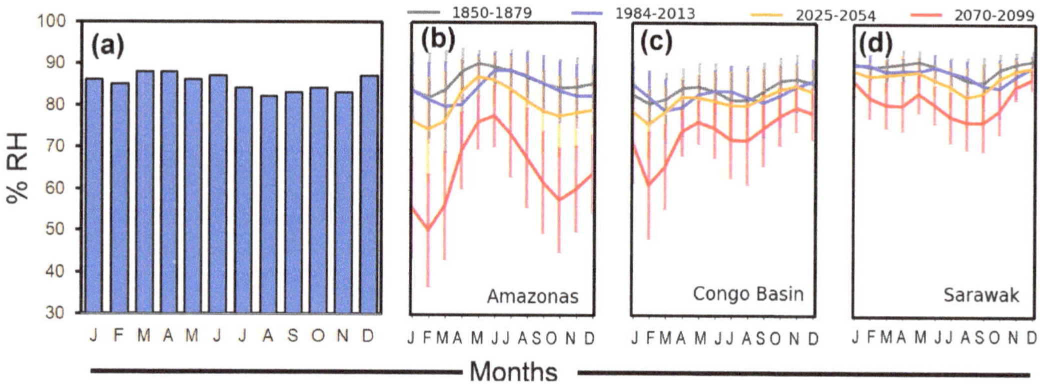

Figure 10. (**a**) Average monthly relative humidity (RH%) in Tefé, Amazonas 1961–1990. The 30-year mean monthly relative humidity (RH%) for the time periods 1850–1879 (grey), 1984–2013 (blue), 2025–2054 (yellow) and 2070–2099 (red) for (**b**) Amazonas, (**c**) the Congo Basin (**d**) and Sarawak. Error bars show one standard deviation from the mean.

Less humid conditions, while seemingly advantageous for the protection of timber heritage from biological damage, would likely mean that wooden objects equilibrated to high humidity would lose water and contract. This could lead to cracks or the disruption of surface coatings [63,88]. In addition, shifts in seasonality can affect the times popular for visits and therefore might require adjustments to site management [89]. The future changes to climate could also mean the loss of some termites in Brazil, with less humid conditions potentially reducing the diversity of termites in tropical forests [90], e.g., a contraction of the range of *Coptotermes formosanus*, though there may be an expansion of other species such as *Mastotermes darwiniensis* [67].

3.4.2. Congo

In the Congo, Kisangani (formerly Stanleyville), the capital of Tshopo province, is a populous and important city. The city is relatively large (1.3 M), but the built heritage is often in poor condition and not part of the standard tourist itinerary [91]. The older colonial buildings are described on a website provided by Jean-Luc Ernst [92]. Iconic for some may be the dilapidated *Villa Regina*, where Katherine Hepburn and Humphrey Bogart stayed while John Huston's *The African Queen* was filmed.

Figure 10c shows that the Congo may become generally less humid over the current century [7], doubtless taking its toll on insects and flora. In terms of precipitation, there has been no significant change over the last century (Figure 8e), and future projected change is highly uncertain, with changes ranging from −50% to +150% for the December–February dry season [93,94]. This uncertainty makes planning future management for timber heritage challenging, emphasising the urgent need for improved climate models for precipitation over Africa.

3.4.3. Sarawak

Sarawak is notable for its Dayak architecture (mainly longhouses with dwelling and communal areas) and many traditional shop houses (which have timber elements). Entirely of wood are the remaining buildings of the Batu Lintang POW camp (e.g., Punjabi Barracks) [95], as featured in Agnes Newton Keith's autobiography *Three Came Home*.

In Borneo, annual rainfall is likely to increase [96], although, as with many regions in the tropics, there is much uncertainty. The seasonal RH in Sarawak seems set to marginally decline (Figure 10d) yet will remain high (~80%). Therefore, surfaces should continue to be damp for long periods under future climates, so here, a biological threat seems likely to persist. Increases in monthly rainfall have been observed in the Limbang River basin (1970–2015) [97], along with increases in short-term intensity [96]. These changes are projected to continue into the future [96], so surface flooding may form an important aspect of management plans.

4. Discussion

Our study shows how global-scale heritage climate pressures can be combined with information from a range of sources to describe the potential deterioration of timber heritage at a local scale. Assessing heritage from several locations allowed us to show that this process is challenging due to (i) limited data and availability; (ii) difficulties in converting environmental pressure to damage; and (iii) translation predictions of future risk to conservation management.

4.1. Data

On a global scale, the use of climate models to calculate heritage climate pressures provides worldwide coverage of the potential threat [7]. This enables researchers to undertake a global analysis, even if there has been little previous research, or limited available data at specific locations. This approach helps address geographical biases in heritage research, addressing issues raised by Simpson et al. [98]. The examples chosen in our study indicate that readily available local data make the translation of heritage pressures from global to local easier. For sites with detailed meteorological records, such as Harmondsworth Barn, it was possible to directly estimate climate pressures and assess potential drivers of deterioration (Figures 2c and 3).

For timber heritage in regions where local data were harder to access, such as the board houses in Freetown and prisoner of war camp in Sarawak, translation from global to local scales required the judicious use of available information. Combining modelled data with information from grey literature, policy documents and archive sources, enabled us to develop a better local understanding of future site threats. However, available data may need careful adjustment, as seen with the board houses in Freetown. Here, the modelled output showed a sharp discontinuity in days of wind-driven rain between 2014 and 2015 (Figure 6c). Improved meteorological monitoring would help resolve such uncertainties, gaps and singularities to better assess risks to timber heritage at specific sites.

4.2. Damage

Quantifying the pressures affecting timber heritage sites provides an indication of the extent of potential damage. However, converting these pressures to damage requires an understanding of scale and process. In our examples, we see that multiple processes can operate individually or synergistically at a site, so focusing on a single damage pathway may not capture the complexity of the threat.

In heritage science, damage functions have commonly been used to convert environmental pressures to damage [12]. However, they are typically developed for historic materials rather than sites. Therefore, damage functions can seem to lack relevance at the site-scale as they include only a limited set of processes, which may not reflect the complexity of conditions at a site. Site-scale environmental pressures can be represented at high temporal and spatial resolution by computational fluid dynamics [99,100]. Such models are computationally expensive to run at the multi-decadal timescales used in this paper, but heritage-specific representations may be possible, especially with increasing computing power.

4.3. Action

The examples in this study also highlight the challenges of translating potential damage into conservation or management action. Future projections are associated with uncertainty, due to the quality of data, understanding of the process and additional unknowns. This can render findings unconvincing if the error is not explicitly addressed. Uncertainties frustrate the decision-making process, as different conservation actions could depend on the balance of such uncertainties. Furthermore, these can lead to distrust of models, potentially stalling action, when heritage managers need to take risky decisions about historic properties.

Protecting heritage against future climate shows parallels with the difficult translation of climate change research to climate change action [101]. This challenge creates a role for narrative in expressing heritage damage in terms of conservation actions and can be important in reducing hesitancy associated with uncertainty. The construction of evidence-based narratives enables researchers to envision a range of possible outcomes to present to stakeholders, such as politicians, funders and the general public. This could represent a powerful mechanism for improving public understanding of timber heritage, in a context relevant to visitors, managers and owners. This approach may also be a useful tool in targeting fieldwork efforts. Comparisons of early field observations with model outputs could act as a preliminary assessment of the need for further work. However, much field data are collected but are not readily accessible due to limited or dispersed publication venues for this form of information. The lack of a dedicated, well-indexed repository limits access to such information.

4.4. Effective Modelling

For models to be effective they need to be usable by those who need them [14]. This means for site-scale conservation, models need to be user-friendly and the output useful to those engaged with site management (e.g., site managers, non-professional heritage practitioners and volunteers). This requires models to be built with the end user requirements prioritised from the start, as otherwise, models of heritage pressures or damage can be impenetrable or highly complex to run. For example, in the case of Iowa, the management of numerous small buildings by local communities could benefit from the development of user-friendly models, with graphical interfaces that reveal local-scale damage, readily interpretable by those caring for the buildings.

Furthermore, when models are isolated from observations and local experience, their effectiveness can be reduced. When enriched with contextual information, the output becomes more specific to the site. As the availability of contextual data to enrich the model is highly variable between sites, flexibility is needed by researchers and conservators in determining what data are acceptable, requiring us to be judicious and potentially creative in utilising available information.

5. Conclusions

Global climate models are a powerful tool in providing a picture of climate across the next century, with outputs that can be tuned to represent a multiplicity of threats to heritage. Heritage is at a fine scale, so it is important to develop approaches to apply global output at the site level. It is challenging, but not impossible, to transfer the focus to local settings. The case studies show that model projections should not be taken in isolation but rather interfaced with a wide variety of local observations and other contextual information to incorporate multiple threats to heritage. At sites with limited information, researchers and practitioners should address the transfer of global to local scale through flexible and imaginative use of disparate quantitative and qualitative sources. This could include, for example, fieldwork observations, oral histories, site descriptions, photographs, films and guidebooks. Weaving together fragmentary understandings into a narrative could effectively communicate threat and uncertainty. Evidence-based narratives provide a mechanism to build trust in modelled outputs, broadening engagement with ideas about the interactions between climate change and heritage. We intend for the current study to promote discussion between heritage researchers, practitioners and managers. Particularly, similar threats facing local heritage in geographically disparate regions could encourage international collaboration between heritage practitioners and policy makers. Effective modelling remains a worthy challenge, yet it is important to remember, no prediction can ever be complete.

Supplementary Materials: The following supporting information can be downloaded at: https://www.mdpi.com/article/10.3390/heritage5040154/s1, Table S1: Copyright information for Figure 1; Figure S1. Scheffer index for Harmondsworth and Scotland; Table S2: Optimum climate conditions for organisms, based on English Heritage (2012) *Practical Building Conservation. Timber*; Figure S2: Modal month with the maximum monthly RH in the American Midwest for the periods (a) 1850–1879, (b) 1984–2013, (c) 2025–2054 and (d) 2070–2099; Figure S3: Isobar map of average annual rainfall (mm) in Sierra Leone sourced from the Sierra Leone Meteorological Agency.

Author Contributions: Conceptualization, P.B.; data curation, J.R.; formal analysis, P.B. and J.R.; investigation, P.B. and J.R.; methodology, P.B. and J.R.; software, J.R.; visualization, P.B. and J.R.; writing—original draft, P.B. and J.R.; writing—review and editing, P.B. and J.R. All authors have read and agreed to the published version of the manuscript.

Funding: This research received no external funding.

Data Availability Statement: Data are available from the links noted in the text.

Conflicts of Interest: The authors declare no conflict of interest.

References

1. Sabbioni, C.; Brimblecombe, P.; Cassar, M. The Atlas Ofclimate Change Impact on European Cultural Heritage. In *Scientific Analysis and Management Strategies*; Anthem Press: London, UK, 2010.
2. Howard, A.J.; Hancox, E.; Hanson, J.; Jackson, R. Protecting the Historic Environment from Inland Flooding in the UK: Some Thoughts on Current Approaches to Asset Management in the Light of Planning Policy, Changing Catchment Hydrology and Climate Change. *Hist. Environ. Policy Pract.* **2017**, *8*, 125–142. [CrossRef]
3. Sesana, E.; Gagnon, A.S.; Ciantelli, C.; Cassar, J.; Hughes, J.J. Climate Change Impacts on Cultural Heritage: A Literature Review. *WIREs Clim. Chang.* **2021**, *12*, e710. [CrossRef]
4. Brimblecombe, P.; Lankester, P. Long-Term Changes in Climate and Insect Damage in Historic Houses. *Stud. Conserv.* **2013**, *58*, 13–22. [CrossRef]
5. Grossi, C.M.; Brimblecombe, P.; Menéndez, B.; Benavente, D.; Harris, I.; Déqué, M. Climatology of Salt Transitions and Implications for Stone Weathering. *Sci. Total Environ.* **2011**, *409*, 2577–2585. [CrossRef]
6. Oguchi, C.T.; Yu, S. A Review of Theoretical Salt Weathering Studies for Stone Heritage. *Prog. Earth Planet Sci.* **2021**, *8*, 1–23. [CrossRef]

7. Richards, J.; Brimblecombe, P. Moisture as a Driver of Long-Term Threats to Timber Heritage. Part I: Changing Heritage Climatology. *Heritage* **2022**, *5*, 1929–1946. [CrossRef]
8. Brimblecombe, P. Refining Climate Change Threats to Heritage. *J. Inst. Conserv.* **2014**, *37*, 85–93. [CrossRef]
9. Brimblecombe, P. Heritage Climatology. In *Climate Change and Cultural Heritage*; Lefevre, R.-A., Sabbioni, C., Eds.; Edipuglia: Bari, Italy, 2010; pp. 57–64.
10. Orr, S.A.; Richards, J.; Fatorić, S. Climate Change and Cultural Heritage: A Systematic Literature Review (2016–2020). *Hist. Environ. Policy Pract.* **2021**, *12*, 434–477. [CrossRef]
11. IPCC. Summary for Policymakers. In *Climate Change 2021: The Physical Science Basis. Contribution of Working Group I to the Sixth Assessment Report of the Intergovernmental Panel on Climate Change*; Masson-Delmotte, V., Zhai, P., Pirani, A., Connors, S.L., Péan, C., Berger, S., Caud, N., Chen, Y., Goldfarb, L., Gomis, M.I., et al., Eds.; Cambridge University Press: Cambridge, UK; New York, NY, USA, 2021; pp. 3–32.
12. Strlič, M.; Thickett, D.; Taylor, J.; Cassar, M. Damage Functions in Heritage Science. *Stud. Conserv.* **2013**, *58*, 80–87. [CrossRef]
13. Choidis, P.; Kraniotis, D.; Lehtonen, I.; Hellum, B. A Modelling Approach for the Assessment of Climate Change Impact on the Fungal Colonization of Historic Timber Structures. *Forests* **2021**, *12*, 819. [CrossRef]
14. Richards, J.; Brimblecombe, P. The Transfer of Heritage Modelling from Research to Practice. *Herit. Sci.* **2022**, *10*, 1–10. [CrossRef]
15. Cassar, M. *Climate Change and the Historic Environment*; Centre for Sustainable Heritage, University College London: London, UK, 2005; Available online: https://discovery.ucl.ac.uk/id/eprint/2082/1/2082.pdf (accessed on 21 August 2022).
16. Brimblecombe, P.; Grossi, C.M.; Harris, I. The Effect of Long-Term Trends in Dampness on Historic Buildings. *Weather* **2006**, *61*, 278–281. [CrossRef]
17. Brimblecombe, P.; Hayashi, M.; Futagami, Y. Mapping Climate Change, Natural Hazards and Tokyo's Built Heritage. *Atmosphere* **2020**, *11*, 680. [CrossRef]
18. Sardella, A.; Palazzi, E.; von Hardenberg, J.; del Grande, C.; de Nuntiis, P.D.; Sabbioni, C.; Bonazza, A. Risk Mapping for the Sustainable Protection of Cultural Heritage in Extreme Changing Environments. *Atmosphere* **2020**, *11*, 700. [CrossRef]
19. Currie, A.; Sterelny, K. In Defence of Story-Telling. *Stud. Hist. Philos. Sci. Part A* **2017**, *62*, 14–21. [CrossRef] [PubMed]
20. Corless, V. The Role of Narrative in Science. Available online: https://www.advancedsciencenews.com/the-role-of-narrative-in-science/ (accessed on 10 August 2022).
21. Nerlich, B.; Koteyko, N.; Brown, B. Theory and Language of Climate Change Communication. *WIREs Clim. Chang.* **2010**, *1*, 97–110. [CrossRef]
22. Pearce, W.; Brown, B.; Nerlich, B.; Koteyko, N. Communicating Climate Change: Conduits, Content, and Consensus. *WIREs Clim. Chang.* **2015**, *6*, 613–626. [CrossRef]
23. Hulme, M. *Why We Disagree about Climate Change: Understanding Controversy, Inaction and Opportunity*; Cambridge University Press: Cambridge, UK, 2009.
24. Matless, D. Climate Change Stories and the Anthroposcenic. *Nat. Clim. Chang.* **2016**, *6*, 118–119. [CrossRef]
25. Wise, M.N. On the Narrative Form of Simulations. *Stud. Hist. Philos. Sci. Part A* **2017**, *62*, 74–85. [CrossRef] [PubMed]
26. Brimblecombe, P. Climate Myths. In *Climate Change and Cultural Heritage*; Lefevre, R.-A., Sabbioni, C., Eds.; Edipuglia: Bari, Italy, 2010; pp. 207–209.
27. Scheffer, T.C. A Climate Index for Estimating Potential for Decay in Wood Structures above Ground. *For. Prod. J.* **1971**, *21*, 25–31.
28. Lisø, K.R.; Hygen, H.O.; Kvande, T.; Thue, J.V. Decay Potential in Wood Structures Using Climate Data. *Build. Res. Inf.* **2007**, *34*, 546–551. [CrossRef]
29. Environmental Protection Agency. Seasonality and Climate Change: A Review of Observed Evidence in the United States; 2021. Available online: https://www.epa.gov/system/files/documents/2021-12/30339_epa_report_climate_change_and_seasonality_v12_release_508.pdf (accessed on 10 August 2022).
30. Angel, J.; Swanston, C.; Boustead, B.M.; Conlon, K.C.; Hall, K.R.; Jorns, J.L.; Kunkel, K.E.; Lemos, M.C.; Lofgren, B.; Ontl, T.A.; et al. Midwest. In *Impacts, Risks, and Adaptation in the United States: Fourth National Climate Assessment, Volume II (Report)*; Reidmiller, D.R., Avery, C.W., Easterling, D.R., Kunkel, K.E., Lewis, K.L.M., Maycock, T.K., Stewart, B.C., Eds.; U.S. Global Change Research Program: Washington, DC, USA, 2018; pp. 872–940.
31. Tourism National Trust. National Trust How Climate Change Will Affect the Future of UK. Available online: https://www.nationaltrust.org.uk/features/how-climate-change-will-affect-the-future-of-uk-tourism (accessed on 17 August 2022).
32. Todd, B. Reconstructing Long-Term Records of UK Drought and Analysis of Variability: 1697–2013. Ph.D. Thesis, University of Liverpool, Liverpool, UK, 2014.
33. Parker, D.E.; Legg, T.P.; Folland, C.K. A New Daily Central England Temperature Series, 1772–1991. *Int. J. Climatol.* **1992**, *12*, 317–342. [CrossRef]
34. Met Office. *MIDAS Open: UK Daily Weather Observation Data, V202107*; NERC EDS Centre for Environmental Data Analysis: Didcot, UK, 2021. [CrossRef]
35. Met Office. *MIDAS Open: UK Hourly Rainfall Data, V202107*; NERC EDS Centre for Environmental Data Analysis: Didcot, UK, 2021. [CrossRef]
36. Met Office. *MIDAS Open: UK Daily Temperature Data, V202107*; NERC EDS Centre for Environmental Data Analysis: Didcot, UK, 2021. [CrossRef]
37. Brown, P.J.; Degaetano, A.T. Trends in U.S. Surface Humidity, 1930–2010. *J. Appl. Meteorol. Climatol.* **2013**, *52*, 147–163. [CrossRef]

38. Feng, Z.; Leung, L.R.; Hagos, S.; Houze, R.A.; Burleyson, C.D.; Balaguru, K. More Frequent Intense and Long-Lived Storms Dominate the Springtime Trend in Central US Rainfall. *Nat. Commun.* **2016**, *7*, 1–8. [CrossRef]
39. Sierra Leone Meteorological Agency. Sierra Leone Meteorological Agency. Available online: https://slmet.gov.sl/ (accessed on 17 August 2022).
40. World Meteorological Organization. WMO Climate Normals. Available online: https://www.ncei.noaa.gov/products/wmo-climate-normals (accessed on 17 August 2022).
41. Climatic Research Unit (University of East Anglia). High-Resolution Gridded Datasets. Available online: https://crudata.uea.ac.uk/cru/data/hrg/ (accessed on 17 August 2022).
42. World Bank Group. Climate Change Knowledge Portal. Available online: https://climateknowledgeportal.worldbank.org/ (accessed on 17 August 2022).
43. Weeks, K.D.; Look, D.W. Exterior Paint Problems on Historic Woodwork. *Preserv. Briefs* **1982**, *10*, 1–12.
44. Sullivan, A.; Leeke, J. Preserving Historic Wood Porches. *Preserv. Briefs* **2006**, *45*, 1–20.
45. Park, S.C. Holding the Line: Controlling Unwanted Moisture in Historic Buildings. *Preserv. Briefs* **1996**, *39*, 1–16.
46. Park, S.C. The Repair and Replacement of Historic Wooden Shingle Roofs. *Preserv. Briefs* **1989**, *19*, 1–12.
47. Basu, P.; Sam, M.A. *Review of the Monuments and Relics Act and Recommendations for New Heritage Legislation for Sierra Leone, Consultation Report*; The Ministry of Tourism and Cultural Affairs and Monuments and Relics Commission, Government of Sierra Leone: Freetown, Sierra Leone, 2016.
48. English Heritage. Harmondsworth Barn. Available online: https://www.english-heritage.org.uk/visit/places/harmondsworth-barn/ (accessed on 17 August 2022).
49. American Gothic. House Center American Gothic House Center. Available online: https://americangothichouse.org/ (accessed on 17 August 2022).
50. Sierra Leone National Tourist Board. Historical and Heritage Sites. Available online: https://ntb.gov.sl/historical-and-heritage-sites/# (accessed on 17 August 2022).
51. English Heritage. *Practical Building Conservation. Timber*; Routledge: Abingdon, UK, 2012.
52. Historic England. The Great Barn, Harmondsworth. Available online: https://historicengland.org.uk/listing/the-list/list-entry/1194332?section=official-list-entry (accessed on 17 August 2022).
53. Society for the Protection of Ancient Buildings. Harmondsworth Barn. *Who Will Save the 'Cathedral of Middlesex'?* Available online: https://web.archive.org/web/20120214143758/http://www.spab.org.uk/cornerstone-magazine/news/harmondsworth-barn/ (accessed on 17 August 2022).
54. Brimblecombe, P. Air Pollution and Architecture: Past, Present and Future. *J. Archit. Conserv.* **2000**, *6*, 30–46. [CrossRef]
55. Andrews, D.P.; Bedford, J.; Bryan, P.G. A Comparison of Laser Scanning and Structure from Motion as Applied to the Great Barn at Harmondsworth, UK. *Int. Arch. Photogramm. Remote Sens. Spat. Inf. Sci.* **2013**, *XL-5/W2*, 31–36. [CrossRef]
56. Frankson, R.; Kunkel, K.E.; Champion, S.M.; Runkle, J. *Iowa State Climate Summary 2022*; NOAA Technical Report NESDIS 150-IA; NOAA/NESDIS: Silver Spring, MD, USA, 2022.
57. Feehan, J. Train Depot Move on Track. Available online: https://www.toledoblade.com/frontpage/2005/06/23/Train-depot-move-on-track.html (accessed on 10 August 2022).
58. Pope, M.K.; Sievert, A.K.; Sievert, S.L. From Pioneer to Tourist: Public Archaeology at Spring Mill State Park. *Int. J. Hist. Archaeol.* **2011**, *15*, 206–221. [CrossRef]
59. The Prairie School Traveler. The Prairie School Traveler. Available online: https://www.prairieschooltraveler.com/html/wi/jumpriver/jumpriver.html (accessed on 10 August 2022).
60. Caba, S. "Gothic" House Being Refurbished. *The Cedar Rapids Gazette*, 20 June 1976.
61. Wurtz, R.L. Constructing Identities in the West Branch Landscape: Herbert Hoover's Life and Legacy as a Common Man, 1935–1992. Master's Thesis, University of Northern Iowa, Cedar Falls, IA, USA, 2017.
62. Lankester, P.; Brimblecombe, P. The Impact of Future Climate on Historic Interiors. *Sci. Total Environ.* **2012**, *417–418*, 248–254. [CrossRef] [PubMed]
63. Bratasz, Ł.; Harris, I.; Lasyk, Ł.; Łukomski, M.; Kozłowski, R. Future Climate-Induced Pressures on Painted Wood. *J. Cult. Herit.* **2012**, *13*, 365–370. [CrossRef]
64. Jacobs, J.L.; Thakur, R. How Advances in Algae-Resistant Roofing Address the Growing Roof Algae Problem. In Proceeding of the Fourth International Symposium on Roofing technology, National Roofing Contractors Association, Gaithersburg, MD, USA, 17–19 September 1997; pp. 99–103.
65. Berdahl, P.; Akbari, H.; Levinson, R.; Miller, W.A. Weathering of Roofing Materials—An Overview. *Constr. Build. Mater.* **2008**, *22*, 423–433. [CrossRef]
66. Ortega-Morales, O.; Montero-Muñoz, J.L.; Baptista Neto, J.A.; Beech, I.B.; Sunner, J.; Gaylarde, C. Deterioration and Microbial Colonization of Cultural Heritage Stone Buildings in Polluted and Unpolluted Tropical and Subtropical Climates: A Meta-Analysis. *Int. Biodeterior. Biodegrad.* **2019**, *143*, 104734. [CrossRef]
67. Buczkowski, G.; Bertelsmeier, C. Invasive Termites in a Changing Climate: A Global Perspective. *Ecol. Evol.* **2017**, *7*, 974–985. [CrossRef]
68. Ludwikowski, J.J.; Peterson, E.W. Transport and Fate of Chloride from Road Salt within a Mixed Urban and Agricultural Watershed in Illinois (USA): Assessing the Influence of Chloride Application Rates. *Hydrogeol. J.* **2018**, *26*, 1123–1135. [CrossRef]

69. Halda, B.J. Doors Number 1, 1989. Available online: https://www.nps.gov/tps/how-to-preserve/tech-notes/Tech-Notes-Doors01.pdf (accessed on 21 August 2022).
70. Powerwash. Roof Cleaning Market Profitability. Available online: https://powerwash.com/spray-tips/roof-cleaning-market-profitability/ (accessed on 10 August 2022).
71. Anderson Pest Solutions. Where Do Termites Live? *Midwestern Termites*. Available online: https://andersonpestsolutions.com/termites/identification/where-do-termites-live/ (accessed on 19 August 2022).
72. Levinson, R.; Berdahl, P.; Asefaw Berhe, A.; Akbari, H. Effects of Soiling and Cleaning on the Reflectance and Solar Heat Gain of a Light-Colored Roofing Membrane. *Atmos. Environ.* **2005**, *39*, 7807–7824. [CrossRef]
73. Thomas, C. Matching New Tiles to Old. Available online: http://www.roofconsult.co.uk/articles/tiling/tips6.htm (accessed on 10 August 2022).
74. Busby, J.W.; Smith, T.G.; White, K.L.; Strange, S.M. Climate Change and Insecurity: Mapping Vulnerability in Africa. *Int. Secur.* **2013**, *37*, 132–172. [CrossRef]
75. Busby, J.W.; Cook, K.H.; Vizy, E.K.; Smith, T.G.; Bekalo, M. Identifying Hot Spots of Security Vulnerability Associated with Climate Change in Africa. *Clim. Chang.* **2014**, *124*, 717–731. [CrossRef]
76. Akinsanola, A.A.; Zhou, W. Projections of West African Summer Monsoon Rainfall Extremes from Two CORDEX Models. *Clim. Dyn.* **2019**, *52*, 2017–2028. [CrossRef]
77. Dosio, A.; Turner, A.G.; Tamoffo, A.T.; Sylla, M.B.; Lennard, C.; Jones, R.G.; Terray, L.; Nikulin, G.; Hewitson, B. A Tale of Two Futures: Contrasting Scenarios of Future Precipitation for West Africa from an Ensemble of Regional Climate Models. *Environ. Res. Lett.* **2020**, *15*, 064007. [CrossRef]
78. Macarthy, J.M. Integrating Climate Change Considerations in Planning for Urban Development in Sierra Leone: The Case of Freetown. Ph.D. Thesis, Newcastle University, Newcastle upon Tyne, UK, 2012.
79. Sierra Leone Heritage. Sierra Leone Heritage. Available online: https://www.sierraleoneheritage.org/ (accessed on 17 August 2022).
80. Basu, P. OBJECT DIASPORAS, RESOURCING COMMUNITIES: Sierra Leonean Collections in the Global Museumscape. *Mus. Anthropol.* **2011**, *34*, 28–42. [CrossRef]
81. Akam, S. Freetown's Wood Homes a Link to Sierra Leone's Past Reuters. Available online: https://www.reuters.com/article/us-sierraleone-architecture-idUSBRE8420IO20120503 (accessed on 2 August 2022).
82. Taylor, E.T.; Kamara, I.S.; Bockarie, A. Rainfall Pattern in Freetown, Sierra Leone: From a Retrospective Viewpoint. In Proceedings of the International Conference ADAPT to CLIMATE, Nicosia, Cyprus, 27–28 March 2014; pp. 1–8.
83. Climate-Data. Climate Data for Cities Worldwide. Available online: https://en.climate-data.org/ (accessed on 17 August 2022).
84. World Bank Group. Sierra Leone Overview. Available online: https://www.worldbank.org/en/country/sierraleone/overview (accessed on 17 August 2022).
85. Corvo, F.; Pérez, T.; Martin, Y.; Reyes, J.; Dzib, L.R.; González-Sánchez, J.; Castañeda, A. Time of Wetness in Tropical Climate: Considerations on the Estimation of TOW According to ISO 9223 Standard. *Corros. Sci.* **2008**, *50*, 206–219. [CrossRef]
86. Bates, H.W. *The Naturalist on the River Amazons, a Record of Adventures, Habits of Animals, Sketches of Brazilian and Indian Life and Aspects of Nature under the Equator during Eleven Years of Travel*; John Murray: London, UK, 1863.
87. Esquivel-Muelbert, A.; Baker, T.R.; Dexter, K.G.; Lewis, S.L.; Brienen, R.J.W.; Feldpausch, T.R.; Lloyd, J.; Monteagudo-Mendoza, A.; Arroyo, L.; Álvarez-Dávila, E.; et al. Compositional Response of Amazon Forests to Climate Change. *Glob. Chang. Biol.* **2019**, *25*, 39–56. [CrossRef]
88. Lee, D.S.-H.; KIM, N.-S.; Scharff, M.; Nielsen, A.V.; Mecklenburg, M.; Fuster-López, L.; Bratasz, L.; Andersen, C.K. Numerical Modelling of Mechanical Degradation of Canvas Paintings under Desiccation. *Herit. Sci.* **2022**, *10*, 1–13. [CrossRef]
89. Brimblecombe, P. Visitor Responses and Climate Change. In *Cultural Heritage from Pollution to Climate Change*; Lefevre, R., Sabbioni, C., Eds.; Edipuglia: Bari, Italy, 2016; pp. 73–80.
90. Woon, J.S.; Boyle, M.J.W.; Ewers, R.M.; Chung, A.; Eggleton, P. Termite Environmental Tolerances Are More Linked to Desiccation than Temperature in Modified Tropical Forests. *Insectes Soc.* **2019**, *66*, 57–64. [CrossRef]
91. Omasombo, J. Kisangani: A City at Its Lowest Ebb. In *Urban Africa. Changing Contours of Surviving in the City*; Abdoumaliq, S., Abdelghani, A., Eds.; CODESRIA Books: Dakar, Senegal; Zed Books: London, UK; University of South Africa Press: Pretoria, South Africa, 2005; pp. 96–119.
92. Ernst, J.-L. Stanleyville, Kisangani. Available online: http://www.stanleyville.be/index.html (accessed on 18 August 2022).
93. Creese, A.; Washington, R.; Jones, R. Climate Change in the Congo Basin: Processes Related to Wetting in the December–February Dry Season. *Clim. Dyn.* **2019**, *53*, 3583–3602. [CrossRef]
94. Saeed, F.; Haensler, A.; Weber, T.; Hagemann, S.; Jacob, D. Representation of Extreme Precipitation Events Leading to Opposite Climate Change Signals over the Congo Basin. *Atmosphere* **2013**, *4*, 254–271. [CrossRef]
95. Ooi, K.G. *Japanese Empire in the Tropics: Selected Documents and Reports of the Japanese Period in Sarawak, Northwest Borneo, 1941–1945*; Ohio University Center for International Studies: Athens, OH, USA, 1998.
96. Sa'adi, Z.; Shahid, S.; Chung, E.S.; Ismail, B.T. Projection of Spatial and Temporal Changes of Rainfall in Sarawak of Borneo Island Using Statistical Downscaling of CMIP5 Models. *Atmos. Res.* **2017**, *197*, 446–460. [CrossRef]
97. Krishnan, M.V.N.; Prasanna, M.V.; Vijith, H. Statistical Analysis of Trends in Monthly Precipitation at the Limbang River Basin, Sarawak (NW Borneo), Malaysia. *Meteorol. Atmos. Phys.* **2019**, *131*, 883–896. [CrossRef]

98. Simpson, N.P.; Clarke, J.; Orr, S.A.; Cundill, G.; Orlove, B.; Fatorić, S.; Sabour, S.; Khalaf, N.; Rockman, M.; Pinho, P.; et al. Decolonizing Climate Change–Heritage Research. *Nat. Clim. Chang. 2022* **2022**, 1–4. [CrossRef]
99. Balocco, C.; Petrone, G.; Maggi, O.; Pasquariello, G.; Albertini, R.; Pasquarella, C. Indoor Microclimatic Study for Cultural Heritage Protection and Preventive Conservation in the Palatina Library. *J. Cult. Herit.* **2016**, *22*, 956–967. [CrossRef]
100. Pineda, P.; Iranzo, A. Analysis of Sand-Loaded Air Flow Erosion in Heritage Sites by Computational Fluid Dynamics: Method and Damage Prediction. *J. Cult. Herit.* **2017**, *25*, 75–86. [CrossRef]
101. Masson, T.; Fritsche, I. We Need Climate Change Mitigation and Climate Change Mitigation Needs the 'We': A State-of-the-Art Review of Social Identity Effects Motivating Climate Change Action. *Curr. Opin. Behav. Sci.* **2021**, *42*, 89–96. [CrossRef]

heritage

Article

Modelling the Alteration of Medieval Stained Glass as a Function of Climate and Pollution: Comparison between Different Methodologies

Aurélie Verney-Carron [1,*], Loryelle Sessegolo [1], Roger-Alexandre Lefèvre [1] and Peter Brimblecombe [2]

1 Univ Paris Est Créteil and Université Paris Cité, CNRS, LISA, F-94010 Créteil, France
2 Department of Marine Environment and Engineering, National Sun Yat-sen University, Kaohsiung 80424, Taiwan
* Correspondence: aurelie.verney@lisa.ipsl.fr

Abstract: Most stained-glass windows installed during the Middle Ages have deteriorated over time due to climate and pollution. To reconstruct their alteration history over the centuries, evaluate the current environmental risk, and predict their alteration in the future, two modelling methodologies have been used. First, based on the short-term exposure of medieval-type glass in different sites, dose–response functions (DRFs) were established. These DRFs correlate relevant environmental factors (temperature, rain quantity, rain pH, relative humidity, and SO_2 concentration) with the response of the material in terms of alteration layer thickness. The second methodology consists of laboratory experiments that aim at parametrising kinetic laws as a function of specific parameters (temperature, rain pH, and relative humidity). These kinetic laws can be extrapolated over long periods, contrary to DRFs. In this study, we compared both methodologies to simulate the alteration of a model stained glass at different European sites or over different time periods. The results highlighted that the kinetic laws were able to closely represent the data, except for the polluted sites where the alteration was underestimated. This indicated that the dependence of the alteration rate on the pollutant concentrations should be included to improve the model.

Keywords: deterioration; cultural heritage; stained-glass windows; dose–response functions; geochemical modelling

Citation: Verney-Carron, A.; Sessegolo, L.; Lefèvre, R.-A.; Brimblecombe, P. Modelling the Alteration of Medieval Stained Glass as a Function of Climate and Pollution: Comparison between Different Methodologies. *Heritage* **2023**, *6*, 3074–3088. https://doi.org/10.3390/heritage6030164

Academic Editor: Artemios Oikonomou

Received: 22 January 2023
Revised: 6 February 2023
Accepted: 10 February 2023
Published: 15 March 2023

1. Introduction

Medieval stained glass deteriorates through environmental exposure. The most ancient stained-glass windows still in place date from the 12th century (e.g., Augsburg Cathedral), and a large number were then installed through to the 13th and 14th centuries in different countries in Europe, mainly in France, Germany, and England [1]. During the Middle Ages, in this geographical zone, glass was produced with siliceous sand and ashes of trees and terrestrial plants (ferns). Thus, most glass pieces have a Si-K-Ca composition with a relatively low SiO_2 content (~50 ± 5 wt.%) but high contents of K_2O (~18 ± 5 wt.%) and CaO (~18 ± 4 wt.%) (see [2] for a review). This type of glass also contains Al_2O_3, Na_2O, MgO, Fe_2O_3/FeO, TiO_2, P_2O_5, and certain metals for coloration. The relatively high variability of the composition of Si-K-Ca glass can be explained by the glassmaking technology and by the local raw materials. In particular, the ash composition depends on the plant species, the part of the tree, the substratum, the harvest period, etc. [1]. By analysing a large set of data, Adlington et al. [1] were able to distinguish different regions of glass production. Due to this variable composition, the glass has poor durability, contrary to Si-Na-Ca glass produced with natron or coastal plants [3].

After installation, stained glass is exposed to the atmosphere and affected by both climatic (rain, temperature, relative humidity) and environmental (e.g., pollution) variation. This exposure leads to the degradation of the glass, limiting the passage of light and the legibility of artistic expression. These alterations are manifested in the form of discontinuous (pits, craters) or continuous alteration layers and deposits of secondary phases (sulphates such as gypsum or syngenite) and carbonates or oxalates [3–17]. The alteration layer is generally depleted in alkalis and alkaline earth elements, but rich in Si, Al, and Fe. Its thickness varies up to 300 μm after six or seven centuries of alteration in most cases. Characterisation studies and laboratory experiments highlight that the alteration layer is formed by interdiffusion or ion exchange between modifier ions of the glass (K^+) and hydrogenated species (H^+) in water. This leads to a hydrated and dealkalinised layer, where local hydrolysis and condensation reactions can contribute to the reorganisation of the layer such that it appears laminated [16,18,19]. A loss of materials caused by dissolution can sometimes be observed [9].

In order to reconstruct the alteration history and predict the degradation of stained-glass windows in the future (in the context of pollution and climate change), it is necessary to understand the alteration mechanisms and to determine the associated kinetics as a function of the climatic and environmental parameters. Several methodologies can be used to achieve this, such as dose–response functions (DRFs) or models based on kinetic laws (KLs), which are detailed below.

First, short-term exposure or laboratory experiments on pristine glass can assess the initial stages of alteration and short-term kinetics in response to selected parameters. Short-term exposure in real atmospheres (in positions sheltered or unsheltered from rain) was applied [20–27] in the context of national European projects (e.g., Multi-Assess, Vidrio) or international programs such as ICP-Materials [28]. The objective of these projects was to compare the alteration of medieval-type glass in different environments (rural, urban, and industrial). Parameters such as temperature; relative humidity; rain quantity and pH; and pollutant (SO_2, NO_2, PM10, etc.) concentrations were monitored. After exposure, the apparent glass alteration rates, corresponding to the K-depleted thickness divided by the exposure time, were calculated. In sheltered positions, the alteration rates of the different model glasses (SC and CL_I, reproducing stained-glass windows from Sainte-Chapelle in Paris and Cologne Cathedral for [22]; and M1 for [24]) ranged between 0.5 and 2.2 $\mu m \cdot a^{-1}$ according to the nine different sites. This suggested that the chemical alteration of the glass was significant even without the direct impact of rain, and that the differences between the sites could induce a four-fold variation in the alteration rate. The temporal evolution also showed that over a year, the rate was not constant or tended to slow down slightly. In unsheltered positions, the alteration thickness can also be measured, but the external surface can be partly dissolved or lost by scaling [29]. The alteration rates of different model glasses (SC, TR, and CL_I, reproducing stained-glass windows from Sainte-Chapelle, Troyes, and Cologne for [22]; M3 for [23,30]; Si-K-Ca for [21,29]; and Ca-K for [27]) after 1 year were between 0.4 and 5.4 $\mu m \cdot a^{-1}$ depending on the composition of the glass and the place of exposure (~25 sites, mainly in Europe). In particular, at the 20 ICP-Materials (International Co-operative Programme on Effects on Materials including Historic and Cultural Monuments) sites, the alteration thickness of a model glass (M3, composition shown in Table 1) varied between 0.8 and 1.9 μm, with an average value of 1.2 μm after three years [23]. The average temperatures at these sites ranged from −0.8 to 24.2 °C, the average RH between 57.5 and 84.3%, the cumulative precipitation over three years from 618 to 5032 mm, the average SO_2 concentration between 0.2 and 35.2 $\mu g \cdot m^{-3}$, and the average NO_2 concentration between 1.4 and 79.1 $\mu g \cdot m^{-3}$. Thus, climate and pollution induced a two- to three-fold variation in the alteration thickness, which was relatively limited. However, this simplified model glass used at the ICP sites was more durable than medieval stained glass. The alterations also slowed down over the six years of exposure [26]. Another glass, M1 (composition shown in Table 1), which was more representative of medieval stained glass, was studied in the ICP-Materials program [31] and the Multi-Assess project [25]. For the 23 ICP-Materials sites [31], the alteration thickness varied from 1.4 to 17.1 μm after six months of exposure and from 1.6 to 22.1 μm after a year of exposure (Table 2). The effect of climate and pollution was thus more pronounced for this glass.

From these data, dose–response functions (DRFs) have been determined. These DRFs correlate relevant environmental factors with the responses of the materials in terms of alterations (e.g., the alteration layer thickness or the percentage of the surface covered by salts for model stained-glass windows) [25,31].

The second methodology consists in laboratory experiments that aim at parametrising kinetic laws and determining the effect of a specific parameter, such as the temperature and pH of a solution to simulate rain [19,32], or the ambient temperature, relative humidity (RH), and SO_2 concentration to simulate atmosphere [33–35]. These kinetic laws can represent inputs for geochemical models. Calculations based on these kinetic laws were undertaken by Verney-Carron et al. [2] to reconstruct the alteration history of medieval stained glass from installation until today. The results were consistent with observations of ancient samples.

The object of this paper was to compare the two methodologies for representing measured data and to simulate theoretical case studies: (i) dose–response functions based on multiple parameters and (ii) kinetic laws based on first-order parameters. To this end, data from the ICP-Materials program ([31], Table 2) as well as a previous case study [36] were used to examine and discuss the advantages and drawbacks of the different models. This comparison provides directions for future research.

Table 1. Chemical composition (in wt.%) of the model glass samples.

Glass	SiO_2	K_2O	CaO	P_2O_5	Na_2O	MgO	Al_2O_3	MnO	Fe_2O_3
M1	48.0	25.5	15.0	4.0	3.0	3.0	1.5		
M3	60.0	15.0	25.0						
SG3	51.3	19.2	16.8	3.8	1.1	4.0	1.8	1.0	1.2

Table 2. Environmental parameters (average temperature T in °C, average relative humidity RH in %, average SO_2 and NO_2 concentrations expressed in μg·m^{-3}, cumulated rainfall amount r in mm, average pH of rain) for 6 and 12 months of exposure (between October 1993 and April or October 1994) [31] and thickness of the leached layer (L_{6U} or L_{12U} in μm) for each site in unsheltered position. The leached layer thickness values are from the ICP-Materials program (data) [31] with an associated uncertainty of 10% [24]. They were also calculated using the dose–response function (DRF, Equations (1) and (2)) and the kinetic laws (KL, Equations (5)–(8)). 'n.a.' means not available.

Site	6-Month Exposure (October 1993–April 1994) Environmental Data						L_{6U}			12-Month Exposure (October 1993–October 1994) Environmental Data						L_{12U}		
	T °C	RH %	SO_2 μg·m^{-3}	NO_2 μg·m^{-3}	r mm	pH	Data μm	DRF μm	KL μm	T °C	RH %	SO_2 μg·m^{-3}	NO_2 μg·m^{-3}	r mm	pH	Data μm	DRF μm	KL μm
Kaperske Hory (Czech Rep.)	1.5	77	20.2	8.5	124.7	4.6	3.9	5.0	2.3	7.3	73	16.7	7.1	1006.7	5.1	7.3	6.4	4.5
Kopisty (Czech Rep.)	3.4	80	52.2	32.9	252.4	4.9	17.1	8.5	2.6	9.8	74	51.1	28.5	1343.5	0.5	22.1	21.4	12.6
Ähtari (Finland)	−6.2	87	2.2	5.4	230.5	4.5	1.4	1.5	2.3	2.0	80	1.3	4	529.1	4.6	4.0	2.9	3.8
Helsinki (Finland)	−2.4	84	9.5	37.8	253.7	4.4	3.4	3.5	2.6	5.0	76	6.8	36.3	621.5	4.4	3.5	5.0	4.3
Waldhof-Langenbrügge (Germany)	2.9	90	11.5	13.3	392.6	4.6	3.9	3.9	3.1	8.9	82	7.3	9.5	723.3	4.5	3.8	6.8	5.0
Aschaffenburg (Germany)	5.4	70	16.2	41.8	403	4.8	4.4	4.5	3.1	11.4	64	11.7	40.2	749	4.8	3.3	3.6	4.5
Bottrop (Germany)	5.8	83	45.8	40.7	439.5	4.7	7.3	8.3	3.3	11.1	79	35.8	37.9	764.6	4.8	7.0	11.8	4.8
Garmisch-Partenkirchen (Germany)	2.2	83	4.7	17.3	391.1	5.2	1.5	2.2	2.8	9.8	80	2.1	10.7	1195.1	5.2	3.6	3.6	4.9
Rome (Italy)	15.3	72	18.5	26.8	662	5.9	5.0	4.6	3.6	19.5	67	13.9	29.1	875.8	5.1	2.3	4.5	5.3
Casaccia (Italy)	10.8	80	4.7	12	n.a.	4.9	1.4	n.a.	n.a.	15.3	74	4.8	11.3	n.a.	4.9	n.a.	3.9	n.a.
Milan (Italy)	9	68	54.2	94.1	504.2	4.2	4.7	11.6	3.6	15.0	68	31.7	85.9	1202.8	4.3	15.3	6.7	6.0
Vredepeel (Netherlands)	5	88	8.8	30.5	451.1	5.8	3.2	3.0	2.9	10.0	83	6.7	27.8	875.2	5.1	10.0	6.8	4.9
Oslo (Norway)	−1.4	76	7.2	59.2	333.6	4.7	1.9	2.9	2.6	6.8	71	5.1	54.9	789.4	4.8	2.9	3.5	4.4
Birkenes (Norway)	−1.4	85	1.1	3.1	1046.6	4.4	2.0	1.4	3.6	4.9	79	0.9	2.3	1646.9	4.4	1.6	2.4	5.4
Stockholm (Sweden)	0.3	78	7.4	26.8	138.7	4.3	4.7	2.9	2.4	6.9	70	5.2	24.8	513.4	4.6	3.2	3.3	4.2
Aspvreten (Sweden)	−0.1	88	2.8	4.5	270.1	4.2	1.9	1.9	2.8	5.9	83	1.8	3.7	585.3	4.4	6.9	3.8	4.5
London (UK)	n.a.	n.a.	n.a.	n.a.	n.a.	n.a.	5.8	n.a.	n.a.	n.a.	n.a.	n.a.	n.a.	n.a.	n.a.	14.1	n.a.	n.a.
Wells (UK)	n.a.	n.a.	n.a.	n.a.	n.a.	n.a.	3.5	n.a.	n.a.	n.a.	n.a.	n.a.	n.a.	n.a.	n.a.	4.3	n.a.	n.a.
Toledo (Spain)	8.3	69	3.9	20.8	362.6	5.8	1.4	1.9	2.6	13.9	58	2.5	19.1	487.3	5.9	1.6	1.4	3.8
Moscow (Russia)	−3.9	80	18.3	26.9	324.1	6.2	5.6	4.5	2.1	4.4	74	17.6	31.3	717.2	6	10.0	6.8	3.6
Lisbon (Portugal)	15.4	68	12.5	37	561.5	5.5	4.3	3.7	3.4	18.2	64	12.5	37	808.2	5.6	6.9	3.7	4.7
Dorset (Canada)	−6.1	81	17.8	n.a.	n.a.	n.a.	2.0	n.a.	n.a.	3.4	81	13.8	n.a.	n.a.	n.a.	1.7	n.a.	n.a.
Steubenville (USA)	2.8	62	51.5	44.3	468.5	n.a.	2.9	n.a.	n.a.	10.6	68	43.8	40.6	1072.4	n.a.	2.6	n.a.	n.a.

2. Materials and Methods

2.1. Data

The ICP-Materials programme (International Co-operative Programme on Effects on Materials including Historic and Cultural Monuments) of the Economic Commission for Europe within the United Nations (UNECE ICP-Materials) is a collaborative project that aims to investigate the effects of air pollution on the degradation of materials and cultural heritage [28]. It has a large network of tests sites and performs regular exposure campaigns involving different materials. The corrosion or soiling rates are evaluated after exposure using parallel meteorological and environmental data. For this, temperature and relative humidity data from local or national meteorological stations are gathered for each site. Rainfall collection is carried out using closed-bucket samplers to measure the pH and the composition. Gaseous pollutants and particulate matter concentrations come either from nearby EMEP net sites (European Monitoring and Evaluation Programme), national or local air quality stations, or passives samplers on site. The results highlighted the evolution of pollution and material responses [37] and allowed the establishment of dose–response functions that relate climate or pollution parameters to the alteration of materials (for example, limestone erosion; the erosion of zinc, aluminium, copper, and various steels; and glass soiling) [38]. The sensitivity of stained-glass windows to the atmosphere led the Institute of Sciences and Technologies in Art of the Academy of Fine Arts (Vienna) to expose two model stained glasses (M1 and M3, Table 1) in sheltered and unsheltered conditions as part of this programme. The first campaign was started between October 1993 and September 1995 and continued until October 1997 and September 2001 to obtain six years of alterations. After exposure, the authors determined the thickness of the altered layer (depleted in K). The evaluation of its thickness after 6 and 12 months of alteration was carried out using nuclear reaction analysis (NRA). Melcher and Schreiner [24] estimated that the uncertainty associated with the procedure of leached depth determination was less than 10%. Several reports (available at https://www.ri.se/en/icp-materials/documents/icp-materials-reports, accessed on 9 February 2023) have gathered the results: Report 27 [31] for two-year exposure in sheltered and unsheltered conditions and dose–response functions, and Reports 48 [39] and 49 [26] characterising glass after three to six years of alteration in sheltered and unsheltered conditions. Woisetschläger and Schreiner [31] characterised the facies of alteration using SEM (scanning electron microscopy). In unsheltered conditions, few crystals were present on the glass surface, as they were regularly rinsed off by the rain. The leached layer appeared to be cracked and flaked off. The authors also developed a dose–response function [31]. For this, they used a statistical evaluation method based on Pearson correlations.

In this paper, we used the set of data concerning the M1 glass [31]. This glass had a high presence of network modifiers (K, Na, Ca, and Mg) and a low amount of network formers (Si, Al, and P), so its composition was relatively close to that of medieval stained glass (Table 1). The model glass was exposed for 6 and 12 months (Table 2) at 23 sites in Europe and North America. Moreover, for this glass, dose–response functions were established: two in unsheltered conditions for 6 and 12 months and one in sheltered conditions for 12 months (see Equations (1)–(3)).

2.2. Dose–Response Functions

Dose–response functions were established by Woisetschläger and Schreiner [31] for the leached layer thickness (L_{DR} in μm) of the M1 glass after six months in unsheltered conditions (U):

$$L_{DR-6U} = exp(-0.13717 + 0.56722\, ln([SO_2]) + 10.2395\, r\, pH) \quad (1)$$

where $[SO_2]$ is the concentration of SO_2 (in $\mu g \cdot m^{-3}$), r is the rain quantity (in mm), and pH is the pH of the rain.

The DRF after 12 months was as follows:

$$L_{DR-12U} = exp(-2.75562 + 0.436\ ln([SO_2]) + 0.04634\ RH + 0.001585\ r\ pH) \quad (2)$$

where RH is the relative humidity. This new parameter was statistically significant in this case.

Melcher and Schreiner [25] also established a dose–response function in sheltered conditions (S) from M1 exposure data in 6 different sites during the Multi-Assess project, which was adapted by Brimblecombe and Lefèvre [36]. After 12 months, the leached layer thickness was expected to be:

$$L_{DR-12S} = -0.64 + (0.03\ RH + 0.04\ [SO_2]) - 0.05\ T \quad (3)$$

where T is the temperature.

2.3. Kinetic Laws

Glass can be altered by two main processes: (1) interdiffusion or ion exchange [40], which leads to the formation of a hydrated and dealkalinised or leached layer that can be reorganised by local hydrolysis and condensation reactions [19]; (2) the dissolution of the glass network, which leads to a loss of material.

Different kinetic laws are associated with both mechanisms. The evolution of the leached layer thickness formed by interdiffusion L generally follows Fick's second law (e.g., [41]):

$$L = 2\sqrt{\frac{D\,t}{\pi}} \quad (4)$$

where D ($m^2 \cdot s^{-1}$) is the diffusion coefficient of the glass modifier elements within the leached layer, and t (s) is the duration of alteration. Therefore, D can be determined under different alteration conditions (T, rain pH, RH) by measuring the leached layer thickness (using SEM or SIMS (secondary ion mass spectrometry) techniques, for example) and knowing the duration of alteration.

The dissolution rate is generally formalised by a first-order law [32,42] that depends on the pH, temperature, and a chemical affinity term based on a silica phase. However, the loss of glass by means other than craters or the scaling of the altered layer was shown to be limited on ancient stained-glass windows (e.g., [9]). Therefore, the alteration rate in the atmosphere is predominantly controlled by diffusion, and dissolution can be neglected.

In ambient conditions, rainfall events alternate with drier periods, where the RH varies. A parametrised geochemical model based on alteration kinetics as a function of climate and pollution must thus consider different water conditions [32,35]. Verney-Carron et al. [2] accounted for three hygric phases: (i) rainfall events; (ii) residual or pore water in the alteration layer when the RH is high; and (iii) water present as vapour. It was also assumed that the alteration layer prevents the direct contact of rain with the pristine glass and limits its dissolution and the loss of materials. Therefore, Equation (5) with suitable parameters is used to determine the simulated alteration thickness using kinetic laws in unsheltered conditions $L_{KL\text{-}U}$:

$$L_{KL-U} = 2\sqrt{\frac{D_r \cdot x_r \cdot t}{\pi}} + 2\sqrt{\frac{D_w \cdot x_w \cdot t}{\pi}} + 2\sqrt{\frac{D_v \cdot x_v \cdot t}{\pi}} \quad (5)$$

where D_r is the diffusion coefficient of the glass modifier elements during rainfall events, which depends on pH and T (see Equation (6)); x_r is the proportion of rainfall time; D_w is the diffusion coefficient of the glass modifier elements when water fills a significant part of the pore network of the altered layer, which also depends on pH and T (Equation (6)); x_w is the fraction of time for which altered layer imbibition takes place; D_v is the diffusion coefficient of the glass modifier elements when the water is in vapour form; and x_v is the fraction of time for which the water is in vapour form.

The parameters D_r and D_w were evaluated by a laboratory study performed at different pH and temperature values on the model stained glass SG3 (composition in Table 1):

$$D = D_0 \times \left(10^{-pH}\right)^n \times exp\left(-\frac{Ea}{RT}\right) \quad (6)$$

where D_0 is an initial constant ($D_0 = 2.4 \pm 0.3 \times 10^{-10}$ m^2·s^{-1}); n is an empirical pH-dependent coefficient ($n = 0.25 \pm 0.02$); and Ea is the activation energy ($Ea = 34.5 \pm 0.3$ kJ·mol^{-1}) [32]. The parameter D_r is calculated using the pH and T values from Table 2 for each site, and D_w is assessed by considering a pH of 8 in the altered layer resulting from interdiffusion and the T value for each site. For D_v, general kinetic laws have yet to be established. Sessegolo et al. [35] have performed some alteration experiments on SG3 glass as a function of T and RH, but complementary data are necessary. However, the authors measured a value of 1×10^{-20} m^2·s^{-1} for the SG3 model glass exposed at 20 °C for nine months. Furthermore, thanks to H isotope tracing on ancient stained-glass windows, Sessegolo et al. [16] measured diffusion coefficients of 3.6×10^{-20} m^2·s^{-1} at 20 °C and 70% RH and 4.9×10^{-20} m^2·s^{-1} at 90% RH. Therefore, in the absence of a complete kinetic law for D_v as a function of T and RH, a unique value of 3.6×10^{-20} m^2·s^{-1} was considered in the calculations.

There is no clear correlation between rainfall (mm) and rain duration as it depends on the intensity of rain. However, based on meteorological data in Paris (617 mm·a^{-1} of rain between 1881 and 2000, corresponding to 548 h·a^{-1} of rain (over 8766 h in total) between 1961 and 1990; Météo-France data), x_r was determined as:

$$x_r = \frac{548/8766}{617} r_{1y} \quad (7)$$

where r_{1y} is the annual rainfall in mm (calculated from the cumulated rainfall over the considered period r in mm).

The duration for which the altered layer is considered wet or soaked by water is dependent on the RH and the pore size in the altered layer (Kelvin's equation). At 12 °C, the pore radius at which water condenses and fills the pore (Kelvin's radius) is 2 nm for 57% RH, 3 nm for 69% RH, 5 nm for 80% RH, 10 nm for 89% RH, and 20 nm for 95% RH. Sessegolo et al. [16] showed that the average pore size of ancient stained-glass windows is around 2 nm. It is difficult to accurately estimate the time required for the altered layer to be sufficiently soaked by water for alteration to occur, but we assumed that x_w is equal to 15% if RH < 70%, 25% if 70 < RH < 80%, and 35% if RH > 80%.

By deduction, the fraction of time for which the glass is exposed to the vapour phase (x_v) is:

$$x_v = 1 - x_w - x_r \quad (8)$$

For sheltered conditions, we could consider that the water is in vapour form 100% of the time, but this would not take into account the imbibition of the alteration layer. In unsheltered conditions, x_w was assumed to vary between 15 and 35% (25% in Paris), but without rain, this percentage could be lower, as the layer is never completely soaked. These two extreme cases were simulated as follows:

$$L_{KL-S} = 2\sqrt{\frac{D_v \cdot x_v \cdot t}{\pi}} \quad (9)$$

where $x_v = 1$; and

$$L_{KL-S} = 2\sqrt{\frac{D_w \cdot x_w \cdot t}{\pi}} + 2\sqrt{\frac{D_v \cdot x_v \cdot t}{\pi}} \quad (10)$$

where $x_v = 1 - x_w$.

3. Results and Discussion

3.1. Leached Layer Thickness in Model Glass

Using environmental data (Table 2), the thickness of the leached layer (L) was calculated based on dose–response functions ($L_{DR\text{-}6U}$ and $L_{DR\text{-}12U}$) and kinetic laws ($L_{KL\text{-}U}$) (Table 2, Figures 1 and 2). The uncertainties in the models (error bars in Figures 1 and 2) were determined by considering an uncertainty for each environmental parameter of ±5%. The minimum and maximum values were obtained, considering that an increase in T, RH, SO_2, and r and a decrease in pH caused an increase in L, and vice versa. For the kinetic laws, the uncertainties for the different assumptions concerning the respective time fractions and the parameterisation of the diffusion laws were difficult to assess and so were not considered here.

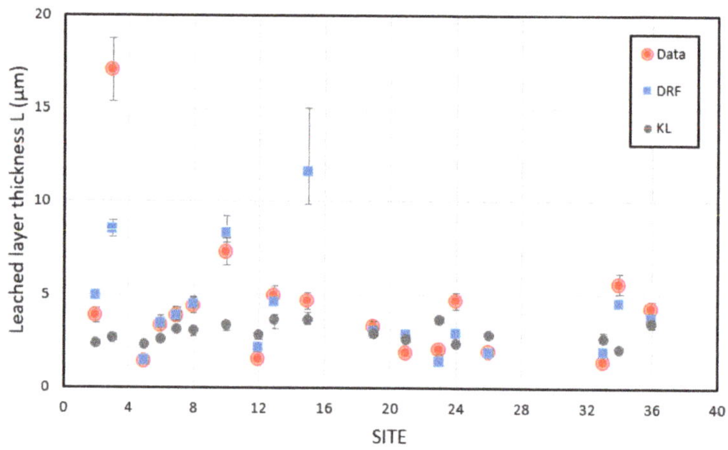

Figure 1. Leached layer thickness (L in µm) after six months of unsheltered exposure in different sites estimated from measurements on model glass (data), based on dose–response functions (DRF) and from kinetic laws (KL).

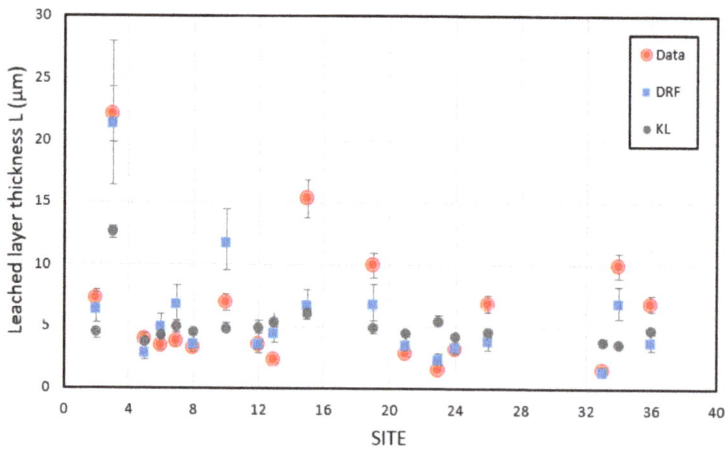

Figure 2. Leached layer thickness (L in µm) after 12 months of unsheltered exposure in different sites estimated from measurements on model glass (data), based on dose–response functions (DRF) and from kinetic laws (KL).

As expected, the values provided by the DRFs at six months (Figure 1) and 12 months (Figure 2) were close to the actual data, as they were established using this set of data.

Furthermore, the values provided by the kinetic laws, which were parameterised in a completely independent way, were also in good agreement. This indicated that despite differences in composition (Table 1), the M1 and SG3 glasses were altered at a similar rate. It also suggested that the kinetic laws determined in the laboratory were able to account for short-term alterations in real conditions and consider the relevant mechanisms. Verney-Carron et al. [2] also showed that the predictions for alterations over the centuries were consistent with observations from ancient stained-glass windows.

The correlation between the data and models was evaluated using Rstudio and calculating Kendall's tau coefficient, because the sample size was small and the variables did not follow a normal distribution. The tau coefficient for the data and the DRFs was 0.68 (with a p-value of 0.0001) for the 6-month series and 0.62 (with a p-value of 0.0004) for the 12-month series. This positive correlation and the low p-values (<0.05) confirmed the ability of the DRFs to fit the data. However, the tau coefficient for the data and the kinetic laws was 0.17 (with a p-value of 0.3370) for the 6-month series and 0.18 (with a p-value of 0.3035) for the 12-month series. In this case, the high p-values highlighted that the kinetic laws produced results of the same order of magnitude as the data but failed to account for the variations in the leached layer thickness as a function of the site. The kinetic laws also underestimated the alteration for highly polluted sites.

The main difference between the models was that the DRFs considered the SO_2 concentration, whereas the KLs only took into account the influence of air pollution via the pH of the rain. However, for polluted sites with high SO_2 concentrations (e.g., sites 3, 10, and 15), the pH was not significantly lower than at the other sites, except for site 3 in the 12-month data (Table 2). Therefore, the pH of the rain did not accurately represent the pollution level of the sites. The SO_2 concentrations in the DRFs allowed us to fit the high L values for sites 3 (partly or totally) and 10 (totally or even more), but produced strange values of L for site 15 (Milan) (Figures 1 and 2).

Unsurprisingly, the DRFs fit the measured data well. For the KLs, the results were close to the data when the L values were relatively low (between 1 and 6 μm); this was encouraging, as they were parameterised on a specific glass altered in laboratory conditions. However, they did not account for the variation as a function of the site or for the high L values, as they did not consider gaseous pollutants other than through their weak and indirect effect on the pH of the rain.

3.2. Long-Term Alteration Rate of Historic Glass in Paris

Ionescu et al. [43] simulated the alteration rate of stone, modern glass, and stained glass in the centre of Paris from 1500 to 2090 using DRFs and climate models. Brimblecombe and Lefèvre [36] extended the study to the different materials (stone, metals, and stained glass) found in Notre-Dame Cathedral between 1325 and 2090. For the past and current periods, they collected historical data (based on proxies or assumptions) and contemporary data (measurements). In particular, they used dendrochronology for temperature and different climate assumptions for precipitation and relative humidity. By using different pollution scenarios, e.g., based on fuel use [44,45]), the authors were able to estimate the concentrations of SO_2, NO_2, O_3, and PM10. For future projections, the authors used outputs of the RCP2.6 and 8.5 scenarios of climate evolution designed by the IPCC (Intergovernmental Panel on Climate Change) AR5 and outputs of the GAINS model for the emission of pollutants. Data concerning temperature, relative humidity, precipitation amount and pH, and SO_2 concentration are presented in the Supplementary Materials (Table S1). This earlier study considered only the alteration of stained-glass windows in sheltered positions, so we proposed here to calculate the evolution of the alteration rate of these ancient stained-glass windows in unsheltered positions using DRFs (Equation (2)) and similar environmental data ([36], Table S1) (Figure 3, solid blue line). For comparison, the alteration rate in sheltered conditions calculated from Equation (3) is also presented (Figure 3, solid grey line).

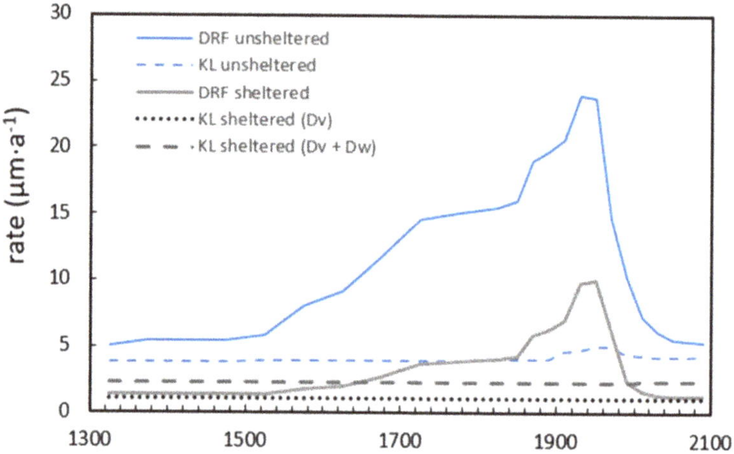

Figure 3. Alteration rate (for the 1st year, in $\mu m \cdot a^{-1}$) of medieval stained-glass windows in Notre-Dame from 1325 to 2090 based on DRFs and KLs. Climate and pollution data used as input are presented in Brimblecombe and Lefèvre [36] and Table S1.

It should be noted that the alteration rate obtained from the DRFs corresponded to the initial alteration over the first year. As these DRFs did not include a time parameter, and the alteration rate was not constant [16,20,23,35], the DRFs could not be used to simulate the alteration over longer time periods. Therefore, the curves could not be read as a cumulative process, as this would lead to very high and overestimated alteration thickness values over the centuries. Rather, the results correspond to an estimation of the environmental risk for stained-glass window alteration as a function of time.

For comparison, it was also possible to estimate this alteration rate for one year using kinetic laws in unsheltered conditions (Equations (5)–(8), Table S1, Figure 3, dashed blue line) and in sheltered conditions (Equations (9) and (10), Table S1, Figure 3, dashed grey lines).

Figure 3 (blue curves) shows that for low levels of pollution, before the increasing use of coal and the Industrial Age, as well as for the 21st century, the alteration rates predicted by the DRFs and KLs in unsheltered conditions were very close. In contrast, during the 20th century, the increase in the alteration rate caused by the pollution was very low for the KLs compared to the DRFs, as only the lower rain pH was considered among the pollution parameters. The large increase in SO_2 resulted in an increase in the alteration rate provided by the DRFs. The sharp drop in the alteration rate starting from the 1950s was then caused by the strong decrease in SO_2 in the atmosphere. For the KLs, a slight effect of the temperature increase was visible during the 21st century. Figure 3 (grey curves) also shows that for sheltered conditions, the results provided by the DRFs and KLs were similar for low-pollution conditions but very different for highly polluted periods.

As the DRFs were parameterised for 6 or 12 months, they could not be extrapolated over the long term, whereas this was feasible with the KLs. Figure 4 shows the evolution of the leached layer thickness of the stained glass windows of Notre-Dame Cathedral (L_{ND}) over the centuries (from 1325 to 2090) calculated from the kinetic laws and using the same climate and pollution data as in Brimblecombe and Lefèvre [36] (blue dashed line); only the situation of 1325 (orange dashed line) is considered for comparison. Equation (5) was adapted to consider the different values of the diffusion coefficient and the time fraction as a function of the environmental data corresponding to the time period:

$$L_{ND} = 2\sqrt{\frac{\sum_{i=1}^{n} D_{ri} \cdot x_{ri} \cdot t_i}{\pi}} + 2\sqrt{\frac{\sum_{i=1}^{n} D_{wi} \cdot x_{wi} \cdot t_i}{\pi}} + 2\sqrt{\frac{\sum_{i=1}^{n} D_v \cdot x_v \cdot t_i}{\pi}} \quad (11)$$

where t_i is the time between each time interval i; and D_{ri}, D_{wi}, x_{ri}, x_{wi}, and x_{vi} are the corresponding values calculated from the environmental data for the given time period. Data and calculations are presented in the Supplementary Materials (Table S1).

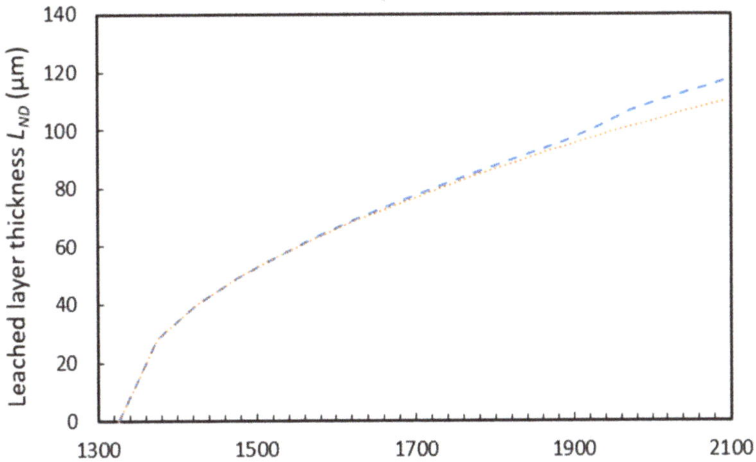

Figure 4. Evolution of the leached layer thickness (in µm) of medieval stained-glass windows in Notre-Dame from 1325 to 2090 predicted by KLs. The blue curve represents the climate and pollution data provided in Brimblecombe and Lefèvre [36] and Table S1. The orange curve represents only the situation of 1325, without changes.

The comparison between the curves showed that the increase in pollution induced a small acceleration in the alteration rate during the 20th century due to the decrease in pH (4.5 between 1910 and 1950 and 4 between 1950 and 1990) and during the 21st century due to the increase in temperature. The results showed that the effect of the temperature predominated over the slight decrease due to the RH and precipitation, but the effect of the RH should be considered more finely in the model, which would require complementary experiments.

3.3. *Advantages and Drawbacks of the Models*

The DRFs were established from data collected on various sites pertaining to the climate and pollution. They correlated different environmentally relevant parameters that were statistically selected, but also incorporated knowledge of the alteration processes. However, they were based on short-term measurements (6 and 12 months), and the alteration rate is not constant and tends to decrease over time [2,16,20,23,35]. For the simulations at Notre-Dame (Figure 3), this alteration rate thus corresponded to the evaluation of a potential threat for stained glass over centuries or decades, and not to a cumulative process, as the alteration thickness does not evolve linearly. Nevertheless, this represents how stained-glass windows have been altered by the environment over time.

A model based on kinetic laws parametrised in the laboratory as a function of specific parameters (T, pH, RH) provides an opportunity to extrapolate results over longer periods. For short-term data, kinetic laws provide good results when pollution is low, but they cannot account for changes at highly polluted sites. It is therefore necessary to improve these kinetic approaches to consider the effect of pollution. The comparison between sheltered and unsheltered conditions indicated that the effect of pollution is significant when the water is in the form of rain, but also when it is in the vapour form (unsheltered and sheltered conditions). In particular, based on laboratory experiments, Boehm [33] highlighted that the SO_2 concentration increases these alterations. The author explained this phenomenon by the higher dissolution of SO_2 in the absorbed water layer, which decreases the pH and increases interdiffusion and crust (gypsum or syngenite) formation. This hypothesis should be confirmed by additional experiments to quantify this effect and to implement it in kinetic laws, e.g., through the pH in the alteration layer for the wet situation in Equation (5).

To be useful, models must be based on physico-chemical principles so that they can help elucidate processes, work at suitable temporal and spatial scales, and have practical applications [46]. Here, the objectives of the stained-glass alteration models were multiple. They were used to reconstruct the alteration history of medieval stained-glass windows. The modelling of long-term alteration requires the understanding of alteration processes and the determination of the associated kinetics. In this sense, the comparison of both kinds of models is fruitful. Because DRFs are statistical models, they cannot be extrapolated over long periods (typically decades or centuries), as the alteration is not constant. However, their comparison to kinetic laws highlighted the key role of pollutants such as SO_2. Geochemical models are based on a more accurate description of the mechanisms and kinetics, but they should include the effects of air pollutants to be more realistic and to account for the variations between sites or environments. Thus, they could anticipate the future degradation of stained-glass windows in the context of climate change and pollution change. The case of Notre-Dame highlighted that an increase in temperature could accelerate the degradation of stained-glass windows. This impact needs to be confirmed by considering more finely the effect of a reduction in RH and precipitation that could counterbalance this increase. However, improved geochemical models could help decision making for the preservation of stained-glass windows. For example, climate data could be evaluated to consider solutions such as protective glazing. Then, the models could have a political application. DRFs have been used in the field of heritage to assess the economic and societal impacts of policies to reduce greenhouse gases and air pollutant emissions [47–51]. In stained-glass window conservation, geochemical models could also have value in simulating the actual process of their alteration as a function of relevant environmental parameters, and they could be appropriate for similar evaluations of climate and pollution scenarios.

4. Conclusions

Two kinds of model can be used to assess the alteration of medieval stained-glass windows. Each has advantages and drawbacks. Dose–response functions based on statistical evaluation can include a large set of climate or pollution parameters to assess the short-term alteration and estimate the risk for medieval stained-glass windows over different time periods. Kinetic laws based on the experimental determination of the alteration rate as a function of specific parameters can be extrapolated over longer time periods but are more limited in the number of environmental parameters they incorporate. The comparison of these models was valuable, as it validated the capacity of kinetic laws to predict alterations in low-pollution situations, but it also underlined the lack of underlying theory. This comparison emphasised the need for future research in the laboratory. Indeed, it would be particularly interesting to conduct experiments to better constrain the effect of SO_2, in order to determine the associated kinetics and build a geochemical model that is able to account for its effect. The objective would be to quantify the effect of SO_2 on the alteration rate under different conditions (concentration, temperature, and RH), and to determine how to implement this effect in a geochemical model. This would require us to understand, for example, if SO_2 has an impact on the pH within the alteration layer or if it favours secondary phase formation, which retroacts on the alteration rate. The model must also better consider the influence of relative humidity.

Supplementary Materials: The following supporting information can be downloaded at: https://www.mdpi.com/article/10.3390/heritage6030164/s1, Table S1: Environmental parameters from 1325 to 2090 in Paris and thickness of the leached layer for medieval stained glass windows (L_{ND} in µm).

Author Contributions: Conceptualisation, A.V.-C.; investigation, A.V.-C., L.S. and P.B.; methodology, A.V.-C., L.S., R.-A.L. and P.B.; writing—original draft preparation, A.V.-C.; writing—review and editing, A.V.-C., L.S., R.-A.L. and P.B.; funding acquisition, A.V.-C. All authors have read and agreed to the published version of the manuscript.

Funding: This research was funded by Agence Nationale de la Recherche, GLAM project (ANR-14-CE22-0007).

Data Availability Statement: All the data presented in this study are given in the article and in the Supplementary Material.

Acknowledgments: Some of the data were collected by members of the Task Force and organisations including supporting organisations of the official UNECE ICP Materials network. These organisations are presented under the 'Acknowledgements' heading on the ICP Materials web page.

Conflicts of Interest: The authors declare no conflict of interest. The funders had no role in the design of the study; in the collection, analyses, or interpretation of data; in the writing of the manuscript; or in the decision to publish the results.

References

1. Adlington, L.W.; Freestone, I.C.; Kunicki-Goldfinger, J.J.; Ayers, T.; Gilderdale Scott, H.; Eavis, A. Regional Patterns in Medieval European Glass Composition as a Provenancing Tool. *J. Archaeol. Sci.* **2019**, *110*, 104991. [CrossRef]
2. Verney-Carron, A.; Sessegolo, L.; Chabas, A.; Lombardo, T.; Rossano, S.; Perez, A.; Valbi, V.; Boutillez, C.; Muller, C.; Vaulot, C.; et al. Alteration of Medieval Stained Glass Windows in Atmospheric Medium: Review and Simplified Alteration Model. *NPJ Mater. Degrad.* **2023**, *under review*.
3. Sterpenich, J.; Libourel, G. Using Stained Glass Windows to Understand the Durability of Toxic Waste Matrices. *Chem. Geol.* **2001**, *174*, 181–193. [CrossRef]
4. Bräutigam, U.; Bürger, H.; Völksch, G. Investigations into Structure and Composition of Gel Layers on Medieval Window Glasses of the Katharinenkirche, Oppenheim (Germany), and the Cathedral St. Gatien, Tours (France). *Glastech. Ber. Glass Sci. Technol.* **1995**, *68*, 29–33.
5. Carmona, N.; García-Heras, M.; Gil, C.; Villegas, M.A. Vidrios y Grisallas Del s. XV de La Cartuja de Miraflores (Burgos): Caracterización y Estado de Conservación. *Bol. Soc. Esp. Ceram. Vidr.* **2005**, *44*, 251–258. [CrossRef]
6. Carmona, N.; Laiz, L.; Gonzalez, J.M.; Garcia-Heras, M.; Villegas, M.A.; Saiz-Jimenez, C. Biodeterioration of Historic Stained Glasses from the Cartuja de Miraflores (Spain). *Int. Biodeterior. Biodegrad.* **2006**, *58*, 155–161. [CrossRef]
7. Garcia-Vallès, M.; Vendrell-Saz, M. The Glasses of the Transept's Rosette of the Cathedral of Tarragona: Characterization, Classification and Decay. *Bol. Soc. Esp. Ceram. Vidr.* **2002**, *41*, 217–224.
8. Gillies, K.J.S.; Cox, A. Decay of Medieval Stained Glass at York, Canterbury and Carlisle. I: Composition of the Glass and Its Weathering Products. *Glastech. Ber.* **1988**, *61*, 75–84.
9. Lombardo, T.; Loisel, C.; Gentaz, L.; Chabas, A.; Verita, M.; Pallot-Frossard, I. Long Term Assessment of Atmospheric Decay of Stained Glass Windows. *Corros. Eng. Sci. Technol.* **2010**, *45*, 420–424. [CrossRef]
10. Marchesi, V.; Messiga, B.; Riccardi, M.P. Window Panes of the Certosa Di Pavia: Chemical Composition, Microstructure and Alteration. *Surf. Eng.* **2005**, *21*, 397–401. [CrossRef]
11. Messiga, B.; Riccardi, M.P. Alteration Behaviour of Glass Panes from the Medieval Pavia Charterhouse (Italy). *J. Cult. Herit.* **2006**, *7*, 334–338. [CrossRef]
12. Perez y Jorba, M.; Dallas, J.P.; Bauer, C.; Bahezre, C.; Martin, J.C. Deterioration of Stained Glass by Atmospheric Corrosion and Micro-Organisms. *J. Mater. Sci.* **1980**, *15*, 1640–1647. [CrossRef]
13. Piñar, G.; Garcia-Valles, M.; Gimeno-Torrente, D.; Fernandez-Turiel, J.L.; Ettenauer, J.; Sterflinger, K. Microscopic, Chemical, and Molecular-Biological Investigation of the Decayed Medieval Stained Window Glasses of Two Catalonian Churches. *Int. Biodeterior. Biodegrad.* **2013**, *84*, 388–400. [CrossRef]
14. Schreiner, M. Deterioration of Stained Medieval Glass by Atmospheric Attack. Part 1. Scanning Electron Microscopic Investigations of the Weathering Phenomena. *Glastech. Ber.* **1988**, *61*, 197–204.
15. Schreiner, M. Deterioration of Stained Medieval Glass by Atmospheric Attack. Part 2. Secondary Ion Mass Spectrometry Analysis of the Naturally Weathered Glass Surfaces. *Glastech. Ber.* **1988**, *61*, 223–229.
16. Sessegolo, L.; Verney-Carron, A.; Saheb, M.; Remusat, L.; Gonzalez-Cano, A.; Nuns, N.; Mertz, J.-D.; Loisel, C.; Chabas, A. Long-Term Weathering Rate of Stained-Glass Windows Using H and O Isotopes. *NPJ Mater. Degrad.* **2018**, *2*, 17. [CrossRef]
17. Lefèvre, R.-A.; Grégoire, M.; Derbez, M.; Ausset, P. Origin Of Sulphated Grey Crusts On Glass In Polluted Urban Atmosphere: Stained Glass Windows Of Tours Cathedral (France). *Glastech. Ber. Glass Sci. Technol.* **1998**, *71*, 75–80.
18. Lombardo, T.; Gentaz, L.; Verney-Carron, A.; Chabas, A.; Loisel, C.; Neff, D.; Leroy, E. Characterisation of Complex Alteration Layers in Medieval Glasses. *Corros. Sci.* **2013**, *72*, 10–19. [CrossRef]

19. Verney-Carron, A.; Sessegolo, L.; Saheb, M.; Valle, N.; Ausset, P.; Losno, R.; Mangin, D.; Lombardo, T.; Chabas, A.; Loisel, C. Understanding the Mechanisms of Si–K–Ca Glass Alteration Using Silicon Isotopes. *Geochim. Cosmochim. Acta* **2017**, *203*, 404–421. [CrossRef]
20. Gentaz, L. Simulation et Modélisation de l'altération Des Verres de Composition Médiévale Dans L'atmosphère Urbaine. Ph.D. Thesis, Université Paris-Est, Marne-la-Vallée, France, 2011.
21. Gentaz, L.; Lombardo, T.; Loisel, C.; Chabas, A.; Vallotto, M. Early Stage of Weathering of Medieval-like Potash-Lime Model Glass: Evaluation of Key Factors. *Environ. Sci. Pollut. Res. Int.* **2011**, *18*, 291–300. [CrossRef]
22. Geotti-Bianchini, F.; Nicola, C.; Preo, M.; Vallotto, M.; Verita, M. MicroIRRS and EPMA Study of the Weathering of Potash-Lime-Silicate Glasses. *Riv. Stn. Sper. Vetro* **2005**, *35*, 49–61.
23. Melcher, M.; Schreiner, M. Evaluation Procedure for Leaching Studies on Naturally Weathered Potash-Lime-Silica Glasses with Medieval Composition by Scanning Electron Microscopy. *J. Non-Cryst. Solids* **2005**, *351*, 1210–1225. [CrossRef]
24. Melcher, M.; Schreiner, M. Leaching Studies on Naturally Weathered Potash-Lime–Silica Glasses. *J. Non-Cryst. Solids* **2006**, *352*, 368–379. [CrossRef]
25. Melcher, M.; Schreiner, M. Quantification of the Influence of Atmospheric Pollution on the Weathering of Low-Durability Potash-Lime-Silica Glasses. *Pollut. Atmos.* **2007**, *49*, 13–22.
26. Melcher, M.; Schreiner, M. *Results from the Multipollutant Programme: Evaluation of the Decay to Glass Samples of Medieval Composition after 3, 4, 5 and 6 Years of Exposure. Part B: Results of the Unsheltered Exposure*; UNECE International Co-Operative Programme on Effects on Materials, Including Historic and Cultural Monuments; Institute of Sciences and Technologies in Art, Academy of Fine Arts: Vienna, Austria, 2004; p. 136.
27. Munier, I.; Lefèvre, R.; Geotti-Bianchini, F.; Verità, M. Influence of Polluted Urban Atmosphere on the Weathering of Low Durability Glasses. *Glass Technol.* **2002**, *43*, 225–237.
28. Tidblad, J.; Kucera, V.; Ferm, M.; Kreislova, K.; Brüggerhoff, S.; Doytchinov, S.; Screpanti, A.; Grøntoft, T.; Yates, T.; de la Fuente, D.; et al. Effects of Air Pollution on Materials and Cultural Heritage: ICP Materials Celebrates 25 Years of Research. *Int. J. Corros.* **2012**, *2012*, e496321. [CrossRef]
29. Gentaz, L.; Lombardo, T.; Chabas, A.; Loisel, C.; Neff, D.; Verney-Carron, A. Role of Secondary Phases in the Scaling of Stained Glass Windows Exposed to Rain. *Corros. Sci.* **2016**, *109*, 206–216. [CrossRef]
30. Melcher, M.; Schreiner, M. Statistical Evaluation of Potash-Lime-Silica Glass Weathering. *Anal. Bioanal. Chem.* **2004**, *379*, 628–639. [CrossRef]
31. Woisetschläger, G.; Schreiner, M. *Evaluation of Decay to Glass Samples after 1 and 2 Years of Exposure*; UNECE International Co-Operative Programme on Effects on Materials, Including Historic and Cultural Monumentsicp Materials; Institute of Chemistry, Academy of Fine Arts: Vienna, Austria, 1998; p. 68.
32. Sessegolo, L.; Verney-Carron, A.; Ausset, P.; Nowak, S.; Triquet, S.; Saheb, M.; Chabas, A. Alteration Rate of Medieval Potash-Lime Silicate Glass as a Function of PH and Temperature: A Low PH-Dependent Dissolution. *Chem. Geol.* **2020**, *550*, 119704. [CrossRef]
33. Boehm, T. The Influence of Temperature, Relative Humidity and SO_2 Concentration on Weathering of Glass. In Proceedings of the 5th ESG Conference, Prague, Czech Republic, 21–24 June 1999; pp. 49–55.
34. Carmona, N.; Villegas, M.A.; Fernández Navarro, J.M. Corrosion Behaviour of R_2O–CaO–SiO_2 Glasses Submitted to Accelerated Weathering. *J. Am. Ceram. Soc.* **2005**, *25*, 903–910. [CrossRef]
35. Sessegolo, L.; Verney-Carron, A.; Valle, N.; Ausset, P.; Narayanasamy, S.; Nowak, S.; Fourdrin, C.; Saheb, M.; Chabas, A. Alteration of Potash-Lime Silicate Glass in Atmospheric Medium: Study of Mechanisms and Kinetics Using ^{18}O and D Isotopes. *J. Non-Cryst. Solids* **2021**, *570*, 121020. [CrossRef]
36. Brimblecombe, P.; Lefèvre, R.-A. Weathering of Materials at Notre-Dame from Changes in Air Pollution and Climate in Paris, 1325–2090. *J. Cult. Herit.* **2021**, *50*, 88–94. [CrossRef]
37. Tidblad, J.; Kreislová, K.; Faller, M.; de la Fuente, D.; Yates, T.; Verney-Carron, A.; Grøntoft, T.; Gordon, A.; Hans, U. ICP Materials Trends in Corrosion, Soiling and Air Pollution (1987–2014). *Materials* **2017**, *10*, 969. [CrossRef]
38. Tidblad, J.; Kucera, V.; Mikhailov, A.A.; Henriksen, J.; Kreislova, K.; Yates, T.; Stöckle, B.; Schreiner, M. UN ECE ICP Materials: Dose-Response Functions on Dry and Wet Acid Deposition Effects After 8 Years of Exposure. *Water Air Soil Pollut.* **2001**, *130*, 1457–1462. [CrossRef]
39. Melcher, M.; Schreiner, M. *Results from the Multipollutant Programme: Evaluation of the Decay to Glass Samples of Medieval Composition after 3, 4, 5 and 6 Years of Exposure. Part A: Results of the Sheltered Exposure*; UNECE International Co-Operative Programme on Effects on Materials, Including Historic and Cultural Monuments; Institute of Sciences and Technologies in Art, Academy of Fine Arts: Vienna, Austria, 2003; p. 167.
40. Doremus, R. Interdiffusion of Hydrogen and Alkali Ions in a Glass Surface. *J. Non-Cryst. Solids* **1975**, *19*, 137–144. [CrossRef]
41. Chave, T.; Frugier, P.; Ayral, A.; Gin, S. Solid State Diffusion during Nuclear Glass Residual Alteration in Solution. *J. Nucl. Mater.* **2007**, *362*, 466–473. [CrossRef]
42. Grambow, B. A General Rate Equation for Nuclear Waste Glass Corrosion. *MRS Online Proc. Libr. Arch.* **1984**, *44*, 15–27. [CrossRef]
43. Ionescu, A.; Lefèvre, R.-A.; Brimblecombe, P.; Grossi, C.M. Long-Term Damage to Glass in Paris in a Changing Environment. *Sci. Total Environ.* **2012**, *431*, 151–156. [CrossRef]
44. Brimblecombe, P. London Air Pollution, 1500–1900. *Atmos. Environ.* **1977**, *11*, 1157–1162. [CrossRef]

45. Brimblecombe, P.; Grossi, C.M. Millennium-Long Recession of Limestone Facades in London. *Environ. Geol.* **2008**, *56*, 463–471. [CrossRef]
46. Richards, J.; Brimblecombe, P. The Transfer of Heritage Modelling from Research to Practice. *Herit. Sci.* **2022**, *10*, 17. [CrossRef]
47. Di Turo, F.; Proietti, C.; Screpanti, A.; Fornasier, M.F.; Cionni, I.; Favero, G.; De Marco, A. Impacts of Air Pollution on Cultural Heritage Corrosion at European Level: What Has Been Achieved and What Are the Future Scenarios. *Environ. Pollut.* **2016**, *218*, 586–594. [CrossRef]
48. Grøntoft, T. Maintenance Costs for European Zinc and Portland Limestone Surfaces Due to Air Pollution since the 1980s. *Sustain. Cities Soc.* **2018**, *39*, 1–15. [CrossRef]
49. Grøntoft, T. Conservation-Restoration Costs for Limestone Façades Due to Air Pollution in Krakow, Poland, Meeting European Target Values and Expected Climate Change. *Sustain. Cities Soc.* **2017**, *29*, 169–177. [CrossRef]
50. Grøntoft, T.; Verney-Carron, A.; Tidblad, J. Cleaning Costs for European Sheltered White Painted Steel and Modern Glass Surfaces Due to Air Pollution Since the Year 2000. *Atmosphere* **2019**, *10*, 167. [CrossRef]
51. Spezzano, P. Mapping the Susceptibility of UNESCO World Cultural Heritage Sites in Europe to Ambient (Outdoor) Air Pollution. *Sci. Total Environ.* **2021**, *754*, 142345. [CrossRef]

heritage

MDPI

Review

Climate Change and Cultural Heritage: Methods and Approaches for Damage and Risk Assessment Addressed to a Practical Application

Alessandra Bonazza [1,*] and Alessandro Sardella [2,3]

[1] National Research Council of Italy, Institute of Atmospheric Sciences and Climate, 40129 Bologna, Italy
[2] Department of Physics and Earth Sciences, University of Ferrara, 44122 Ferrara, Italy; a.sardella@isac.cnr.it
[3] National Research Council of Italy, Institute of Atmospheric Sciences and Climate, 88046 Lamezia Terme, Italy
* Correspondence: a.bonazza@isac.cnr.it; Tel.: +39-0516399576

Abstract: In the last 20 years, research on the observed and projected impacts of climate change on cultural heritage has led to significant developments regarding damage quantification and risk assessment, which unfortunately are not yet exhaustively transferred to practical applications and to the sector of policy and decision making. One of the major reasons for this still lacking alignment remains with the inadequate handover of quantitative data, which is a prerequisite for the development of measures and strategies for the mitigation of the impacts and risk reduction. In this paper, we focus on the methods and approaches put in place for the production of projections providing quantitative assessments of climate change-induced impacts in the near and far future (up to the 21st century) on outdoor built heritage mainly constituted by stone and stone-like materials. Our critical study found that different approaches have been applied for quantifying slow cumulative damage due to the ongoing variations of climate and air pollution parameters and to risk assessment caused by hydrometeorological extreme events induced by variations of temperature and precipitation. There is clear evidence that efforts are still needed for directing research to provide concrete solutions and tools addressed to meet the requirements of stakeholders and to solve the existing challenges in the field: selected effective models and tools are illustrated. The discussion is structured in order to highlight the driving role of research in supporting the definition of priorities for heritage managers and the development of strategies by decision and policy makers for the prevention and safeguarding of cultural heritage at risk.

Keywords: outdoors built heritage; stone; stone-like materials; slow cumulative damage; extreme events; damage function; vulnerability; projections; downscaling; policy-decision-makers; user-driven approach

Citation: Bonazza, A.; Sardella, A. Climate Change and Cultural Heritage: Methods and Approaches for Damage and Risk Assessment Addressed to a Practical Application. *Heritage* **2023**, *6*, 3578–3589. https://doi.org/10.3390/heritage6040190

Academic Editor: Claudia Pelosi

Received: 28 February 2023
Revised: 27 March 2023
Accepted: 7 April 2023
Published: 10 April 2023

1. Introduction

The risks on cultural heritage imposed by climate change have gained increasing attention during the last 20 years and several efforts have been made in order to assess the projected impacts on different building materials and heritage categories, both for outdoors and indoors [1,2]. In spite of the state of advancements unquestionably achieved in the research, the safeguarding of cultural heritage from climate-induced hazards still suffers from a lack of integration of measures purposely dedicated in the national plans for adaptation to climate change and disaster risk reduction and management [3]. On this aspect, the recent report "Strengthening Cultural Heritage Resilience for Climate Change" (2022) of the EU Open Method of Coordination (OMC) expert group of Member States stresses that only 12 out of the 28 countries participating mentioned the presence of cultural heritage in climate change policies, while merely 7 are the countries with available plans to coordinate climate change and cultural heritage (i.e., Ireland, Greece, Italy, Cyprus, Slovenia, Finland and Sweden) [4]. Additionally, the last report of the Intergovernmental

Panel on Climate Change (IPCC) highlights that cultural policies are still limited, although the integration of culture into policy and planning is recognized as a key step for the development of sustainable and resilient cities [5]. One of the reasons for the still ineffective handover of scientific results to policy and decision-makers in the field of cultural heritage protection is surely an inefficient transfer of research outputs into concrete tools and solutions addressed to meet the stakeholder needs and to solve the existing challenges at a territorial level [6,7].

A most impending requirement from policy and decision-makers is undoubtedly the availability of quantitative data of the observed and projected impacts for different scenarios on cultural and natural heritage, which are fundamental for establishing thresholds of acceptable risk and for setting up strategies of adaption and mitigation. Additionally, the need for improved knowledge about the scale and rates of damage on cultural heritage (both tangible and intangible) and the lack of a coherent methodology for its assessment are claimed as being still existing gaps [4]. Initiatives addressed to bridge these gaps will surely contribute to supporting the correct planning of mitigation and adaptation measures in different countries and, consequently, to define the priorities of intervention and the appropriate allocation of resources for their implementation.

The current article addresses methods and approaches mainly put in place for the development of projections providing impact evaluations in the near and far future of climate change on outdoor cultural heritage, both regarding slow cumulative damage due to ongoing climate/air pollution changes and to the risks associated with extreme hydrometeorological events linked to changes in temperature and precipitation. Focus is given to the methodological approaches applied to attempt a quantification of the damage and to the development of risk indicators in the field of protection and management of built heritage mainly in stone and stone-like materials. Quantifying and/or ranking the damage and risk continue to represent a challenge for the scientific community as they require a selection of prioritized parameters and atmospheric forces, namely, a limitation of the field of reliability and applicability (in our specific case heritage building materials, and cultural heritage categories), an awareness of the impossibility of comprising all the aspects [8], and the establishment of a dose-response link possibly on the basis of experimental work in the laboratory and by performing long-term field exposure tests.

The contents are provided with the additional objective to support heritage managers and non-technical experts in prioritizing climate and pollution parameters to monitor and select a more adequate time frequency and space scale of their measurement, in order to adequately support the methods and approaches for damage quantification and risk assessment. Major focus is given to the methods and approaches addressed to assist policy-decision-makers and operational bodies in dealing with setting up and putting into practice measures for the protection of cultural heritage in danger.

2. Dealing with Projected Impact and Risk on Cultural Heritage: Methods and Approaches

2.1. *Slow Cumulative Damage Due to Ongoing Variations of Climate/Air Pollution Parameters*

Research on the climate change impact on cultural heritage started by focusing on, and has been more exhaustively dedicated to up to now, to the evaluation of the impacts of gradual changes of climate and air pollution parameters on cultural heritage both outdoors and indoors [1,6,9–11]. It is within this framework that we can count on the higher number of efforts aiming at developing projections up to 2100 of the quantitative evaluations of damage at a European and Mediterranean level. Of major interest for this article is the research conducted on the damage processes of subaerial outdoor built heritage, specifically: soiling/blackening and surface recession of carbonate stones (namely, marble and compact limestone) due to air pollution and rain (both clean and acid); biological degradation; decohesion/fracturing caused by salt crystallization and thermoclastism.

This focus is motivated by:

1. Science-based evidence that monumental complexes, archaeological sites and historic buildings are likely to continue to undergo the effects of these damage processes in the near and far future, particularly in urban and coastal areas [1,2,9,11,12];

2. The availability of studies on the quantification of damage on heritage building materials by the development and application of damage functions in combination with outputs from climate projections [13–18].

Table 1 reports the available key equations utilized for damage quantification related to the processes taken into consideration (i.e., the outdoors, stone and stone-like materials), with materials for which the function and, therefore, the evaluation is valid and a list of the climate and pollution parameters recommended to be monitored.

Table 1. Key equations mainly utilized for the damage quantification of heritage building materials (i.e., stone and stone-like materials) exposed outdoors.

Damage Process	Damage Function/Risk Expression	Valid For	Climate/ Pollution Parameters
Surface recession	■ Lipfert (1989) [19]; Bonazza et al; (2009) [13] $L = 18.8 \cdot R + 0.016 \cdot [H^+] \cdot R + 0.18 \cdot (V_{dS} \cdot [SO_2] + V_{dN} \cdot [HNO_3])$ L = surface recession per year ($\mu m \cdot year^{-1}$); 18.8 = intercept term based on the solubility of $CaCO_3$ in equilibrium with 330 ppm CO_2 ($\mu m \cdot m^{-1}$); R = precipitation ($m \cdot year^{-1}$); 0.016 = constant valid for precipitation pH in the range 3–5; $[H^+]$ = hydrogen ion concentration ($\mu mol \cdot l^{-1}$) evaluated from rain yearly pH; 0.18 = conversion factor from ($cm \cdot s^{-1}$) ($\mu g \cdot m^{-3}$) to µm; V_{dS} = deposition velocity of SO_2 ($cm \cdot s^{-1}$); $[SO_2]$ = SO_2 concentration ($\mu g \cdot m^{-3}$); V_{dN} = deposition velocity of HNO_3 ($cm \cdot s^{-1}$)and $[HNO_3]$ = HNO_3 concentration ($\mu g \cdot m^{-3}$). ■ Kucera et al. (2007) [20] $R = 3.95 + 0.0059 \cdot [SO_2] \cdot RH_{60} + 0.054 Rain \cdot [H^+] + 0.078 \cdot [HNO_3] \cdot RH_{60} + 0.0258 \cdot PM_{10}$ R = surface recession per year ($\mu m \cdot year^{-1}$); $[SO_2]$ = SO_2 concentration ($\mu m \cdot m^{-3}$); RH_{60} = is the measured relative humidity when RH = 60% otherwise 0; Rain = amount of rainfall (mm) and $[H^+]$ = H^+ concentration (0.0006–0.13 $mg \cdot l^{-1}$); [HNO3] = HNO_3 concentration ($\mu m \cdot m^{-3}$); PM_{10} = particulate matter concentration ($\mu g \cdot m^{-3}$).	Marble and limestone with porosity lower than 25%	• Rain amount • Rain pH • Temperature (T) • Relative humidity (RH) • Sulphur dioxide (SO_2) • Nitric acid (HNO_3) • Carbon dioxide (CO_2) • Particulate matter (PM)
Soiling/ Blackening	■ Kucera (2005) [21] $R = R_0 \exp(-k_s \cdot PM_{10} \cdot t)$)R = reflectance after time t; t = time; R_0 = initial value of reflectance; k_s = rate constant for blackening and PM_{10} = particulate matter concentration = 10 ($\mu g \cdot m^{-3}$). ■ Brimblecombe and Grossi (2009) [16] $-dR/dt = (R_0 - R_p)\, Vd_{EC} \cdot EC/\tau$ dR = rate of change in reflectance of the material (clean stone); *t* = time; R_0 = reflectivity of the clean stone; R_p = final reflectance of the crust; Vd_{EC} = deposition velocity of elemental carbon; EC = elemental carbon concentration ($\mu g \cdot m^{-3}$); τ = folding density (surface concentration of elemental carbon required to reduce the reflectivity by a factor e). ■ Brimblecombe and Grossi (2009) [16] $R_t = (R_o - R_c) \cdot exp(-k_s t) + Rc$ R_t = rate of reduction in reflectance; R_0 = initial reflectance of the clean stone; R_C = reflectance of the deposited material; t = time; k_s = soiling constant.	Carbonate stones in general (sedimentary and metamorphic), mortars	• Rain amount • Temperature (T) • Relative humidity (RH) • Sulphur dioxide (SO_2) • Particulate matter (PM) • Carbon fractions of particulate matter (PM): elemental carbon (EC) * and organic carbon (OC) *
Biodeterioration/ Biomass accumulation	■ Gòmez-Bolea et al. (2012) [15] $B = \exp(-0.964 + 0.003P - 0.01T)$ B = biomass accumulation ($mg \cdot cm^{-2}$); P = annual precipitation (mm); T = annual mean temperature (°C).	Siliceous stones	• Rain amount • Temperature (T)

Table 1. *Cont.*

Damage Process	Damage Function/Risk Expression	Valid For	Climate/ Pollution Parameters
Thermoclastism	▪ Bonazza et al., (2009) [14] $\sigma_T = E \cdot \alpha \cdot (daily\Delta T_{air} + 20°C)/(1 - v)$ σ_T = maximum thermal stress (MPa); E = Young's modulus (GPa); α = thermal expansion coefficient (K^{-1}); $\Delta T_{air} = T_{airmax} - T_{airmin}$ (°C); ν = Poisson's ratio.	Marble	• Surface temperature • Temperature (T)
Salt crystallization	▪ Evaluation based on cycles per year/season of temperature and relative humidity (Sabbioni et al, 2010; Grossi et al., 2011; Menendez 2018) [9,12,22]	Porous stones in general	• Relative humidity (RH) • Temperature (T)

* rarely available from air quality monitoring networks. Specific aerosol monitoring campaigns are necessary in proximity of the sites.

Among the listed equations, those employed for surface recession and biomass accumulation offer a direct quantification of the damage, while the functions provided for soiling/blackening, thermoclastism and decohesion caused by salt crystallization require the establishment and adoption of the acceptable thresholds of damage and/or formulation of risk expressions determining the frequency of events likely to cause deterioration.

The majority of the European-based projections for the near (2021–2050) and far future (2071–2100) for the deterioration processes listed have been produced in the framework of the EC Noah's Ark Project by applying the Global and Regional Hadley climate models (i.e., a grid resolution of 295 × 278 km and 50 × 50 km, respectively) under the A2 scenarios (i.e., IPCC SRES Emission Scenarios used in TAR and FAR) [9]. Ciantelli et al. 2018 [18] provided downscaled projections covering the Panamanian isthmus for surface recession, biomass accumulation and deterioration due to salt crystallization at a grid resolution of 25 × 25 km, by using the climate model EC-Earth for the middle-future period (2039–2068). The available downscaled analyses are mainly at a local level and are focused on case studies [12,17,22–24].

Undoubtedly, downscaling in the resolution remains a still-existing gap in dealing with projections of the slow cumulative damage processes induced by climate changes [25].

Additionally, projections produced by the application of damage functions such as those listed in Table 1 do not account for the different rates of vulnerability or for the exposure of cultural heritage sites, and they consider the whole area investigated as constituted by the building materials for which the functions are valid. On the other hand, in spite of the recognized limits, these equations find concrete current examples of practical application from the actors and institutions in charge of the protection and management of cultural heritage, such as the Italian Risk Map of Cultural Heritage system coordinated by the *Direzione Generale Sicurezza del Patrimonio Culturale* of the Italian Ministry of Culture, which exploits climate and air pollution data from a monitoring station network [26]. In addition, they continue to be applied for site-specific analyses and for substantially improving evidence-based scientific data in support of the measures and policies of air pollution reduction and climate change mitigation, with benefits for cultural heritage [27].

The recommended/suggested optimal frequency of the measurements for all climate and pollution parameters listed in Table 1 for each damage process is daily (i.e., averages and for specific cases, such as thermoclastism, maximum and minimum values), in order to obtain data for representative evaluations of the monthly, seasonal and yearly values. These values are the most commonly used in dose-response functions for slow cumulative damage processes, of which the rate of degradation is in general subtle and can be evidenced only over time.

Air quality networks with free and open-access data play, in this framework, a very important role, even though they do not always offer all the required parameters (for example, elemental carbon is quite difficult to find in spite of its recognized driving role in blackening), as well as measurements at a proper distance of the heritage site under study. Nevertheless, the accessibility and the certainty in a standard method of measurement make the use of climate and air pollution data from these sources highly recommendable.

Data and products from Copernicus services, specifically, Climate Change (C3S) and Atmosphere Monitoring (CAMS) also represent a significant source of data for the damage assessment of cultural heritage by providing climate and pollution data at 10 × 10 km of spatial resolution [28]. The Copernicus services ensure coverage of remote areas or those not accessible (such as areas under armed conflict) and where there is the lack of an in situ environmental monitoring network. The examples of exploitation of data from the CAMS and C3S services for cultural heritage protection are still sporadic and the potential offered is still far from being fully explored.

2.2. *Risks Associated with Extreme Hydrometeorological Events Related to Climate Change*

Research has only recently started to focus on the development of projections of the impacts and risks imposed on cultural heritage by extreme events linked to climate change. Basically, the methodological approach builds on the concept of risk as the combination of three components: hazard, vulnerability and exposure. European and Mediterranean-based projections up to 12 × 12 km in spatial resolution for climate-induced extreme hazards under different scenarios are available. They are produced by applying individual regional climate models and ensemble climate simulations to reduce the uncertainties and provide outputs and tools for practical solutions in response to the challenges faced by the cultural heritage community in protecting and managing cultural heritage at risk [29–31]. The pursued approach goes beyond the analysis of damage for a single material or materials group, as adopted for the damage quantification caused by ongoing variations of climate and air pollution parameters (Section 2.1). It instead embraces the complexity of diverse categories of cultural and natural heritage by attempting to include the criticalities that increase its vulnerability by a physical and managerial point of view [32]. Vulnerability assessment still remains a complex issue and the selection of the more appropriate method to be employed continues to be under debate. Empirical and analytical methods have been applied and the majority of the evaluations available for cultural heritage are hazard-oriented (such as for flooding, and fire) and are sporadically combined with climate projections of the likelihood of an increase or decrease in a hazard for a comprehensive evaluation of the risk [30,33–35].

Empirical methods lend themselves to a more direct practical application being based on the analysis of observed damage, expert opinion and, consequently, a score assignment. The application of this method ensures a full understanding of the critical factors which influence the vulnerability in the field at an operational level from the experience gained by non-technical users, such as owners and managers, who are actively involved in the overall assessment. The adoption of a more accessible and comprehensive method for vulnerability evaluation entails an improved potential for practical application.

Figure 1 sets out the overall concept underlying the pursued approach in the framework of the Interreg Central Europe Projects ProteCHt2save and STRENCH [36,37] for the risk assessment of cultural heritage exposed to climate extreme events, while explaining the methods and tools applied for the hazard mapping and vulnerability assessment, with the final aim of setting preparedness strategies for the resilience of cultural heritage at a local level.

For the hazard analysis, the methodology focuses on events linked to climate change associated with precipitation and temperature extreme variations, such as heavy rain, flash and large basin floods, and prolonged drought periods.

The elaboration of hazard maps at a territorial level linked to a hydrometeorological extreme event has been conceived as a key step to identify the hazard prone areas in Europe and the Mediterranean Basin, that are exposed to calamitous events (i.e., a flood of a large basin, flash flood, heavy rain, etc.). The identification of the hazard-prone areas, together with the vulnerability assessment carried out at a local and building scale, allow the users to set up mitigation and preparedness measures in order to increase the resilience of diverse categories of cultural and natural heritage, among them archaeological sites, small, ruined villages, monumental complexes, historic buildings, and cultural landscapes.

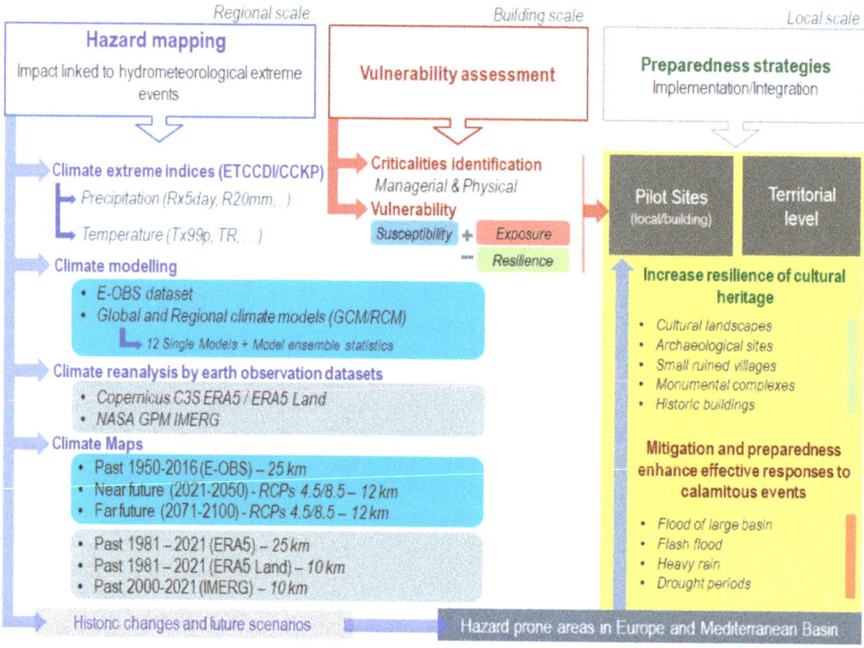

Figure 1. Overall concept of the methodology applied for the risk assessment of cultural heritage categories exposed to climate-induced hydrometeorological extreme events in the framework of the Interreg Central Europe ProteCHt2save and STRENCH. For hazard mapping section, blue boxes refer to data and maps deriving from climate models, while grey boxes refer to data and products from earth observation domain (Copernicus and NASA).

The methodology applied for mapping historic and future climate change referring to extreme variations in precipitation and temperature, basically comprises the following steps:

1. Search and selection of appropriate climate extreme indices and climate variables among the 27 indices defined by the Expert Team on Climate Change Detection (ETCCDI), whose definition can be found at the Climdex Project web site [38], but also among the indices defined by the Climate Change Knowledge Portal (CCKP) [39] (Table 2);
2. Computation and elaboration of selected indices to produce maps and analyses of their historical changes by using:
    a. climate data (T, P) from E-OBS observational dataset, from 1950 to present with a spatial resolution of 25 × 25 km [40] (Table 3);
    b. products provided by the EU Earth Observation program, Copernicus (ERA5 and ERA5 Land form C3S), and NASA (GPM-IMERG), providing climate data and reanalysis at a spatial resolution of 10 × 10 km [41] (Table 3).
3. Computation and elaboration of high-resolution maps of their future projection by using numerical climate model simulations. Twelve different combinations of GCM/RCM ensembles based on the EUROCORDEX initiative (with a resolution ~12 km) have been produced [40] (Table 3). Future projections cover two 30-year future periods, namely, the near future (2021–2050) and the far future (2071–2100), with respect to a historic reference (i.e., 1975–2005). The projections are available under the

two emission scenarios of RCP4.5 (stabilizing) and RCP8.5 (pessimistic). The use of an ensemble approach has been proved to reduce the uncertainty in climate change projections, particularly at a regional level, and it is widely used in climate change impact research, giving more reliable results than individual models [42].

Table 2. List of climate extreme indices selected as the most representative for the extreme events taken into account in the projects computed for the STRENCH WGT.

Extreme Event	Index	Definition and Description
Heavy rain	R20 mm	Very heavy precipitation days *. Number of days in a year with precipitation larger or equal to 20 mm/day.
Heavy rain	R95pTOT	Precipitation due to extremely wet days *. The total precipitation in a year cumulated over all days when the daily precipitation is larger than the 95th percentile of the daily precipitation on wet days. A wet day is defined as having a daily precipitation ≥ 1 mm/day. A threshold based on the 95th percentile selects only 5% of the most extreme wet days over a 30-year-long reference period.
Flooding	Rx5 day	Highest 5-day precipitation amount *. Yearly maximum of cumulated precipitation over consecutive 5-day periods.
Flooding	CWD	Consecutive wet days *. Seasonal maximum number of consecutive days with RR ≥ 1 mm.
Drought	CDD	Maximum number of consecutive dry days *. Maximum length of a dry spell in a year, that is, the maximum number in a year of consecutive dry days with a daily precipitation smaller than 1 mm/day.
Drought	CDD5	5 days of consecutive dry days **. Seasonal number of events of >5 consecutive dry days with a daily precipitation smaller than 1 mm/day.
Extreme heating	Tx90p	Extremely warm days *. Percentage of days in a year when the daily maximum temperature is greater than the 90th percentile. A threshold based on the 90th percentile selects only 10% of the warmest days over a 30-year-long reference period.
Extreme heating	su30	Strong summer days *. Seasonal count when TX (daily maximum) > 35 °C.
Extreme heating	HWI	Heat waves index **. Seasonal count of days TX >5°C above the monthly average for 5+ days.
Extreme heating	Tx99p	Hot days **. Seasonal N° days above average 99th percentile of TX (on basis of 1986–2005)
Extreme heating	TR	Tropical nights *. Seasonal count of days when TN (daily minimum temperature) > 20 °C.

* [38]. ** [39].

Table 3. Climate dataset, numerical products and re-analyses used for computing the selected climate extreme indices with an indication of time aggregation and resolution.

	E-OBS	C3S ERA5	C3S ERA5Land	NASA GPM IMERG	GCM/RCM Future Projection
R20 mm	✓	✓	✓	✓	✓
R95pTOT	✓	✓	✓	✓	✓
Rx5 day	✓	✓	✓	✓	✓
CWD		✓	✓	✓	

Table 3. *Cont.*

	E-OBS	C3S ERA5	C3S ERA5Land	NASA GPM IMERG	GCM/RCM Future Projection
CDD	✓	✓	✓	✓	✓
CDD5		✓	✓	✓	
Tx90p	✓				✓
su30			✓		
HWI		✓	✓		
Tx99p		✓	✓		
TR			✓		

E-OBS = historical observations for the 30-year-periods of 1987–2016 and 1951–1980. A 25 km resolution from 1950. C3S ERA5 = seasonal. An ~31 km – 0.25° resolution, from 1981. C3S ERA5 Land = monthly, seasonal, and yearly. Resolution ~9 km resolution, from 1981. NASA GPM IMERG = seasonal. A 10 km resolution, from 2000. GCM/RCM Future projections = 2021–2050 and 2071–2100 (reference period 1976–2005) under RCP4.5/RCP8.5. A 12 km resolution.

The final result is the production of climate maps at a territorial level showing historic and future changes and the likelihood of an increase/decrease in climate extremes, with the aim to evaluate the hazard-prone areas in Europe and the Mediterranean Basin.

The likelihood of an increase and decrease in a hazard subsequently needs to be integrated with the vulnerability ranking of the heritage site for a risk assessment (Figure 1).

Risk Mapping Tool for Cultural Heritage Protection

Recently developed in the framework of the Interreg Central Europe Projects ProteCHt2save and STRENCH [36,37], and based on the methodological approach above explained, the "Risk mapping tool for Cultural Heritage protection" [43] has the major objectives of supporting policy and decision-makers in the management of cultural heritage at risk from climate change induced by hydrometeorological extreme events, and fostering the inclusion of dedicated measures for cultural heritage safeguarding in national disaster risk-reduction plans, in the framework of a transnational perspective.

The tool combines, for the first time, the outputs from 12 climate models, a historic data set and data from the Earth Observation domain. Past and future projections of purposely-selected climate extreme indices (e.g., maps and time series) with a high spatial resolution and addressed to the safeguarding of cultural heritage are provided for the first time in a unique point of access. An user-driven approach has been adopted since the beginning in order to foster the use by non-technical experts and stakeholders of the cultural heritage field.

By accessing the "Risk mapping tool for Cultural Heritage protection" (shortened to "STRENCH WGT" in Figure 2) and applying the tools available in its "Maps" section, users can directly identify hazard prone areas and download the related historic and future maps [43]. The historic time series at specific locations of the climate extreme indices listed in Table 2 can be also visualized by using the products of Copernicus (from 1981 to present) and NASA (from 2000) (Figure 1).

The "Risk mapping tool for Cultural Heritage protection" also includes a methodology for a vulnerability ranking at a building scale, considering the vulnerability as a result of the interaction among the susceptibility, exposure and resilience. Starting from these requirements, a hierarchy tree is introduced including various branches (referred to as criterion or sub-criterion) which help in conceptualizing the evaluation. The vulnerability is ranked from 0 (low) to 1 (high). Details of the procedure are given in [44]. By using this methodology and following a guided procedure, users can then rank the vulnerability of the site under investigation. The overall aim is to enable the assignment of the values for each criterion or sub-criterion necessary for the evaluation of the three identified requirements.

Figure 2. Guided procedure for the application of "Risk mapping tool for Cultural Heritage protection".

The diagram in Figure 2 serves as a guideline to users, including non-technical experts, for the hazard and vulnerability assessment of cultural heritage categories at a local scale (i.e., a case study analysis) by using the "Risk mapping tool for Cultural Heritage protection" and related tools based on the methodological approach outlined in Figure 1.

The procedure requires the active participation of users who operate in an interactive way and is required as a first step to provide a general overview of a case study concerning the geographical location and the main environmental features.

The second step is to focus on an in depth study and description of the cultural heritage category that needs to be protected against one or more environmental hazards linked to climate change.

Then, the assessment of its vulnerability by applying the vulnerability tool/methodology integrated in the "STRENCH WGT" can be performed.

The subsequent step foresees an investigation of the main risks impacting the site and the execution of a detailed research of the past calamitous events that have occurred at the site, while also considering the protective and recovery measures put in place during and after those events in order to highlight the good and bad practices of safeguarding, and to determine the still-existing gaps needing to be overcome. Having once collected all this information, it is then possible to analyze the past and future changes of hazards by selecting the proper climate extreme indices (Table 2). All the procedures are then finally addressed to identify strategies of preparedness and prevention for resilience strengthening of the sites under consideration.

The data and results obtained though the testing of the "Risk mapping tool for Cultural heritage protection" for the 14 case studies in Central Europe involved in the ProteCHt2save (seven sites) and STRENCH (seven sites) projects are available online [43–46]. This activity was addressed in the improvement of the tool on the basis of the active involvement and feedback on use from the partners responsible for the case studies under the guidance of CNR-ISAC, the lead partner of both projects.

3. Concluding Remarks and Future Directions

The safeguarding of cultural heritage, including built heritage such as historic centers, archaeological sites, monumental complexes, and ruined villages, but also landscapes, historic parks and gardens, requires a holistic and multidisciplinary approach in order to identify all the critical parameters and factors that can put it in danger in a changing environment. A paradigm shift is necessary for the proper management of this important heritage. Ambitious policy choices increasingly supported by scientific research are essential for the implementation of strategies and measures to tackle the slow cumulative damage due to ongoing variations in climate and air pollution; however, the impact of short-term events, and emergency situations due to hydrometeorological extreme events, cannot be ruled out. The preparedness, prevention and protection of cultural heritage must be planned carefully, taking into account all the peculiarities of a territory, while analyzing the vulnerability of the cultural heritage assets included in it, and estimating the exposure to a potentially dangerous event in a short- and long-term perspective.

In order to ameliorate the communication between the research sector and practical application in the policy-decision-making process, research focused on the assessment of the climate change impacts on cultural heritage should be, first of all, based on a proper identification of the environmental (e.g., climate and air pollution) parameters to be monitored with a definition of an appropriate spatial and temporal resolution. Adopting the continuous environmental monitoring of prioritized climatic parameters in proximity to the heritage site, and/or planning specific checking with a monthly or seasonal frequency, can support determining the risks as a consequence of climate change effects. It is suggested that cultural heritage managers, owners and non-technical experts in charge of the protection of cultural heritage be reminded that dealing with the identified climate drivers causing deterioration is also dependent upon a comprehension of the vulnerability of a heritage asset, and the environmental context in which it is located. This will allow the scientific community to further explore the potentialities and tailor the damage functions and models with a user-driven approach.

The obtained results in terms of the observed and projected impacts should be translated into pragmatic guidelines for stakeholders, including urban planners, conservation practitioners, cultural heritage owners and managers. The understanding of the synergic effects of related climate change impacts and an improved practical use of outputs will assist in establishing priorities in relation to the protection needs of tangible heritage objects and assets.

Author Contributions: Conceptualization: A.B. and A.S.; methodology: A.B. and A.S.; formal analysis: A.B. and A.S; investigation: A.B. and A.S.; resources: A.B., writing—original draft preparation: A.B. and A.S., writing—review and editing: A.B. and A.S.; visualization: A.S., supervision: A.B.; funding acquisition: A.B. and A.S. All authors have read and agreed to the published version of the manuscript.

Funding: This research was funded by the Interreg Central Europe Projects ProteCHt2save "Risk assessment and sustainable protection of Cultural Heritage in changing environment, ProteCHt2save, Project index number CE1127"and the "STRENgthening resilience of Cultural Heritage at risk in a changing environment through proactive transnational cooperation–STRENCH, Project index number CE1665".

Data Availability Statement: The data presented in this study are available after registering at https://www.protecht2save-wgt.eu/, (accessed on 28 February 2023).

Acknowledgments: Authors wish to thank all colleagues and partners involved in the consortia of Interreg Central Europe Projects ProteCHt2save and STRENCH.

Conflicts of Interest: The authors declare no conflict of interest.

References

1. Sesana, E.; Gagnon, A.S.; Ciantelli, C.; Cassar, J.; Hughes, J.J. Climate change impacts on cultural heritage; A literature review. *WIREs Clim. Chang.* **2021**, *12*, e710. [CrossRef]
2. Bonazza, A. Sustainable Heritage and Climate Change. In *Routledge Handbook of Sustainable Heritage*, 1st ed.; Fouseki, K., Cassar, M., Dreyfuss, G., Ang Kah Eng, K., Eds.; Routledge: London, UK, 2022; pp. 263–271. [CrossRef]
3. Daly, C.; Purcell, C.E.; Donnelly, J.; Chan, C.; MacDonagh, M.; Cox, P. Climate change adaptation planning for cultural heritage, a national scale methodology. *J. Cult. Heritage Manag. Sustain. Dev.* **2021**, *11*, 313–329. [CrossRef]
4. European Commission Directorate-General for Education, Youth, Sport and Culture. *Strengthening Cultural Heritage Resilience for Climate Change: Where the European Green Deal Meets Cultural Heritage*; Publications Office of the European Union: Luxembourg, 2022; Available online: https://data.europa.eu/doi/10.2766/44688 (accessed on 28 February 2023). [CrossRef]
5. IPCC Sixth Assessment Report. Impacts, Adaptation and Vulnerability. Available online: https://www.ipcc.ch/report/ar6/wg2/ (accessed on 28 February 2023).
6. Bonazza, A.; Maxwell, I.; Drdácký, M.; Vintzileou, E.; Hanus, C.; Ciantelli, C.; De Nuntiis, P.; Oikonomopoulou, E.; Nikolopoulou, V.; Pospíšil, S.; et al. *Safeguarding Cultural Heritage from Natural and Man-Made Disasters. A Comparative Analysis of Risk Management in the EU.* ©; European Union: Maastricht, The Netherlands, 2018; p. 207. ISBN 978-92-79-73945-3. [CrossRef]
7. Sesana, E.; Gagnon, A.S.; Bertolin, C.; Hughes, J. Adapting Cultural Heritage to Climate Change Risks: Perspectives of cultural heritage experts in Europe. *Geosciences* **2018**, *8*, 305. [CrossRef]
8. Richards, J.; Brimblecombe, P. The transfer of heritage modelling from research to practice. *Herit. Sci.* **2022**, *10*, 17. [CrossRef]
9. Sabbioni, C.; Brimblecombe, P.; Cassar, M. *The Atlas of Climate Change Impact on European Cultural Heritage—Scientific Analysis and Management Strategies*; Anthem Press: London, UK, 2010; p. 146. ISBN 978927909800-0. [CrossRef]
10. Leissner, J.; Kilian, R.; Kotova, L.; Jacob, D.; Mikolajewicz, U.; Brostrom, T.; Ashley-Smith, J.; Schellen, H.L.; Martens, M.; Van Schijndel, J.; et al. Climate for Culture: Assessing the impact of climate change on the future indoor climate in historic buildings using simulations. *Herit. Sci.* **2015**, *3*, 38–52. [CrossRef]
11. Fatoric, S.; Seekamp, E. Are cultural heritage and resources threatened by climate change? A systematic literature review. *Clim. Chang.* **2017**, *142*, 227–254. [CrossRef]
12. Grossi, C.; Brimblecombe, P.; Menéndez, B.; Benavente, D.; Harris, I.; Déqué, M. Climatology of salt transitions and implications for stone weathering. *Sci. Total Environ.* **2011**, *409*, 2577–2585. [CrossRef]
13. Bonazza, A.; Messina, P.; Sabbioni, C.; Grossi, C.M.; Brimblecombe, P. Mapping the impact of climate change on surface recession of carbonate buildings in Europe. *Sci. Total Environ.* **2009**, *407*, 2039–2050. [CrossRef]
14. Bonazza, A.; Sabbioni, C.; Messina, P.; Guaraldi, C.; De Nuntiis, P. Climate change impact: Mapping thermal stress on Carrara marble in Europe. *Sci. Total Environ.* **2009**, *407*, 4506–4512. [CrossRef]
15. Gómez-Bolea, A.; Llop, E.; Ariño, X.; Saiz-Jimenez, C.; Bonazza, A.; Messina, P.; Sabbioni, C. Mapping the impact of climate change on biomass accumulation on stone. *J. Cult. Herit.* **2009**, *13*, 254–258. [CrossRef]
16. Brimblecombe, P.; Grossi, C.M. Millennium-long damage to building materials in London. *Sci. Total Environ.* **2009**, *407*, 1354–1361. [CrossRef]
17. Brimblecombe, P.; Lefvre, R. Weathering of materials at Notre-Dame from changes in air pollution and climate in Paris. *J. Cult. Herit.* **2021**, *50*, 88–94. [CrossRef]
18. Ciantelli, C.; Palazzi, E.; von Hardenberg, J.; Vaccaro, C.; Tittarelli, F.; Bonazza, A. How can climate change affect the UNESCO cultural heritage sites in Panama? *Geosciences* **2018**, *8*, 296. [CrossRef]
19. Lipfert, F.W. Atmospheric damage to calcareous stones: Comparison and reconciliation of recent experimental findings. *Atmos. Environ.* **1989**, *23*, 415–429. [CrossRef]
20. Kucera, V.; Tidblad, J.; Kreislova, K.; Knotkova, D.; Faller, M.; Reiss, D.; Snethlage, R.; Yates, T.; Henriksen, J.; Schreiner, M.; et al. UN/ECE ICP materials dose-response functions for the multi-pollutant situation. *Water Air Soil Pollut. Focus* **2007**, *7*, 249–258. [CrossRef]
21. Kucera, V.; MULTI-ASSESS. Model for Multi-Pollutant Impact and Assessment of Threshold Levels for Cultural Heritage. EU 5FP RTD Project, Contract Number: EVK4-CT-2001-00044. Deliverable 0.2, Publishable Final Report. 2005. Available online: http://www.corr-institute.se/icp-materials/web/page.aspx?refid=35 (accessed on 28 February 2023).
22. Menéndez, B. Estimators of the impact of climate change in salt weathering of cultural heritage. *Geosciences* **2018**, *8*, 401. [CrossRef]
23. Smith, B.J.; McCabe, S.; McAllister, D.; Adamson, C.; Viles, H.A.; Curran, J.M. A commentary on climate change, stone decay dynamics and the 'greening' of natural stone buildings: New perspectives on 'deep wetting'. *Environ. Earth Sci.* **2011**, *63*, 1691–1700. [CrossRef]
24. Orr, S.A.; Young, M.; Stelfox, D.; Curran, J.; Viles, H. Wind-driven rain and future risk to built heritage in the United Kingdom: Novel metrics for characterising rain spells. *Sci. Total Environ.* **2018**, *640–641*, 1098–1111. [CrossRef]
25. Vyshkvarkova, E.; Sukhonos, O. Climate Change Impact on the Cultural Heritage Sites in the European Part of Russia over the past 60 Years. *Climate* **2023**, *11*, 50. [CrossRef]
26. Gaddi, R.; Cacace, C.; Bucchianico, A.D.M.D. The risk assessment of surface recession damage for architectural buildings in Italy. *J. Cult. Herit.* **2022**, *57*, 118–130. [CrossRef]
27. Spezzano, P. Mapping the susceptibility of UNESCO World Cultural Heritage sites in Europe to ambient (outdoor) air pollution. *Sci. Total Environ.* **2021**, *754*, 142345. [CrossRef] [PubMed]

28. Bonazza, A.; Bonora, N.; Ducke, B.; Spizzichino, D.; Recchia, A.P.; Taramelli, A. Copernicus in support of monitoring, protection and management of cultural and natural heritage. *Sustainability* **2022**, *14*, 2501. [CrossRef]
29. Sardella, A.; Palazzi, E.; von Hardenberg, J.; Del Grande, C.; De Nuntiis, P.; Sabbioni, C.; Bonazza, A.P. Risk mapping for the sustainable protection of cultural heritage in extreme changing environments. Special Issue Assessing the Impact of Climate Change on Urban Cultural Heritage. *Atmosphere* **2020**, *11*, 700. [CrossRef]
30. Bonazza, A.; Sardella, A.; Kaiser, A.; Cacciotti, R.; De Nuntiis, P.; Hanus, C.; Maxwell, I.; Drdácký, T.; Drdácký, M. Safeguarding cultural heritage from climate change related hydrometeorological hazards in Central Europe. *Int. J. Disaster Risk Reduct.* **2021**, *63*, 102455. [CrossRef]
31. Kotova, L.; Leissner, J.; Winkler, M.; Kilian, R.; Bichlmair, S.; Antretter, F.; Moßgraber, J.; Reuter, J.; Hellmund, T.; Matheja, K.; et al. Making use of climate information for sustainable preservation of cultural heritage: Applications to the KERES project. *Herit. Sci.* **2023**, *11*, 18. [CrossRef]
32. Cacciotti, R.; Kaiser, A.; Sardella, A.; De Nuntiis, P.; Drdcký, M.; Hanus, C.; Bonazza, A. Climate change-induced disasters and cultural heritage: Optimizing management strategies in Central Europe. *Clim. Risk Manag.* **2021**, *32*, 100301. [CrossRef]
33. Figueiredo, R.; Romao, X.; Pauperio, E. Component-based flood vulnerability modelling for cultural heritage buildings. *Int. J. Disaster Risk Reduct.* **2021**, *61*, 102323. [CrossRef]
34. Miranda, F.N.; Ferreira, T.M. A simplified approach for flood vulnerability assessment of historic sites. *Nat. Hazards* **2019**, *96*, 713–730. [CrossRef]
35. Gandini, A.; Egusquiza, A.; Garmendia, L.; San-Jose, J.T. Vulnerability assessment of cultural heritage sites towards flooding events. *IOP Conf. Ser. Mater. Sci. Eng.* **2018**, *364*, 012028. [CrossRef]
36. Interreg Central Europe ProteCHt2save. Available online: https://programme2014-20.interreg-central.eu/Content.Node/ProteCHt2save.html (accessed on 28 February 2023).
37. Interreg Central Europe Strench. Available online: https://programme2014-20.interreg-central.eu/Content.Node/STRENCH.html (accessed on 28 February 2023).
38. Climdex. Available online: www.climdex.org (accessed on 28 February 2023).
39. Climate Change Knowledge Portal. Available online: https://climateknowledgeportal.worldbank.org (accessed on 28 February 2023).
40. European Climate Assessment & Dataset. Available online: http://www.ecad.eu/download/ensembles/ensembles.php (accessed on 28 February 2023).
41. Copernicus Programme. Available online: https://www.copernicus.eu/en (accessed on 28 February 2023).
42. Euro-CORDEX Simulations. Available online: https://euro-cordex.net/imperia/md/content/csc/cordex/20180130-eurocordex-simulations.pdf (accessed on 28 February 2023).
43. Risk Mapping Tool for Cultural Heritage Protection. Available online: https://www.protecht2save-wgt.eu/ (accessed on 28 February 2023).
44. Cacciotti, R.; Drdácký, M.; with the contribution of all partners. Strench Project Deliverable 1.2.2 "Definition of a methodology for ranking vulnerability of cultural heritage". 2020. Available online: https://programme2014-20.interreg-central.eu/Content.Node/STRENCH/CE1665-STRENCH-D.T1.2.2-Vulnerability-ranking.pdf (accessed on 7 April 2023).
45. Sardella, A.; with the contribution of all partners. Strench Project Deliverable D.T2.2.1 "Testing of the WebGIS tool for landscape protection". 2022. Available online: https://programme2014-20.interreg-central.eu/Content.Node/STRENCH.html (accessed on 7 April 2023).
46. Sardella, A.; with the contribution of all partners. Strench Project Deliverable D.T2.2.2 "Testing of the WebGIS tool for ruined hamlet protection". 2022. Available online: https://programme2014-20.interreg-central.eu/Content.Node/STRENCH.html (accessed on 7 April 2023).

heritage

Article

Mass-Transfer Air Pollution Modeling in Heritage Buildings

Morten Ryhl-Svendsen * and Signe Hjerrild Smedemark †

Institute for Conservation, Royal Danish Academy, 1435 Copenhagen, Denmark;
signe.hjerrild.smedemark@hum.ku.dk
* Correspondence: mrsv@kglakademi.dk; Tel.: +45-4170-1929
† Current address: Department of Nordic Studies, and Linguistics, University of Copenhagen, 2300 Copenhagen, Denmark.

Abstract: Two simple mass-balance models for estimating the concentration of air pollutants inside buildings are presented for pollutants originating from outdoors or generated indoors. The models can be used to establish average pollution loads on heritage objects inside buildings and assist in risk assessment for conservation. The models can be run with a minimum of data, either based on fixed conditions or as a Monte Carlo simulation based on plausible intervals of the input factors. Input data can be obtained by simple measurements or based on the literature. A museum storage hall in Denmark was used as a test site for demonstrating the models. They were evaluated with regard to the prediction of the indoor/outdoor concentration ratio for ozone and nitrogen dioxide and the build-up concentration of indoor generated organic acids. The pros and cons of such models were discussed, where the main reservation is related to shortcomings when real buildings are more complicated than the single-zone structure of the models. A strength of the models is the easy adaption to an indoor environment and, despite being semi-quantitative at times, the simplicity of the models, which allows for practical everyday use in air quality management of heritage buildings.

Keywords: mass-balance; emission; deposition velocity; ozone; nitrogen dioxide; organic acids; indoor-outdoor ratio; indoor air pollution; Monte Carlo simulation

Citation: Ryhl-Svendsen, M.; Smedemark, S.H. Mass-Transfer Air Pollution Modeling in Heritage Buildings. *Heritage* **2023**, *6*, 4768–4786. https://doi.org/10.3390/heritage6060253

Academic Editors: Peter Brimblecombe and Jenny Richards

Received: 15 May 2023
Revised: 9 June 2023
Accepted: 10 June 2023
Published: 12 June 2023

1. Introduction

Air pollution is recognized to cause damage to a wide range of heritage materials, e.g., by direct oxidation or by conversion into acid at contact [1–3]. Pollutants may be present in outdoor air or have local, indoor sources. The main gaseous outdoor pollutants known to cause damage to heritage collections are ozone, nitrogen dioxide, and various sulfur compounds, and may have both natural and anthropogenic sources [4–6]. The most critical indoor air pollutants known to engage in material damage are formic acid and acetic acid (in this paper, "organic acids" is used to refer to formic acid and acetic acid collectively) [4–7]. Wood, a common heritage material, as well as a widely used construction material, is known to emit organic acids and is one of the main indoor sources of this type of compound in indoor air [8–10].

Outdoor pollutants will, other things being equal, become reduced to a fraction of the outdoor level when the compounds infiltrate a building, either directly through forced ventilation systems, open windows, etc., or by natural ventilation at tortuous paths bypassing closed but leaky doors, or through small leaks, cracks, and holes in the building envelope. If, on the other hand, indoor pollution sources are present, the indoor level can exceed the outdoor level by many orders of magnitude [11]. Several publications collate typical pollution levels in museums and other heritage buildings [6,12–15], and guidelines on pollution levels for the preservation of heritage objects are given, for example, by the American Society of Heating, Refrigerating and Air-Conditioning Engineers (ASHRAE) [16].

In order to assess the risk associated with air pollution for a heritage collection, the pollution levels must be known. This can be obtained either by measurement or by calculation based on other available data. Even though air pollution monitoring solutions are becoming increasingly cheap and available [17], air quality monitoring for a range of pollutants, and maybe in several locations at one time, is often more comprehensive than what most museums have the resources for. Instead, air pollution models are a useful and cost-effective mean to evaluate conservation conditions for heritage buildings without having to initiate large and expensive monitoring campaigns. Model input can be based on easily accessed outdoor pollution data from public air pollution monitoring network and for indoor-generated compounds by data from the literature on material's emissions.

As will be described below, models for indoor air pollution are available in many levels of complexity. However, while advanced models are suitable for detailed research and for gaining a fundamental understanding of air quality mechanisms, they may not be feasible for everyday long-term surveying. In that case, a sufficient and useful estimate of the pollution load in a heritage building can be obtained by more simple mathematical expressions based on basic yet decisive factors. Besides being easy to use and understand, the output provides sufficient information necessary for assessing the risk of material damage. As damage to heritage objects caused by air pollutants depends on the dose rather than its momentary concentration in the air [6,18,19], steady-state calculations, rather than modeling dynamic situations, are fully comprehensive for predicting conservation conditions in heritage environments.

In this paper, we demonstrate such models, which provide an effective and easy estimation of gaseous air pollution levels to be used in practice for conservation risk assessment in buildings containing heritage collections. Table 1 lists the nomenclature used throughout the paper.

Table 1. Nomenclature used in this paper: Symbols and their units.

Symbol	Description	Unit
A	Surface area	m^2
C_i	Indoor concentration of a pollutant in the air	$\mu g\ m^{-3}$
C_o	Outdoor concentration of a pollutant in the air	$\mu g\ m^{-3}$
I/O	Ratio between indoor and outdoor pollution concentration	Dimensionless
L	Loading: The surface-to-volume ratio of objects or materials in a room. $L = A/V$	$m^2\ m^{-3}$
n	Air exchange rate (exchange of air with ambient)	$hour^{-1}$
Q	Air flow rate (e.g., through a filter unit)	$hour^{-1}$
S	Surface removal rate. $S = v_d(A/V)$	$hour^{-1}$
SER	Area-specific emission rate of a pollutant from a material	$\mu g\ m^{-2}\ h^{-1}$
V	Room volume	m^3
v_d	Deposition velocity of a pollutant in the air onto a surface	$m\ hour^{-1}$

2. Indoor Air Pollution Models

In its most basic form, the level of outdoor pollutants can be approximated for indoor environments by the "100/10/1" rule-of-thumb by Tetreault [6] (pp. 35–38). It suggests, as a generalization, that outdoor pollutants will be decimated from a 100% presence outdoors to about 10% inside a building and again down to about 1% of the outdoor level inside smaller indoor enclosures such as display cases, cabinets, etc. As simple as it may sound, the "100/10/1" rule provides a fair idea of which order of magnitude one may expect for air pollution inside a museum building if the outdoor pollution level is known. The ratio between the three locations, outdoors, room, and confined enclosure, reflects a mass-balance based on several factors (building geometry and characteristics, weather, use of the building, etc.), which control the transfer of air between the different zones.

2.1. Mass-Balance at Steady-State

Mass-balance models are based on the fundamental principle of mass conservation and can be expressed as an ordinary differential equation that balances sources and sinks [20,21] (pp. 57–65):

$$\frac{dC_i}{dt} = C_{Sources} - C_{Sinks} \quad (1)$$

where C_i is the concentration at steady-state, t is time, $C_{sources}$ the influx, and C_{sinks} is the loss.

Such expressions are fast to compute and can provide a basic means to model air pollution in buildings or rooms that can be treated as a single zone, e.g., one room surrounded by ambient environment, and assumes well-mixed air (uniform distribution of pollutants throughout a zone). The mass-balance calculates the concentration at steady-state conditions.

2.2. Indoor-Outdoor Ratio (I/O)

Weschler et al. [22] presented a steady-state mass-balance model to determine the ozone concentration that one can expect inside a building if the outdoor level is known. This relation is also known as the I/O (indoor/outdoor) ratio. Several parameters must be known to use the I/O model. It treats the building (or room) as a single zone and uses the outdoor concentration together with the air change rate (n), the deposition velocity of the pollutant onto an internal surface (v_d), the ratio between the surface area of materials in the zone and the volume of the zone (A/V), in order to estimate the concentration of ozone indoors:

$$\frac{C_i}{C_o} = \frac{n}{n + \left(v_d \times \frac{A}{V}\right)} \quad (2)$$

The surface-to-volume ratio (A/V) is also called the loading of material (L). The deposition velocity expresses the rate at which the pollutants react on surfaces. It is defined as the flux of the pollutant toward a surface divided by its concentration in the air and is by this a mass-transfer coefficient with the unit of velocity [23]. The factor $v_d(A/L)$ from Equation (2) expresses the loss of pollutants by surface uptake (sorption reactions). This is often referred to as the surface removal rate (S) [24], and is comparable to an air exchange rate with which it shares unity (number of room volumes removed per unit time). For a room with several materials present ($material_1$, $material_2$, $material_3$, …), the total surface removal rate S is the sum of each material's surface removal rate (S_1, S_2, S_3, …):

$$\frac{C_i}{C_o} = \frac{n}{n + (S_1 + S_2 + S_3 + \ldots)} \quad (3)$$

The I/O model was put forward only for ozone [22], but since then, it has been used for other pollutants as well, such as nitrogen dioxide and sulfur dioxide [25]. The model assumes that the air pollutants are removed only through deposition onto surfaces or by ventilation. According to Spicer et al. [26], this is a reasonable assumption for reactive gaseous outdoor air pollutants, although pollutants, such as nitrogen dioxide, are known to engage in heterogeneous reactions and be re-released as nitrous and nitric acid [27,28]. Reactions in the air are also disregarded, although, for example, the production of nitrogen dioxide by atmospheric reactions involving ozone and nitrogen oxide may take place indoors [29].

In heritage studies, the I/O model has been known as the "IMPACT model" due to its use in an online I/O calculator aimed for museums and other heritage buildings, developed through the European research project "Innovative Modelling of Museum Pollutants and Conservation Thresholds" (IMPACT) [25]. The online IMPACT calculator is currently not available. However, the I/O model (Equation (2)) can easily be set up in Microsoft Excel or similar spreadsheet software. Deposition velocities are available in the literature for a range of pollutants and materials, for example [30].

2.3. Indoor Air Pollution (IAP)

A steady-state balance similar to that of Equations (2) and (3) can be expressed for pollutants generated within the indoor zone (the building, the room) [31,32]:

$$C_i = \frac{SER \times L}{n + S} \quad (4)$$

The IAP model assumes the outdoor concentration $C_o = 0$ and treats the building as a single zone. The indoor generation of pollutants happens as emission from materials, and in the case of museum environments, especially the release of organic acids is considered harmful for a range of heritage materials [4–6]. The area-specific emission rate *SER* of organic acids has been measured for samples of heritage wood objects [10], and the IAP model has previously been applied to museum storage environments [31,32]. Smedemark and Ryhl-Svendsen [32] provided a general step-by-step guide to its use based on easily measured building characteristics and data collated from the literature. This approach is used in the present study as well (see Section 4 below).

2.4. Monte Carlo Simulations

The factors which the air pollution models depend on may be difficult to estimate precisely. Some parameters may vary widely between sites or within the same building over time. It can, therefore, be difficult to choose one specific value for each factor without giving rise to a large uncertainty in the calculated pollution level. Instead, it may be desirable to model within a given interval (for example, expected variation in the air exchange rate or uncertain estimation of surface areas of objects). In the case of the I/O ratio and IAP models, some parameters remain constant and are easy to measure (e.g., room volume), while others may vary within a given interval, influenced continuously by climate or by sudden impacts caused by human activity (Figure 1).

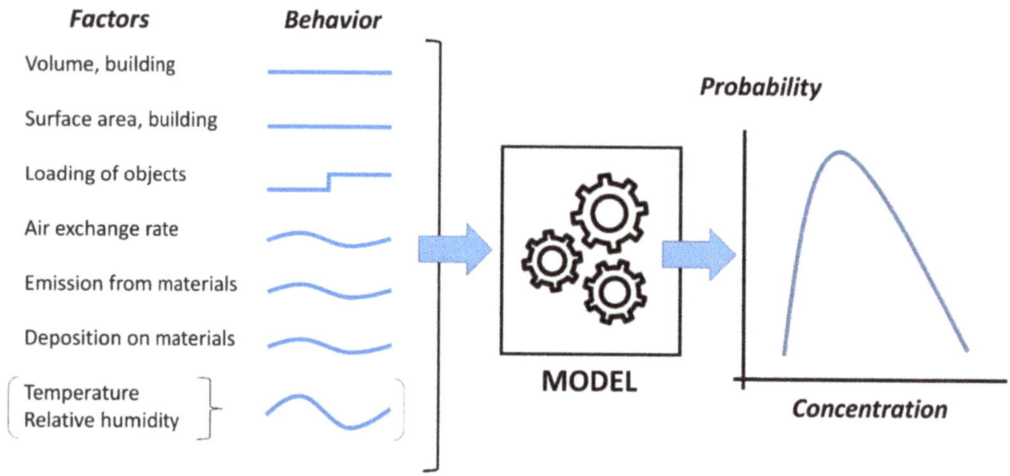

Figure 1. Principal sketch of factors influencing air pollution modeling and their possible behavior. Some factors remain constant over time (straight line), some may change due to sudden events (jagged line) or by more or less continuous variations (curved lines). Temperature and humidity are not directly included in the models. However, indirectly, they influence other factors such as emission rate. Together they result in a range of more or less probable outputs (concentrations).

A Monte Carlo simulation is a statistical tool that can be used to model an output that depends on several random variables. The parameters are defined as an interval, and a probability distribution is assigned to each variable. The results are obtained through repeated random sampling over numerous model runs. Monte Carlo simulations have previously been used in several indoor air pollution studies [32–36].

2.5. Other Computational Simulations

It must be stressed that the simple models of Equations (2)–(4) do not take into account all the variables which affect indoor air pollution levels. Chemical reactions may take place in the air or on surfaces, which re-release new compounds [37,38]. This is especially relevant for nitrogen oxides [27,28]. While detailed chemical models exist [39,40], they are much more complex than what is necessary for day-to-day air quality assessment for heritage buildings and will not be dealt with here. Although such models have indeed been applied to heritage environments, this has mainly been for short-term continuous monitoring periods with a focus on time-resolved measurements [41–43].

A number of non-constant factors influence emission and deposition rates, e.g., temperature and relative humidity [10,30]. Air exchange may vary, and the air movement and turbulence in a room create variations and gradients in the indoor environments, which are not reflected in the simple steady-state models as they assume a perfect mixing of air within a space. Instead, fluid simulations (computational fluid dynamics, CFD) can be used to explore such situations [44]. Again, although the use of CFD indeed has a place also for heritage environment research, it requires expert know-how to use and may primarily be applied for exploring special situations where airflow within a room must be visualized [45], or as an integrated part of designing buildings together with climate and energy consumption simulations [46].

Particle pollution loads can be described by expressions similar to those of the gaseous compound models, e.g., I/O ratio [47,48]. However, other, more specific mechanisms do also influence particle infiltration and deposition (among other things, particle size, aerodynamic properties, gravity, and thermophoresis-related phenomena) [49–51]. These conditions cannot be given the attention they deserve within the limits of this paper. Therefore, particle modeling will not be dealt with.

3. Case Study

3.1. The National Museum Storage Facility

For the demonstration of the application of mass-balance models, we take as an example a newly built Danish storage facility. Located in the town Vinge, about 40 km outside Copenhagen, the building contains the largest heritage storage facility in Northern Europe. The facility, designed by Gottlieb Paludan Architects and MOE Consulting Engineers, opened in May 2022, and will house the main part of the collections of The National Museum of Denmark and the Danish Royal Library [52].

The entire facility has about 25,000 m^2, which, besides storage areas, includes an indoor truck bay, service areas, pest disinfection chambers, workshops, and offices. The library part of the storage facility is mechanically cooled for the keeping of chemically unstable collections (e.g., acid paper), while the main hall for museum objects is without active temperature control. This part of the storage area is designed following the low-energy concept previously developed and put in use in several Danish storage buildings [53,54]. We focus here on the National Museum's part of the facility, which is the main, unheated storage hall (Figures 2 and 3).

3.2. The Building

The storage building has a well-sealed building envelope with a low air change rate. The exact air exchange rate is not known (not yet measured) but assumed to be on the order of 0.5 per day (0.02 h^{-1}). There are no windows, and entrances are fitted with double-door airlocks in order to reduce the impact of the outdoor climate.

The storage hall contains eight adjacent storage sections, separated by rolling gates (Figure 2). Each section measures 1000 m^2 with a volume of 9400 m^3 (total 8000 m^2 and 75,200 m^3). Despite the division in eight sections (due to fire sectioning), the eight closely connected areas can be regarded as one large, combined storage volume, which shares the internal distribution of air, and by this, a uniform climate and air quality.

Figure 2. A view down the central aisle that connects the storage sections, of the still empty storage hall. Photo by the authors.

Figure 3. A compact-shelves section, about half-full, with wooden furniture. Photo by the authors.

A ventilation system is in place for humidity control by mechanical dehumidifiers (desiccant type). It runs entirely by internal recirculation and without intake of ambient air.

The relative humidity is maintained at a moderate level (set-point 50% RH, limits 40–60% RH). The ventilation loop is only running when dehumidification is needed, and is VAV (variable air volume) controlled, with a maximum recirculation rate of 0.3 room volume per hour. The recirculated air passes through a combined particle and chemical air filter (Camfil City-Flo combination bag filter, particle filter grade F7 with broad-spectrum carbon media; Camfil, Stockholm, Sweden, https://www.camfil.com/ (accessed on 1 May 2023)). The hall is unheated and has been designed to always remain below 20 °C, however, the actual thermal performance has still to be validated by a full year of normal operation.

3.3. Ambient Conditions

The facility is located in a rural area near the small city of Frederikssund, and a few km from Roskilde Fjord. Danish climate is within a temperate climate zone (Köppen classification Dfb), and the ambient pollution level of the area is about 60 $\mu g\ m^{-3}$ ozone and 6 $\mu g\ m^{-3}$ nitrogen dioxide annual average (data from the national air pollution monitoring program at station Risø, 15 km away) [55].

3.4. Collection and Interior

Work started in 2022 to move in collection objects, however, this operation will take years due to derivative tasks, such as cleaning, pest disinfection procedures, and documentation along the moving process. At the time of writing (Spring 2023), the facility is loosely estimated 20% full. The collection contains a large variety of cultural history objects of many types of materials (wood, metals, painted objects, plastics, etc.). However, the vast majority are wooden objects. For simplicity in the present modeling study, we regard the collection items to be a uniform quantity of wooden objects (e.g., furniture, see Figure 3).

The storage is equipped with mobile compact shelves, which allow for storing objects at a high storage capacity (Figure 3). The shelves are made of powder-coated and galvanized steel. Other main materials present in the storage areas are concrete (walls and ceiling) and epoxy paint (floor). In Table 2, the surface area and loading (ratio of surface area to room volume) are given for each class of materials.

Table 2. Distribution of materials in one section of the storage hall. Materials are assumed to be equally distributed in all eight sections.

Material	Area [m^2]	Loading [$m^2\ m^{-3}$]
Wall and ceiling (concrete)	4300	0.5
Floor (syntetic paint)	1000	0.1
Shelves (metal)	1000	0.1
Objects (wood), now	6500	0.7
Objects (wood), when full	28,000	3.0

4. Methods

The performance of the storage hall with regard to indoor air quality was demonstrated first for outdoor pollutants by the I/O model, followed by a prediction of indoor air pollution levels by the IAP model. Both models predict the pollution levels at steady-state conditions, however, we demonstrate how the steady state will vary at different air exchange rates. The IAP model was then used as the basis for a series of Monte Carlo simulations taking possible variations of the input parameters into account.

Several of the parameter values were estimates. The air exchange rate of the storage hall has not yet been validated by measurement. However, based on previous measurements, probable values were assumed [13]. Data on heritage materials' emission rates, as well as deposition rates for organic acids, are scarce. The input values, many of which are assumptions, were mainly taken from previous model studies [31,32]. Deposition velocity data for ozone and nitrogen dioxide were given by Grøntoft and Raychaudhuri [30].

The dimensions of the storage area, its interior, ventilation system, and the loading and nature of objects, were measured and observed at visits onsite, and by consulting architectural drawings and technical descriptions of the building. The estimation of objects' loading is subject to some uncertainty, as the museum objects are all different in size and complex in shape (object types include furniture, household utensils, musical instruments, wooden sculptures, timber objects for buildings, etc.). For one typical shelf section, we measured the projected surface area of all objects in great detail. From this, we calculated the average object surface-area per running meter shelf, which was then multiplied up to the full storage capacity.

4.1. I/O Model

The I/O ratio was calculated by Equation (2) for the storage hall when full, assuming an air exchange rate of 0.02 h^{-1}, loading of materials as given in Table 2, and for each material the deposition velocities given in Table 3. From this, I/O ratios and the distribution of the pollution loss on the different material surfaces were calculated for ozone and nitrogen dioxide. For ozone and nitrogen dioxide, it was modeled how the I/O depends on the air exchange rate. Calculations were performed in Microsoft Excel using Equation (3), at steps of 0.02 h^{-1} within the interval $0 < n < 1$ h^{-1} (50 steps).

Table 3. Deposition velocity values at 50% relative humidity. Converted from cm s^{-1} in the original source [30].

Material	Ozone Deposition Velocity (v_d) [m h^{-1}]	Nitrogen Dioxide Deposition Velocity (v_d) [m h^{-1}]
Fine concrete	0.0612	0.0360
Brick	0.4320	0.2268
Wood-work surface treated	0.0198	0.0108
Metal	0.0050	0.0036
Synthetic floor covering	0.0202	0.0108

4.2. IAP Model

The concentration of organic acids was calculated by Equation (4) for the storage hall in three general scenarios:

- Sparsely filled with museum objects (as today) at a loading (L) of 0.7 m^2 m^{-3}
- Half-filled storage at L = 1.5 m^2 m^{-3}
- Full storage at L = 3 m^2 $m^{-3.}$

Only emission from wood was taken into account. The emission of organic acids directly affects the concentration in the air. As the emission rate is influenced by temperature (the higher temperature, the higher the emission rate) [10,56,57], the IAP modeling was carried out for the two extremes, a high summer and a low winter storage temperature. The model input is given in Table 4.

Table 4. Area-specific emission rate of wood (general), loading of objects, air exchange rate, and surface removal rate for the storage hall. *SER* data from [32].

Scenario	*SER* Winter [µg m^{-2} h^{-1}]	*SER* Summer [µg m^{-2} h^{-1}]	L [m^2 m^{-3}]	n [h^{-1}]	S [h^{-1}]
Sparsely filled storage (today)	50	200	0.7	0.02	2
Half full storage	50	200	1.5	0.02	2
Full storage	50	200	3	0.02	2
Full storage with filter recirculation	50	200	3	0.30 *	2

* Air exchange and filter removal combined.

Following this, it was modeled for each scenario how the organic acid concentration varied with a change in the air exchange rate (by ventilation or internal filtration). Calculations were performed in Microsoft Excel by Equation (4) at steps of 0.02 h^{-1} within the interval $0 < n < 10$ h^{-1} (500 steps).

4.3. Monte Carlo Simulation

A Monte Carlo simulation was used to model the indoor air pollution level following Equation (4). The simulation modeled the concentration based on the probability within a defined range for each of the four parameters. For simplicity, a uniform probability distribution was assigned between the minimum and maximum value of each variable of the pollution mass-balance. It may be that some factors are distributed differently (e.g., air exchange rate). However, more data are needed in order to verify this. The four parameters were considered independent.

The simulation was made by 1000 repetitions of Equation (4) where the value of each input parameter was selected randomly with a linear probability distribution using the SLUMP function (random number generator) in Microsoft Excel. The input data for the area-specific emission rate (*SER*) from wood were taken from a study on the emission from three heritage wood object samples [10] in winter and summer temperature conditions. Qualified estimates of possible ranges of *L*, *n*, and *S* were based on previous model studies [31,32]. Model input conditions (parameter intervals) are given in Table 5.

Table 5. Monte Carlo simulation input data. *SER* ranges from [10]. Intervals of *L*, *n*, and *S* were estimated based on [31,32].

Input Factor	Interval
SER, winter [$\mu g\ m^{-2}\ h^{-1}$]	39–108
SER, summer [$\mu g\ m^{-2}\ h^{-1}$]	145–303
L, low: Sparsely filled storage [$m^2\ m^{-3}$]	0.4–0.9
L, high: Full storage [$m^2\ m^{-3}$]	3–5
n [h^{-1}]	0.01–1
S [h^{-1}]	0.2–2

As the organic acid emission from heritage collections depends on temperature, one simulation was made for winter and one for a summer scenario. Furthermore, a simulation was first conducted for the present conditions (sparsely filled storage), and then repeated for a filled storage room. Thus, four scenarios were simulated:

A. Storage hall as now, sparsely loaded with objects, winter temperature.
B. Storage hall as now, sparsely loaded with objects, summer temperature.
C. Storage hall at full capacity, filled with objects, winter temperature.
D. Storage hall at full capacity, filled with objects, summer temperature.

A reservation must be made: Data for temperature-dependent emission rates are scarce, and it has not been possible to find input values for the exact temperature range in the hall. Our summer input relates to tests conducted at a standard room test temperature of 23 °C. It is, therefore, possible that the summer conditions will be overestimated, and the resulting summer concentration must therefore be considered an absolute worst-case scenario.

4.4. Pollution Measurements at Site

The concentration of ozone, nitrogen dioxide, and organic acids (acetic acid and formic acid) were measured shortly after the storage facility was taken into use. Sampling was conducted by the use of passive samplers (badge type) supplied and analyzed by The Swedish Environmental Research Institute IVL (Gothenburg, Sweden, www.ivl.se (accessed on 1 May 2023)). Sampling was carried out in duplicates for an exposure period of one month. Indoor samplers were mounted with metal clips on a shelf-end, half-ways between the floor and the ceiling. Outdoor samplers were mounted at 2 m height under a rain screen on a pole placed on an open grass lawn. Conditions were not yet representative of normal storage operation (empty storage hall, start-up adjustment of the ventilation system, increased traffic through docking gate, etc.), but can be regarded as background screening. Follow-up measurements will be made in the coming years as the storage is gradually being filled.

5. Results

5.1. I/O Model Results

For a constant air exchange rate of 0.02 h^{-1} and a fully loaded storage room (Table 2) the I/O ratio was for ozone 0.19 and for nitrogen dioxide 0.29. If all collection objects were removed from the model (bare room with only concrete walls, ceiling, and floor) the I/O ratio increased to 0.37 for ozone and 0.50 for nitrogen dioxide. The fate of pollutants which enter the storage is shown in Figure 4, where the distribution of the materials the pollutants deposit onto are shown, together with the fraction leaving the room again via the air exchange. For ozone, 53% will deposit on the heritage objects (wood), while for nitrogen dioxide, only 14% ends up onto the collection. The rest is lost on building surfaces or removed again by ventilation.

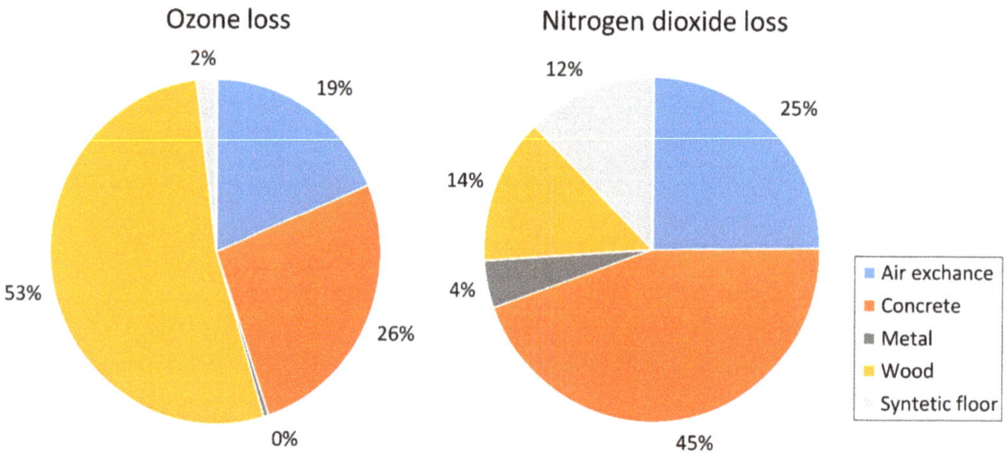

Figure 4. The fate of ozone and nitrogen dioxide in a full storage room (28,000 m^2 of wooden objects). The charts show the percentage distribution between sorption on different materials and removal by ventilation.

As the I/O ratio strongly depends on the air exchange rate, it is shown in Figure 5 how the I/O ratio for ozone and nitrogen dioxide will increase if the ventilation rate increases, and vice versa. This is illustrated for air changes up to 1 h^{-1}. The surface removal rates for the storage room, as calculated on the basis of Tables 2 and 3, were rather low, for ozone 0.1, and for nitrogen dioxide 0.05 h^{-1}. As an imaginary example of an indoor environment with a higher surface removal rate, Figure 5 also shows the I/O ratio for a room with $S = 3\ h^{-1}$, which is a mid-range condition for typical occupied buildings such as homes, shops, offices, and museum exhibitions, having a variety of highly sorptive surfaces, such as carpets and furniture textiles [24].

5.2. IAP Model Results

The span in organic acid concentrations between summer and winter conditions is shown in Figure 6, assuming an air exchange rate of 0.02 h^{-1} (1/2 room volume per day) for the situation of the storage hall today (sparsely filled), and as it will gradually become filled up (input data from Table 4). For the full storage, the effect of adding the current filtration system (0.3 h^{-1} clean air delivery rate) was also tested.

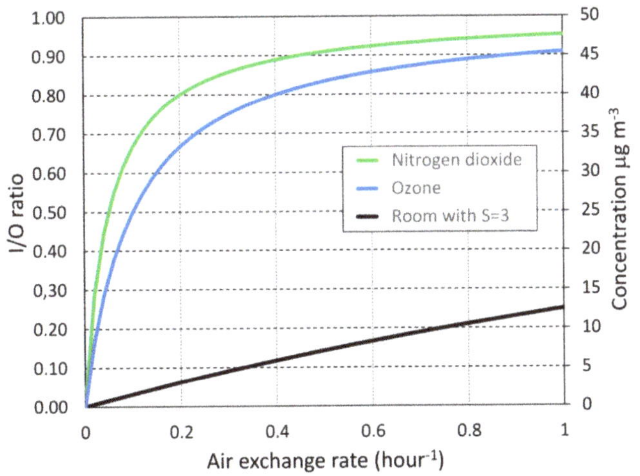

Figure 5. The I/O ratio of ozone and nitrogen dioxide in the storage hall when full as a function of the air exchange rate. Also shown is a room with a surface removal rate $S = 3\ h^{-1}$.

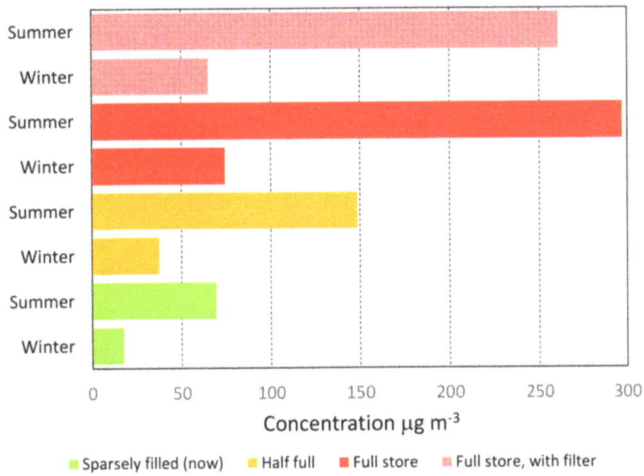

Figure 6. Organic acid concentration at winter and summer temperature inside the storage hall at different degrees of loading with wooden objects. At the top, in red, is shown the full storage with and without air filtration turned on.

In Figure 7 it is shown how the steady-state concentration of organic acids changes if the air exchange rate varies. This is illustrated for air changes up to $10\ h^{-1}$, which covers a range from solely naturally ventilated buildings up to a high rate of forced ventilation (with new air and/or by internal filtration). The models are shown for winter and summer temperature, and at each instance for the three general scenarios: Sparsely filled with wooden museum objects (as today) at a loading (L) of $0.7\ m^2\ m^{-3}$, a half-filled storage room at $L = 1.5\ m^2\ m^{-3}$; and a full storage at $L = 3\ m^2\ m^{-3}$.

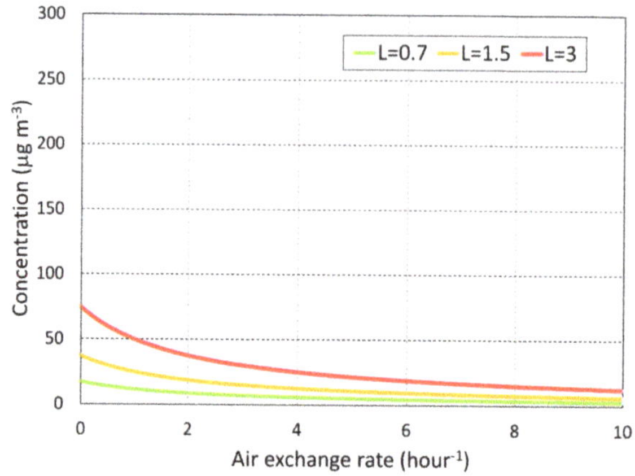

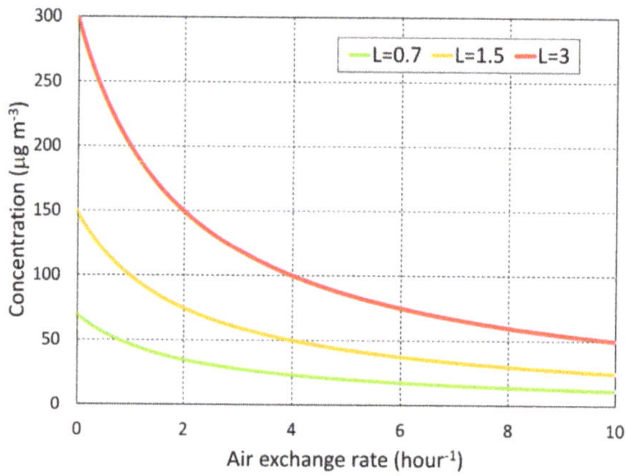

Figure 7. The concentration of organic acids as a function of the air exchange rate at winter temperature (**top**) and at summer temperature (**bottom**). For each scenario: Full room ($L = 3$ m^2 m^{-3}), half-full room ($L = 1.5$ m^2 m^{-3}); sparsely filled room ($L = 0.7$ m^2 m^{-3}).

5.3. Monte Carlo Simulation Results

The distribution of probable organic acid concentrations is given in Figure 8 below. The output from each of the 1000-iterations model runs was collected into pillars of intervals of 50 µg m^{-3} in a probability density histogram (0–50, 51–100, 101–150, a.s.o.). The simulation output presents the likely span of the indoor air pollution level for: (A) Sparsely filled storage room in winter, (B) sparsely filled storage in summer, (C) full storage in winter, (D) full storage in summer. The sparsely filled storage reflects the status of the facility at the time of writing. For each simulation, the median, average, minimum, and maximum 95% confidence interval bounds are shown in Table 6.

Table 6. The minimum and maximum 95% confidence interval bounds, median and average concentration [µg m^{-3}] for each Monte Carlo simulation (1000 iterations).

Simulation	5th Percentile	Median	Average	95th Percentile
A	15.3	29.3	30.9	51.5
B	53.3	88.5	91.9	144
C	56.0	115	123	215
D	197	360	377	648

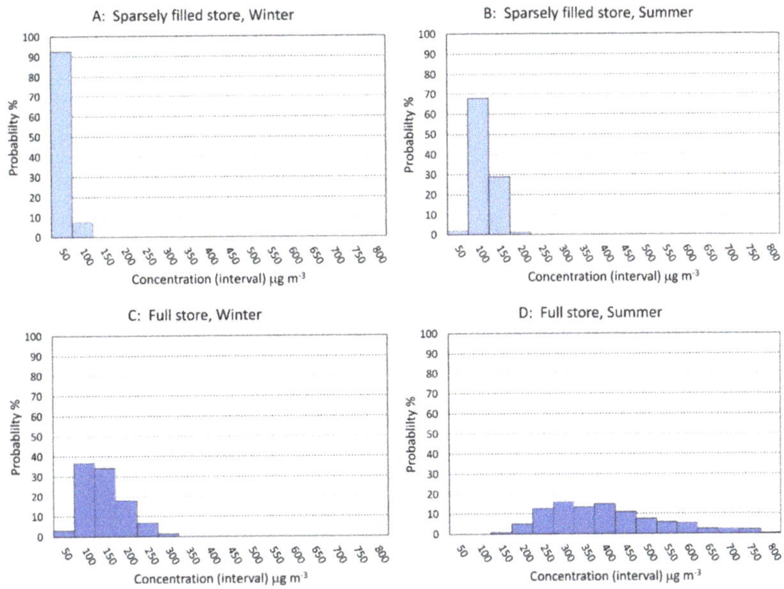

Figure 8. The probability distribution of pollution levels for the storage building as today (sparsely filled) in winter (**A**) and summer (**B**) conditions, and when filled with objects in winter (**C**) and summer (**D**) conditions.

In order to establish the least necessary number of simulation iterations, the result (average concentration) and the spread (standard deviation) of repeating a simulation ten times were compared for the modeling scenario C at 10, 100, 1000, and 2000 iterations. At 1000 and 2000 iterations, the results were very similar (average 123.4 and 122.7, both with a standard deviation of 1), which indicated that 1000 iterations were enough to obtain stable and reliable results (Figure 9).

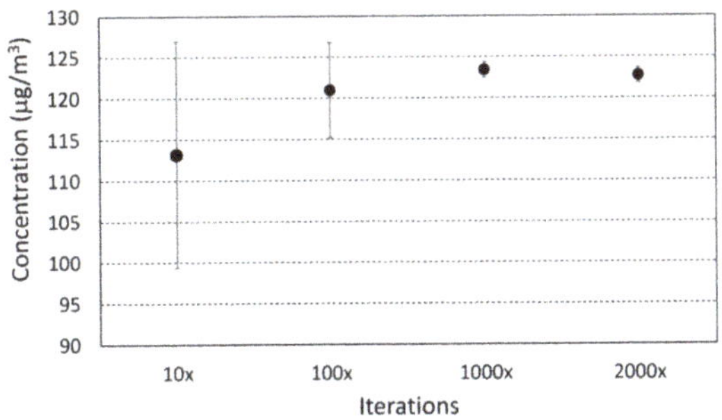

Figure 9. The average and standard deviation of 10 repeated simulations of scenario C at 10, 100, 1000, and 2000 iterations.

5.4. Pollution Measurements

The measured concentrations are reported in Table 7. Each value is the average of a duplicate measurement. Organic acids were only measured indoors.

Table 7. Passive sampling results for air pollutants indoors and outdoors. The indoor-to-outdoor ratio is given for ozone and nitrogen dioxide.

Location	Ozone	Nitrogen Dioxide	Organic Acids
Indoor concentration [$\mu g\ m^{-3}$]	1	0.2	12
Outdoor concentration [$\mu g\ m^{-3}$]	52	5.1	n.a.
I/O ratio [dimensionless]	0.02	0.04	n.a.

6. Discussion and Conclusions

6.1. Background Measurements

The pollution measurements by passive sampling, which were performed at the beginning of the operation of the building, revealed a low concentration of organic acids indoors (12 $\mu g\ m^{-3}$), and a low ingress of outdoor pollutants (I/O ratio = 0.02–0.04) (Table 7). The I/O model. However, predicted higher ratios for ozone 0.19, and for nitrogen dioxide 0.29, even had the storage room been full of objects. On the other hand, the low organic acid concentration reflects well a storage almost without wooden objects.

6.2. Outdoor Pollutants

While we acknowledge that the initial pollution measurement was done at a time when normal operation routines within the building had not yet come fully into force, the difference in outdoor pollutants between real measurements and the model calls for a closer examination. As displayed in Figure 5, the I/O ratio is highly influenced by air change. Although the air exchange rate of the hall is unknown, it is unlikely that it was much lower than our estimate of 0.02 h^{-1}. A more likely explanation is that the single-zone model has limitations if real buildings have a more complicated geometry than just one room. Perhaps the ingress of air had a more tortuous path into the storage hall than just through the nearest doors so that adjacent rooms became secondary zones between the storage area and ambient air. In that case, the indoor pollution level would become a fraction of the levels in the adjacent zones, which were a fraction of ambient, etc. However, in order to investigate such behavior in detail, air exchange measurements using several tracer gases at the same time are required [57], and the ignored influence of internal airflow between adjacent areas is a general weakness of single-zone models.

Another plausible explanation for the low I/O ratio is the additional effect of air filtration. Assuming filtration was on full-time, the current filtration system would deliver 0.3 room volumes of clean air per hour. This can be added to Equation (3) as a contributing loss (Q):

$$\frac{C_i}{C_o} = \frac{n}{n + (S_1 + S_2 + S_3 + \ldots) + Q_{filter}} \tag{5}$$

(input parameters: $n = 0.02$; $S_{total\text{-}ozone} = 0.1$; $S_{total\text{-}NO2} = 0.05$; $Q_{filter} = 0.3$).

Which, in that case, lowers the I/O ratio to 0.05 for both ozone and nitrogen dioxide, a level close to the real-life measurements.

An important feature of the I/O model is the illustration of the distribution of pollution loss. The collection received 53% of the indoor ozone and 14% of the nitrogen dioxide. Had the collection contained larger fractions of sorptive materials, such as textiles, this would have received an even higher part of the pollution deposition. This clarifies a dilemma, which is that storing many collection items will help clean the air. However, it will happen at the cost of the harmful pollutants' deposition onto the collection. Deliberate exploitation of exposing large surface areas is indeed an efficient pollution control strategy, as long as the sacrificial material is something other than the collection, e.g., sorptive wall covering [58]. In the planning of such actions, the I/O model will be a useful tool for estimating the effect of applying pollution-scavenging materials in rooms.

As previously mentioned, another shortcoming of the I/O model is the ignoring of atmospheric reactions. Studies in museum buildings have demonstrated how ozone and nitrogen oxide may react indoors and produce nitrogen dioxide beyond what is already introduced by ventilation, sometimes to a level of I/O > 1 [29]. In an American museum gallery, three weeks of measurements showed a loss distribution for ozone where surface uptake accounted for 62%, reactions with nitrogen oxide 31%, nitrogen dioxide 2%, and sorption on occupants for 5% [43]. In other words, for a model, which only accounts for surface and air exchange removal, the result may be underestimated by 30–40%, and this must be taken into account, especially for situations with a high influx of outdoor pollutants (high air exchange rate).

6.3. Indoor-Generated Pollutants

The IAP model, as run either as a single calculation based on fixed conditions or by the Monte Carlo simulation based on plausible intervals of each factor, offered a fair estimate of how the indoor pollution level will react to the continuous filling up of the storage area with wooden objects. For the storage still only sparsely filled (L = 0.7 m^2 m^{-3}), the most probable concentration levels were within well-defined intervals (0–50 in winter and 50–100 μg m^{-3} in summer). For a full storage room (L = 3 m^2 m^{-3}), the spread of the most probable concentration range was much wider, especially in summer (between about 200–500 μg m^{-3}), due to the wider summer intervals of the decisive factor *SER*.

In general, the difficulties in measuring L and n, as well as the little data available on *SER*, add uncertainty to the model. Besides scarce *SER* values for heritage materials as such, data are lacking the actual temperature interval of the storage facility. We used data measured at 10–23 °C, as this is what was available in the literature. However, the storage hall is intended to be cooler (at least below 20 °C), although this has not yet been verified by a full year of measurement. This winter (2022–2023), the lowest indoor temperature was 11 °C, and we still have to observe a summer period at normal operation. The uncertainty of the temperature level may, therefore, lead to an overestimation of the concentration level in summer. In any case, the model predicts how the organic acid level varies between the seasons, and while we still need to test the exact values by measurement when normal operation has been set up, the trend has been laid out. Previous monitoring in another storage building of The National Museum convincingly illustrated the same behavior [57], which we, based on the modeling, also expect to observe in the new storage hall (Figure 10).

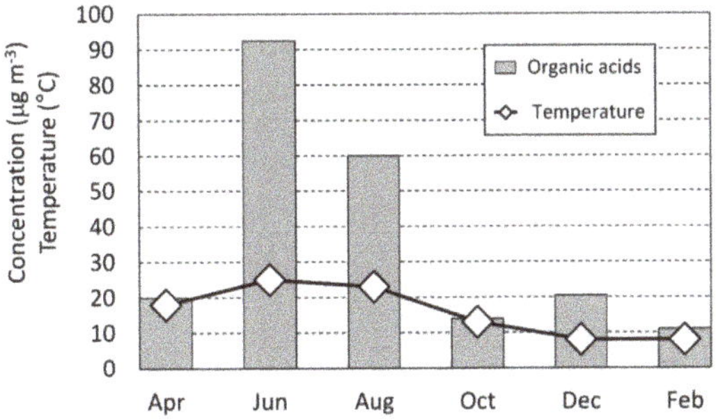

Figure 10. Organic acid concentration and temperature during one year in the storage room of the National Museums of Denmark's Music Museum. Data from [57] (converted from ppb).

6.4. Constant versus Dynamic Conditions

The static nature of steady-state models is a challenge when applied to dynamic environments. Although we assume a number of factors to be constant, they may, as discussed in the introduction, in fact, be of a dynamic nature (Figure 1). Examples are the emission rate, which may vary with the annual seasons due to the influence of temperature, or the air exchange rate, which may vary due to changes in weather and peoples' use of the building. The dynamic nature of an indoor environment can be demonstrated by a continuous series of instant ozone measurements performed at yet another of the National Museum's storage buildings (Figure 11). When measured at a resolution of 1-h intervals, it was revealed how the indoor level varied sometimes more than 10 μg m^{-3} per day, in a pattern closely following the outdoor ozone level. Over the twelve days shown in Figure 11, the average I/O ratio was 0.39. However, on a short-term basis, it cycled between 0.25 and 0.70. This would rarely be reflected in a steady-state simple I/O modeling, as the input values often are long-term average values (e.g., passive sampling measurements carried out over periods of weeks or longer).

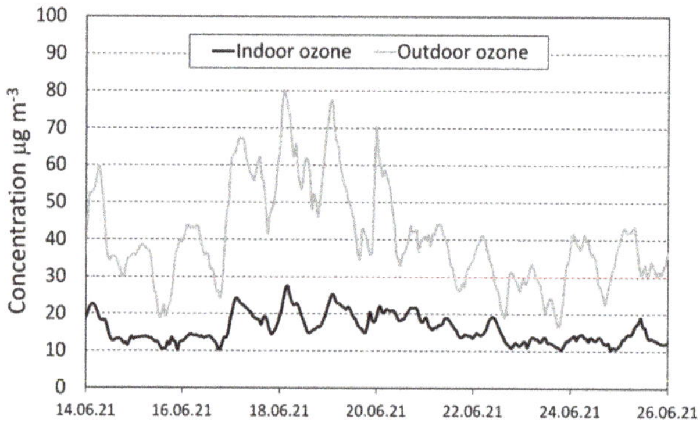

Figure 11. Twelve days of continuous ozone measurements inside and outside The National Museum of Denmark's storage facility Ørholm (unpublished data by the authors).

Although Monte Carlo simulations reflect the possible outcomes at steady-state conditions, the spread of results also reflects the possible numerical extent of the short dynamic variations (however, not necessarily distributed in the same way). Real dynamic condition modeling would require a large amount of data, and all factors should be available in time-resolved high resolution in order to show their mutual influence on the concentration over time. However, as argued previously, for conservation risk management, this is an unnecessary level of detail when long-time average concentration is sufficient for dose calculations for the assessment of materials damage risks.

6.5. Practical Implications and Perspectives

For our example, the National Museum's storage hall in Vinge, modeling showed that outdoor pollutants are efficiently retarded by the low air exchange of the building. Furthermore, the additional air filtration rate of 0.3 h^{-1} provides sufficient control. The indoor generated organic acids may be expected in a level up to about 300 (plausible range 200–500) μg m^{-3} in summer, once the storage hall has been filled with objects. However, should summer temperature turn out to be much lower than 20 °C, the concentration will be correspondingly smaller as well. Wintertime levels are expected to be less than 100 μg m^{-3}.

If desired, the organic acid concentration can be decreased by filtration. However, for the indoor generated pollutants, the flow rate of the current filtration system is inadequate. According to Figure 7, the filter recirculation rate (clean air delivery rate) should be 2 h^{-1} to halve the level at current conditions or 4 h^{-1} to lower it to one-third. In any case, a decision to increase filtration should be based on assessing the potential risk for the actual collection and the question of whether there are particularly susceptible objects present. Even for the least probable high concentrations predicted by the Monte Carlo simulation, the level is below the maximum average level of 1000 μg m^{-3} recommended by ASHRAE for a general museum collection [16].

As the museum storage hall is still just beginning to be filled up, it will take some time before true steady-state conditions enter into force. Our planned follow-up monitoring of the pollution levels over the coming years will reveal how well the models actually predicted the reality. More data are needed in order to improve input for the models. The authors encourage readers to share data from pollution monitoring in heritage buildings, as well as any experiences in using the I/O and IAP models, in order to refine them and validate their use.

The models presented here are practical for assessing the impact of indoor and outdoor pollution loads in a building. Even for fast and overall screening, it will be easy to establish the order of magnitude estimates, from which it can be decided whether a problem needs to be addressed in more detail. In our example, the modeling revealed a need to investigate further the air exchange between adjacent indoor areas, and to establish the overall air exchange rate. Steady-state models allow, with a minimum of input data, to assess indoor conditions based on basic properties or as the result of initiated control actions (e.g., increased ventilation, applying filters, etc.), and by this contribute as a tool for better practice in air quality management in heritage buildings.

Author Contributions: Conceptualization, methodology, formal analysis, investigation, software, data curation, writing—review and editing, M.R.-S. and S.H.S.; writing—original draft preparation, M.R.-S. All authors have read and agreed to the published version of the manuscript.

Funding: This research received no external funding.

Data Availability Statement: The model input data are available in tables in the article.

Acknowledgments: We thank the management of the National Museum of Denmark for permission to use the storage buildings as the subject of our research. We especially thank conservator Lars Aasbjerg Jensen, National Museum of Denmark, for providing information and data about the building, and for practical support during site visits.

Conflicts of Interest: The authors declare no conflict of interest.

References

1. Baer, N.S.; Banks, P.N. Indoor air pollution: Effects on cultural and historic materials. *Int. J. Mus. Manag. Curatorship* **1985**, *4*, 9–20.
2. Brimblecombe, P. The composition of museum atmospheres. *Atmos. Environ. Part B Urban Atmos.* **1990**, *24B*, 1–8. [CrossRef]
3. Graedel, T.E.; McGill, R. Degradation of materials in the atmosphere. *Environ. Sci. Technol.* **1986**, *20*, 1093–1100. [CrossRef]
4. Blades, N.; Oreszczyn, T.; Bordass, B.; Cassar, M. *Guidelines on Pollution Control in Museum Buildings*; Museums Association: London, UK, 2000.
5. Hatchfield, P. *Pollutants in the Museum Environment-Practical Strategies for Problem Solving in Design, Exhibition and Storage*; Archetype Publications Ltd.: London, UK, 2002.
6. Tétreault, J. *Airborne Pollutants in Museums, Galleries and Archives: Risk Assessment, Control Strategies and Preservation Management*; Canadian Conservation Institute: Ottawa, ON, Canada, 2003.
7. Paterakis, A.B. *Volatile Organic Compounds and the Conservation of Inorganic Materials*; Archetype Publications Ltd.: London, UK, 2016.
8. Gray, V.R. The Acidity of Wood. *J. Wood Sci.* **1958**, *1*, 58–64.
9. Clarke, S.G.; Longhurst, E.E. The Corrosion of Metals by Acid Vapours from Wood. *J. Appl. Chem.* **1961**, *11*, 435–443. [CrossRef]
10. Smedemark, S.H.; Ryhl-Svendsen, M.; Schieweck, A. Quantification of formic acid and acetic acid emissions from heritage collections under indoor room conditions. Part I: Laboratory and field measurements. *Herit. Sci.* **2020**, *8*, 58. [CrossRef]
11. Smedemark, S.H. The dynamics and control of indoor air pollution in repositories without mechanical ventilation for cultural heritage collections. A literature review. *e-Preservation Sci.* **2018**, *15*, 17–29.
12. Grzywacz, C.M. Using Passive Sampling Devices to Detect Pollutants in Museum Environments. In *ICOM Committee for Conservation, Proceedings of the 10th Triennial Meeting, Washington, DC, USA, 22–27 August 1993*; Bridgland, J., Ed.; James & James: London, UK, 1993; pp. 610–615.
13. Ryhl-Svendsen, M.; Smedemark, S.H. Environmental conditions for Danish storage buildings: Reviewing 20 years of air quality surveys. *Stud. Conserv.* **2023**. (*submitted*).
14. Ryhl-Svendsen, M. Indoor air pollution in museums: Prediction models and control strategies. *Stud. Conserv.* **2006**, *51* (Suppl. S1), 27–41. [CrossRef]
15. Dabanlis, G.; Loupa, G.; Tsalidis, G.A.; Kostenidou, E.; Rapsomanikis, S. The Interplay between Air Quality and Energy Efficiency in Museums, a Review. *Appl. Sci.* **2023**, *13*, 5535. [CrossRef]
16. ASHRAE. Museums, Galleries, Archives, and Libraries. In *ASHRAE Handbook-HVAC Applications*; American Society of Heating, Refrigerating and Air-Conditioning Engineers: Atlanta, GA, USA, 2019; Chapter 24.
17. Grzywacz, C.M. *Monitoring for Gaseous Pollutants in Museum Environments*; (Tools for Conservation); The Getty Conservation Institute: Los Angeles, CA, USA, 2006.

18. Brimblecombe, P. The balance of environmental factors attacking artifacts. In *Durability and Change. The Science, Responsibility, and Cost of Sustaining Cultural Heritage. Dahlem Workshop, Berlin, Germany, 1992*; Krumbein, W.E., Brimblecombe, P., Cosgrove, D.E., Staniforth, S., Eds.; John Wiley & Sons: New York, NY, USA, 1994; pp. 67–79.
19. Brimblecombe, P. Pollution Studies. In *ENVIRONMENT Leather Project Deterioration and Conservation of Vegetable Tanned Leather*; Larsen, R., Ed.; Protection and Conservation of the European Cultural Heritage, Research Report No. 6 of European Commission; The Royal Danish Academy of Fine Arts: Copenhagen, Denmark, 1996; pp. 25–32.
20. Abdalla, T.E.; Peng, C. Evaluation of housing stock indoor air quality models: A review of data requirements and model performance. *J. Build. Eng.* **2021**, *43*, 102846. [CrossRef]
21. Pepper, D.W.; Carrington, D. *Modeling Indoor Air Pollution*; Imperial College Press: London, UK, 2009.
22. Weschler, C.J.; Shields, H.C.; Naik, D.V. Indoor Ozone Exposures. *J. Air Pollut. Control. Assoc.* **1989**, *39*, 1562–1568. [CrossRef]
23. Nazaroff, W.W.; Gadgil, A.J.; Weschler, C.J. Critique of the use of deposition velocity in modeling air quality and exposure. In *ASTM Special Technical Publication: Modeling of Indoor Air Quality and Exposure*; Nagda, N.L., Ed.; ASTM STP 1205; American Society for Testing and Materials: Philadelphia, PA, USA, 1993; pp. 81–104.
24. Weschler, C.J. Ozone in indoor environments: Concentration and chemistry. *Indoor Air* **2000**, *10*, 269–288. [CrossRef] [PubMed]
25. Blades, N.; Kruppa, D.; Cassar, M. Development of a web-based software tool for predicting the occurrence and effect of air pollutants inside museum buildings. In *ICOM Committee for Conservation, Proceedings of the 13th Triennial Meeting, Rio de Janeiro, Brazil, 22–27 September 2002*; Verger, I., Ed.; James & Jamens: London, UK, 2002; pp. 9–14.
26. Spicer, C.W.; Kenny, D.V.; Ward, G.F.; Billick, I.H. Transformations, Lifetimes, and Sources of NO_2, HONO, and HNO_3 in Indoor Environments. *J. Air Waste Manag. Assoc.* **1993**, *43*, 1479–1485. [CrossRef]
27. Febo, A.; Perrino, C. Prediction and experimental evidence for high air concentration of nitrous acid in indoor environments. *Atmos. Environ.* **1991**, *25A*, 1055–1061. [CrossRef]
28. Katsanos, N.A.; De Santis, F.; Cordoba, A.; Roubani-Kalantzopoulou, F.; Pasella, D. Corrosive effects from the deposition of gaseous pollutants on surfaces of cultural and artistic value inside museums. *J. Hazard. Mater. A* **1999**, *64*, 21–36. [CrossRef] [PubMed]
29. Brimblecombe, P.; Blades, N.; Camuffo, D.; Sturaro, G.; Valentino, A.; Gysels, K.; Van Grieken, R.; Busse, H.-J.; Kim, O.; Wieser, M. The indoor environment of a modern museum building, The Sainsbury Centre for Visual Arts, Norwich, UK. *Indoor Air* **1999**, *9*, 146–164. [CrossRef]
30. Grøntoft, T.; Raychaudhuri, M.R. Compilation of tables of surface deposition velocities for O_3, NO_2 and SO_2 to a range of indoor surfaces. *Atmos. Environ.* **2004**, *38*, 533–544. [CrossRef]
31. Ryhl-Svendsen, M. The role of air exchange rate and surface reaction rates on the air quality in museum storage buildings. In *Museum Microclimates, Proceedings of the Copenhagen Conference, Copenhagen, Denmark, 19–23 November 2007*; Padfield, T., Borchersen, K., Eds.; National Museum of Denmark: Copenhagen, Denmark, 2007; pp. 221–226.
32. Smedemark, S.H.; Ryhl-Svendsen, M. Determining the level of organic acid air pollution in museum storage rooms by mass-balance modelling. *J. Cult. Herit.* **2022**, *55*, 309–317. [CrossRef]
33. McVoy, G.R. Monte Carlo Simulation Techniques Applied to Air Quality Impact Assessment. *J. Air Pollut. Control Assoc.* **1979**, *29*, 843–845. [CrossRef]
34. Johnson, M.; Lam, N.; Brant, S.; Charron, D.; Gray, C.; Pennise, D. Modeling indoor air pollution from cookstove emissions in developing countries using a Monte Carlo single-box model. *Atmos. Environ.* **2011**, *45*, 3237–3243. [CrossRef]
35. Kruza, M.; Shaw, D.; Shaw, J.; Carslaw, N. Towards improved models for indoor air chemistry: A Monte Carlo simulation study. *Atmos. Environ.* **2021**, *262*, 118625. [CrossRef]
36. Dai, X.; Liu, J.; Zhang, X. Monte Carlo simulation to control indoor pollutants from indoor and outdoor sources for residential buildings in Tianjin, China. *Build. Eviron.* **2019**, *165*, 106376. [CrossRef]
37. Weschler, C.J.; Shields, H.C.; Naik, D.V. Indoor chemistry involving O_3, NO, and NO_2 as evidence by 14 months of measurements at a site in Southern California. *Environ. Sci. Technol.* **1994**, *28*, 2120–2132. [CrossRef] [PubMed]
38. Weschler, C.J.; Shields, H.C. Potential reactions among indoor pollutants. *Atmos. Environ.* **1997**, *31*, 3487–3495. [CrossRef]
39. Nazaroff, W.W.; Cass, G.R. Mathematical modeling of chemically reactive pollutants in indoor air. *Environ. Sci. Technol.* **1986**, *20*, 924–934. [CrossRef]
40. Carslaw, N. A new detailed chemical model for indoor air pollution. *Atmos. Environ.* **2007**, *41*, 1164–1179. [CrossRef]
41. Druzic, J.; Adams, M.; Tiller, C.; Cass, G.R. The measurement and model predictions of indoor ozone concentrations in museums. *Atmos. Environ.* **1990**, *24A*, 1813–1823. [CrossRef]
42. Drakou, G.; Zerefos, C.; Ziomas, I.; Gantis, V. Numerical simulation of indoor air pollution levels in a church and in a museum in Greece. *Stud. Conserv.* **2000**, *45*, 85–94.
43. Pagonis, D.; Price, D.J.; Algrim, L.B.; Day, D.A.; Handschy, A.V.; Stark, H.; Miller, S.L.; de Gouw, J.; Jimenez, J.L.; Ziemann, P.J. Time-resolved measurements of indoor chemical emissions, deposition, and reactions in a university art museum. *Environ. Sci. Technol.* **2019**, *53*, 4794–4802. [CrossRef]
44. Li, Y.; Nielsen, P.V. CFD and ventilation research. *Indoor Air* **2011**, *21*, 442–453. [CrossRef] [PubMed]
45. Grau-Bové, J.; Mazzei, L.; Strlic, M.; Cassar, M. Fluid simulations in heritage science. *Herit. Sci.* **2019**, *7*, 16. [CrossRef]
46. Ascione, F.; Bellia, L.; Capozzoli, A. A coupled numerical approach on museum air conditioning: Energy and fluid-dynamic analysis. *Appl. Energy* **2013**, *103*, 416–427. [CrossRef]

47. Grau-Bové, J.; Strlič, M. Fine particulate matter in indoor cultural heritage: A literature review. *Herit. Sci.* **2013**, *1*, 8. [CrossRef]
48. Mašková, L.; Smolík, J.; Ondráček, J.; Ondráčková, L.; Travnickova, T.; Havlica, J. Air quality in archives housed in historic buildings: Assessment of concentration of indoor particles of outdoor origin. *Build. Eviron.* **2020**, *180*, 107024. [CrossRef]
49. Grau-Bové, J.; Mazzei, L.; Malki-Ephstein, L.; Thickett, D.; Strlič, M. Simulation of particulate matter ingress, dispersion and deposition in a historical building. *J. Cult. Herit.* **2016**, *18*, 199–208. [CrossRef]
50. Camuffo, D. *Microclimate for Cultural Heritage: Measurements, Risk Assessment, Conservation, Restoration, and Maintaince of Indoor and Outdoor Monuments*, 3rd ed.; Elsevier: Amsterdam, The Netherlands, 2019; pp. 197–234.
51. Morawska, L.; Salthammer, T. Fundamentals. In *Indoor Environment, Airborne Particles and Settled Dust*; Morawska, L., Salthammer, T., Eds.; Wiley-WCH Verlag GmbH & Co. KGaA: Weinheim, Germany, 2003; pp. 1–46.
52. Fællesmagasinet (In Danish: The Joint Storage Facility). Danish Agency for Culture and Palaces. Available online: https://slks.dk/omraader/slotte-og-ejendomme/bygge-og-udviklingsprojekter/faellesmagasinet (accessed on 1 May 2023).
53. Janssen, H.; Christensen, J.E. Hygrothermal optimisation of museum storage spaces. *Energy Build.* **2013**, *56*, 169–178. [CrossRef]
54. Ryhl-Svendsen, M.; Jensen, L.A.; Larsen, P.K.; Bøhm, B.; Padfield, T. Ultra-low-energy museum storage. In *Proceedings of the ICOM-CC 16th Triennial Conference, Lisbon, Portugal, 19–23 September 2011*; Bridgland, J., Ed.; ICOM Committee for Conservation: Paris, France, 2011; p. 8.
55. Department of Environmental Sciences, Aarhus University. Air Quality Data. Available online: https://envs.au.dk/en/research-areas/air-pollution-emissions-and-effects/air-quality-data (accessed on 1 May 2023).
56. Smedemark, S.H.; Ryhl-Svendsen, M. Indoor air pollution in archives: Temperature dependent emission of formic acid and acetic acid from paper. *J. Pap. Conserv.* **2020**, *21*, 22–30. [CrossRef]
57. Ryhl-Svendsen, M.; Jensen, L.A.; Larsen, P.K. Air quality in low-ventilated museum storage buildings. In *ICOM Committee for Conservation, Proceedings of the 17th Triennial Conference Preprints, Melbourne, Australia, 15–19 September 2014*; Bridgland, J., Ed.; International Council of Museums: Paris, France, 2014; p. 8.
58. Ryhl-Svendsen, M.; Clausen, G. The effect of ventilation, filtration and passive sorption on indoor air quality in museum storage rooms. *Stud. Conserv.* **2009**, *54*, 35–48. [CrossRef]

heritage

MDPI

Article

Practical Use of Damage Functions for Environmental Preventive Conservation and Sustainability—Examples from Naturally Ventilated Buildings

David Thickett

English Heritage Trust, Rangers House, London SE108QX, UK; david.thickett@english-heritage.org.uk

Abstract: This work explores the potential of using damage functions to assess cultural heritage environments. Changes caused by dimensional variation due to fluctuations in relative humidity are assessed using two accessible functions, and a third is discussed. The risk of mould growth is assessed from a time series of temperature and RH data. The results of previous studies comparing predictions from four functions to observed mould formation are reviewed, and the practical aspects of using the functions are described. Two situations related to metal and stone risk are described, comparing environments for display and assessing new or refitted buildings for storage. The use of functions to improve sustainability and their combination with performance models to predict carbon footprints are discussed.

Keywords: damage function; mould; corrosion; plastic deformation; carbon footprint

Citation: Thickett, D. Practical Use of Damage Functions for Environmental Preventive Conservation and Sustainability—Examples from Naturally Ventilated Buildings. *Heritage* **2023**, *6*, 2633–2649. https://doi.org/10.3390/heritage6030139

Academic Editors: Jenny Richards and Peter Brimblecombe

Received: 30 January 2023
Revised: 22 February 2023
Accepted: 25 February 2023
Published: 1 March 2023

1. Introduction

A large proportion of cultural heritage collections exist outside of air-conditioned spaces. These include historic houses, religious buildings, and archaeological sites, amongst many others. Whilst attempts have been made to install air conditioning in such buildings, without a good understanding of the building envelope performance, many of these have resulted in very poor control. The standards and guidelines for preventive conservation have, on the whole, been written for large institutions, explicitly or implicitly assuming air conditioning. They specify bands for temperature and relative humidity (RH), and sometimes for maximum pollution concentrations and light levels [1–3]. The assessment of compliance is fairly straightforward. Unfortunately, such an approach is extremely energy-intensive and has a large carbon footprint. These concerns recently prompted a major alteration to a modern art museum in Berlin that is under construction [4]. Assessing monitored environmental data in naturally ventilated buildings is more complex. The environment is complex, with the temperature, RH, and both gaseous and particulate pollution levels varying over several time periods and frequently with strong seasonal variations [5,6]. In naturally lit buildings, the light, UV, and NIR levels also vary strongly. Damage functions are well-suited to assess environmental data in such situations. The term originated in engineering and does not translate directly to material cultural heritage, although it is widely used. In engineering, the materials and their processing are generally well-characterised, and the required properties and failure levels are well-known. The materials forming cultural heritage are almost never fully characterised (possibly with the exception of some iconic objects), and their processing methods are only roughly known. Their response to the environment depends on their exact composition—copper responds differently to bronze, and the response of wood depends on the species and cut. Whilst historical techniques can provide some indication, only close analysis (often instrumental) can provide the necessary information. This issue is compounded for portable collections, which often have several potential forming techniques and a long history of materials masquerading as other materials. The values ascribed to cultural heritage objects are complex, and deterioration may affect different values in very different ways. The superimposition of a value function on top of a damage function has been suggested as a way to overcome this [7]. Determining such value functions requires significant social science research. Perhaps the term response function is better suited for the numerical description of changes in cultural heritage objects. It is probably best to consider the results of such functions in terms of the magnitude of risk and not exact change, as the materials are very variable, and small differences can lead to very different amounts of change. For example, the main deterioration reaction in archaeological iron is the transformation of iron chlorides into akageneite. This reaction generates a volume increase that can lever layers of wrought iron apart. The extent to which this occurs depends on the size of the void in which the iron chlorides are located. The amount of akageneite generated is a function of the RH, but the same amount in two objects can cause no or a catastrophic loss of material [8–10].

Response functions can be determined through various routes.

A property or series of properties of an object or surrogate can be measured through aging under various levels in laboratory experiments. The change in properties is related mathematically to the levels using a variety of models. The complexity of cultural heritage objects means there are many issues with surrogates and there are ethical issues to using objects [6,11]. The slow deterioration rates of many objects under ambient conditions mean some form of acceleration is generally required, and the issues of relating deterioration rates at higher temperatures, RH values, pollution concentrations, or light doses to those under ambient conditions are well-recognised [12,13]. The complexity of environments in naturally ventilated buildings also generates a need for many levels to be produced. The dimensional change and mould functions developed by the Image Permanence Institute (IPI) in their eClimateNotebook software are examples of this approach. The dimensional change function maps the RH onto the United States Forest Products Laboratory isotherms for an average wood, with a user-selectable equilibrium time (the data are averaged over this period) [14]. The mould function slightly adapts published microbiological growth studies on dried beans, grains, and grasses and relative growth studies on agar plates using observations in archival collections [14].

With sensitive methods to determine changes in properties, similar work can be undertaken in the actual environments of interest. This requires long exposures and multivariate methods to generate the response functions. Most functions determined in this way have been for outdoor exposure, where the deterioration rate is much faster [15]. However, improvements in analytical methods and their in situ application has allowed a limited number of indoor functions to be generated [16–18]. The functions developed are only valid under the range of conditions the materials were exposed to.

The modelling of the physical, chemical, or combined properties of a material or materials in an object and their response to environmental conditions provides a third general approach. Capturing sufficiently the complexity of the objects is a significant challenge. This approach requires a degree of expertise, and the models are difficult to make available for non-expert users. The wide number of possible environmental conditions limits the utility of providing a series of scenarios, and some way of inputting the user's actual data is the best outcome. This has only been developed with one model for two materials and a limited number of geometries so far [19,20]. Many other models have been or are being developed [21–25]. It is possible to model the decay processes themselves, and the ECOS model for salt activity has been used quite widely. It does have some limitations, however, as recently described [26].

There has been a reluctance in the conservation field to accept modelling results to make environmental decisions [27]. There are legitimate concerns regarding surrogates or models capturing the complexity of cultural heritage objects and the impact of natural aging and especially burial. Measurements of deterioration rates to verify modelled data are therefore important. A number of examples are included in this work.

The use of response functions to answer a number of preventive conservation questions in different scenarios will be illustrated through a series of case studies. The examples were selected to illustrate the different types of questions faced in preventive conservation and are limited by the available damage functions. The work is structured according to the three main deterioration routes observed: physical damage (mechanical), biological damage, and chemical damage, with one or more examples given for each.

Materials science effectively dictates the value of preventive conservation, since its outcomes define criteria applied by practitioners. The conservation and care of collections increasingly uses specified values and ranges for variables such as relative humidity (RH), temperature, and light to delineate safe criteria for the storage and display of materials, as well as for identifying regions of risk. These values form the preventive conservation evidence that is used to argue for resources and design specifications within the heritage sector. Their value in practice is related to the strength of the data used to derive them. Understanding structure and decay mechanisms and rates of change is an essential factor for deriving preservation standards. This paper focuses on the individuality of objects and what this may mean for setting standards.

2. Physical Damage to Rigid Hygroscopic Organic Materials

The expansion and contraction of materials such as wood through interaction with water vapour is acknowledged as a major cause of deterioration. In most situations, the response to RH dominates this interaction. Temperature has an effect on isotherms, but this is much smaller. The response is complicated by the slow response of thick materials to environmental RH changes, induced moisture gradients, and creep and fatigue. The physical properties and water vapour uptake and transport alter with aging. There is evidence of objects acclimatising to their environments, as demonstrated by various approaches [19,28,29]. Three response functions are readily accessible, HERIe [20], IPI [30], and Climate toolbox [31].

Probably the most common question is as follows: is the monitored environment acceptable, or does it require modification? The ruins of St Augustine's Abbey, Canterbury, UK contain a small, modern museum and visitor centre built in 1998. A contemporary (1960s) panel painting with a view of the Abbey ruins is part of the collection displayed (Figure 1).

The building has thermostatically controlled underfloor heating. The painting was in poor condition when displayed on opening. Several years of modifications to the underfloor heating set points eventually yielded an acceptable winter RH in the space, above 35%. In the recent unprecedentedly hot summer in the UK, the RH dropped below 35% on four short occasions (less than 6 h), reaching as low as 28% on one of those occasions. The RH data (Rotronic Hydroclip probe) were input into the HERIe mechanical damage module, using the parameters for a 5 mm oak panel with soft gesso and water ingress from both faces (the lacquer only remains as small patches on the painted surface). The panel was clearly re-used carpentry from an unknown source and date. The results are shown in Figure 2.

The modelled strain did not exceed the elastic limit for gesso (0.002) for the majority of the data. There was a slight exceedance for a short period at the lowest RH, generating a risk index of 0.08.

Another common task is to compare two environments, often prior to a potential object move or when deciding where an object is displayed, if several suitable locations are available. A common approach to assess environments is to set a percentage time in certain limits specification [32,33]. Kenwood house, London, UK has many rooms used to display objects. RH data (Rotromic hygroclip II probe) from two rooms (Figure 3) were assessed from June 2020 to September 2020 using a 40–65% RH range. Room A was within the range for 2.25% of the period, and room B for 2.47% of the period. Despite these very similar metrics, the RH in room B clearly reached much higher values, and the drops in RH value were also larger. The data were also fed into the HERIe mechanical damage model for 5 mm oak, tangentially cut, and with water ingress from one side. Two oak cabinets are displayed in the abovementioned rooms, with approximately the same 5 mm thickness of wood. The limit was set to 0.02% strain. Acoustic emission monitoring was undertaken over the period described. Two Physical Acoustics Micro II systems were used, each with two WD sensors in anti-correlation mode [34]. The energy detected was proportional to the increase in crack area occurring in the wood [35]. Acoustic emission energy was only detected from the cabinet in room B, and the energy of each event is included in Figure 3. The acoustic emission events occurred rapidly after the RH dropped from a high value. These corresponded with the points at which the HERIe model exceeded a strain of 0.004.

Figure 1. St Augustine's Abbey Museum; the painting is on the far left.

Given the large scale of many collections (the English Heritage Trust cares for over one million objects) and the limited conservation time, it is not unusual for small instances of deterioration to take some time to be observed on display. When deterioration is observed, a frequent question is: when did that event happen? Response functions can give an indication of this. In November 2008, an existing crack in a Cattaro panel painting displayed at Rangers House, London, UK was observed to have extended by approximately 2 mm (Figure 4).

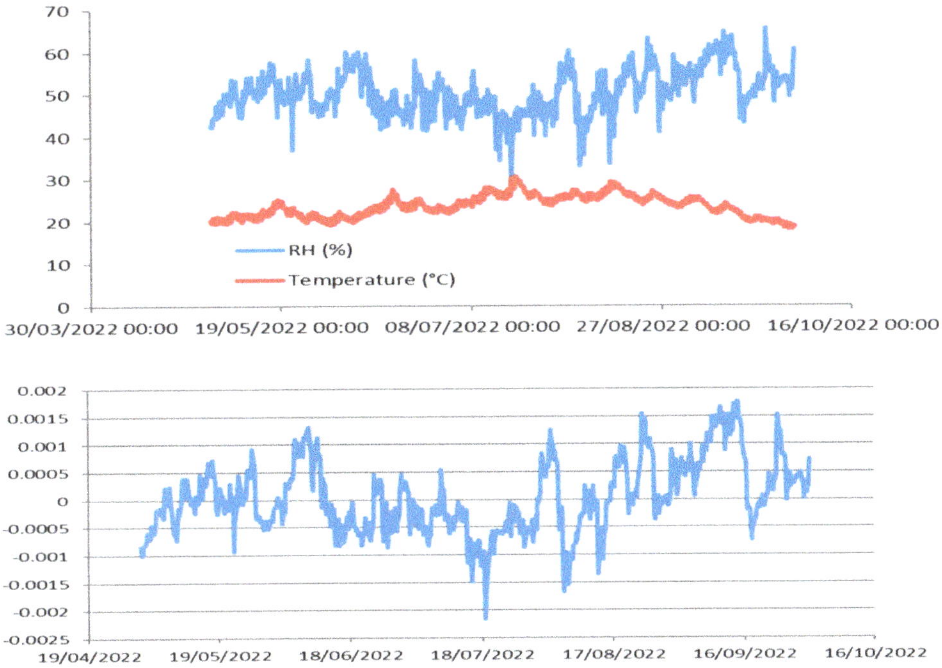

Figure 2. Strain calculations from HERIe for 5 mm oak panel with soft gesso at St Augustine's Abbey Museum.

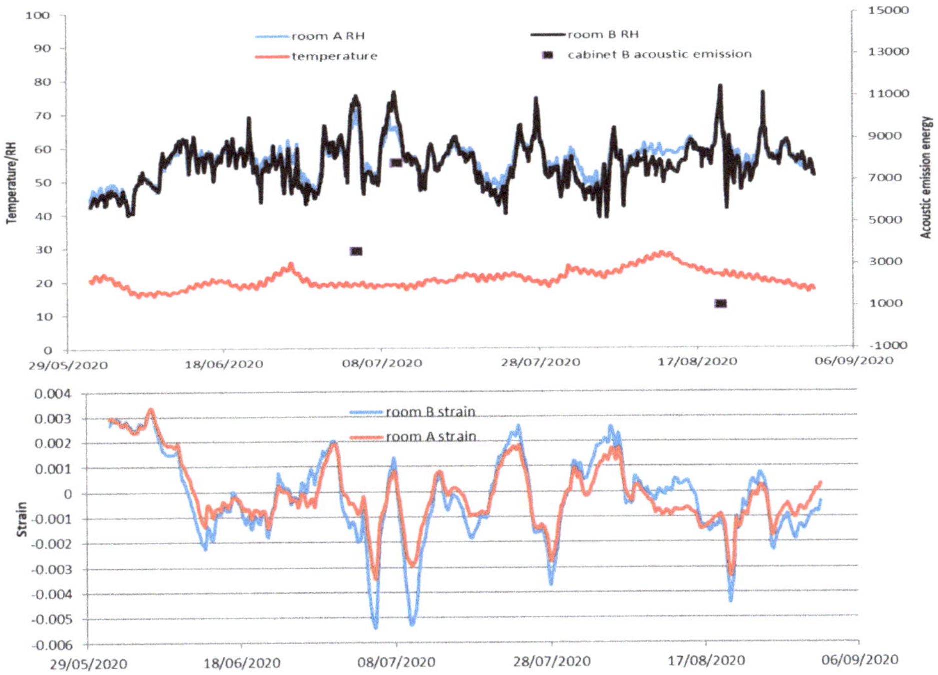

Figure 3. RH values in two rooms at Kenwood House, acoustic emission energy from oak cabinets in those rooms, and strain calculations from HERIe.

Figure 4. Annunciation by Lorrenzo Marini da Cattaro, b. 1478.

RH data from January to November that year were input into IPI climate notebook, and the % dimensional change index was calculated. This was the only response function available at that date. An equilibrium time of 3 days was selected from unpublished measurements of changes in panel painting dimensions. This figure is consistent with the dimensional changes measured by Wilk et al. [36]. The zero dimension change is set at the value calculated from the average RH over the period used. The RH and dimensional change index from the IPI software are shown in Figure 5.

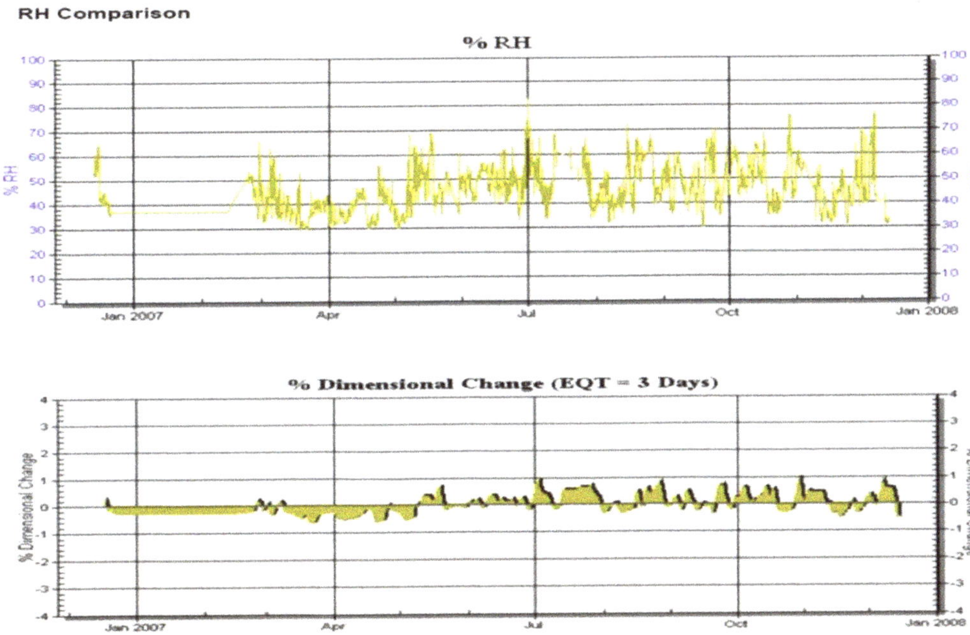

Figure 5. RH values from Rangers House and dimensional change calculations from IPI climate notebook.

As can be seen, two instances reached an index of 1, in early July and late October. IPI documentation states that: "%DC numbers identify areas of significant fluctuations in the environment, <1% the wood will go back to its actual form". This seems to indicate a value above 1 may risk plastic deformation.

3. Biological Damage—Mould Growth

Mould growth is a significant problem, and outbreaks consume significant resources and have a large carbon footprint to remediate [37]. Future climate models also predict an increasing risk of mould growth in many areas. The material has a strong influence on the likelihood of mould growth, as does the deposited dust (which can have a lower water activity than the material itself [38]) and nutrients from gaseous pollution or contamination from prior use or handling. Environmentally, the temperature and RH, the time of exposure, and (probably) the ventilation rate are known to contribute, although the effect and mechanism of ventilation are not clear. Four response functions for mould growth were compared to observed mould and, in particular, outbreaks in more than fifty instances [38]. The functions produced by the IPI and Wufi were found to best correlate with observed mould outbreaks and growth. Many historic environments are damp and encourage mould growth, but using the building enhances its chance of long-term preservation. A series of chalk tunnels beneath Dover Castle, UK were used as a hospital and combined operations centre during World War II and are dressed with collections from those dates, but not original to the tunnels. These are accessioned as props and not museum objects. However, the extreme environment means understanding the mould risk is essential to managing the site, and very significant mould outbreaks caused long closures in 2002 and 2006 (Figure 6).

After remediation in 2006, a number of strategies were suggested to reduce the risk of mould in the future. The additional heating of the air was suggested, updating the 1940s heating elements to warm air circulated by a circulation system installed from a decommissioned battleship. Monitored temperature and RH data (ACR SR002 logger) from Gun Operations in the middle-level Casemate of the tunnels from 2007 were assessed with the IPI mould function (Figure 7).

Figure 6. Mould on the underside of plotting table in Gun Operations in 2002.

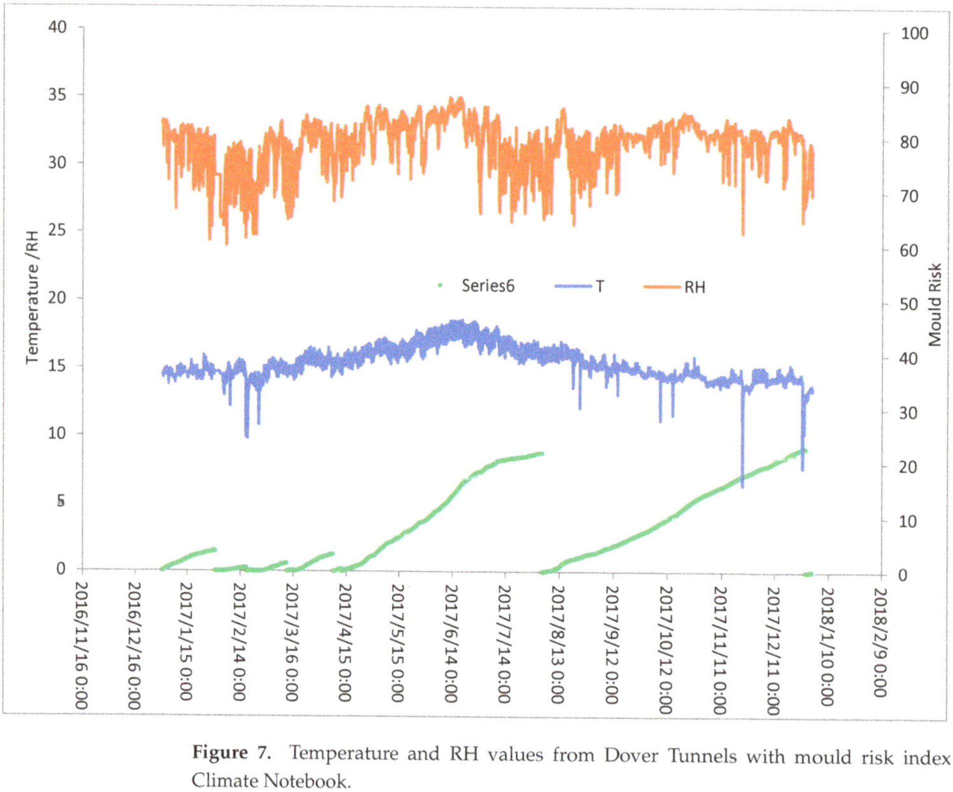

Figure 7. Temperature and RH values from Dover Tunnels with mould risk index from IPI Climate Notebook.

The mould risk was highest in the summer months. The large number of drops in RH was caused by the ventilation system bringing outside air into the tunnels. The distribution of mould risk would have made heating an inefficient strategy, as its strongest effect would be in the colder winter months. The ventilation system was upgraded to remove dead areas and controlled with an adaptive system so that it brought external air into the tunnels when the external absolute humidity was below that measured internally. When this was not the case (approximately 32% of the time), it recirculated the internal air. The inlet filtration was upgraded to remove complex hydrocarbon pollution from the adjacent ferry port.

4. Chemical Damage—Metal Corrosion

Metal corrosion is strongly affected by the presence of pollutant gases. For most metals, the RH also has a strong synergistic effect with pollution. Many metals have a critical RH, above which the corrosion rate increases dramatically [39]. Depending on the pollutant concentrations present, this is frequently between 60 and 80% RH. Many locations housing cultural heritage metals reach such values at some point in the year. In the UK, this is often over the summer/autumn. With the exception of silver, temperature has much less impact on corrosion rates.

A copper alloy sculpture, shown in Figure 8, was proposed for loan to Aplsey House, London, UK.

Figure 8. Copper alloy sculpture of Queen Victoria.

It could be accommodated in two rooms. The corrosion risk in the two rooms was assessed from temperature and RH data (Rotronic hydroclip II) and nitrogen dioxide, ozone, hydrogen chloride, sulfur dioxide, and hydrogen sulfide concentrations (diffusion tubes) [40,41]. The pollution data were collected from 30-day exposures, and average temperatures and RH values were calculated from the 60 min of measured data. The response function for copper developed by Thickett et al. was used with the data to estimate monthly corrosion rates over the proposed display period [16]. Whilst resembling bronze and catalogued as such, portable XRF analysis (Bruker Tracer III/IV) indicated a relatively pure copper composition, over 98%, with the colouration presumably derived from patination. The results are shown in Figure 9.

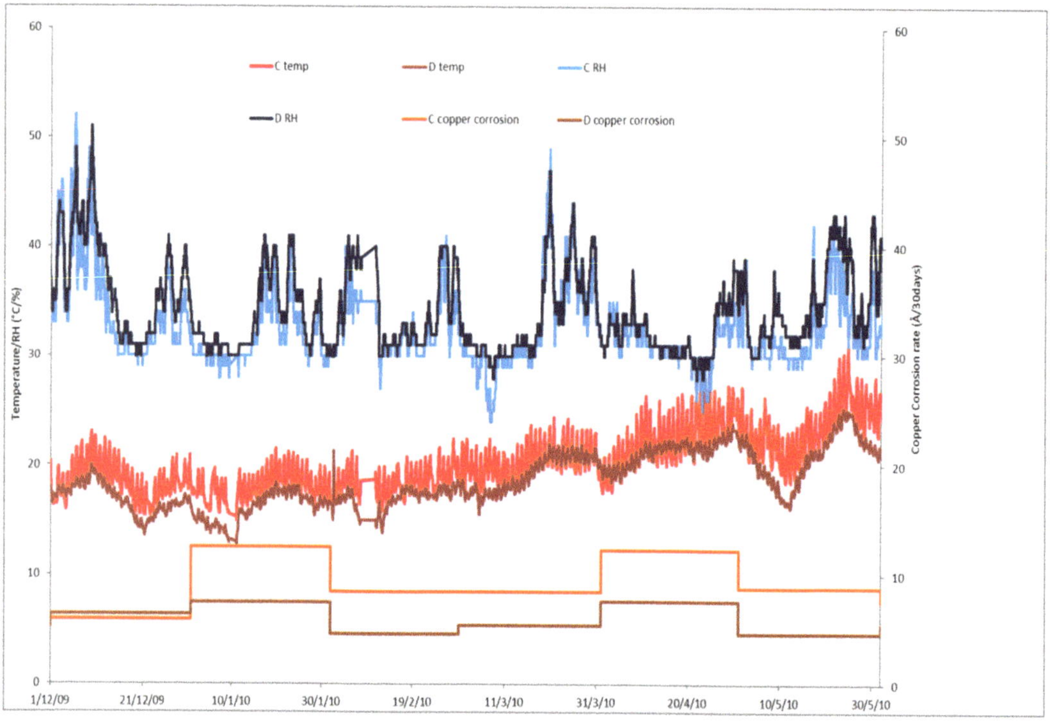

Figure 9. Temperature and RH values from two rooms at Apsley House and calculated monthly copper corrosion rates.

Solely considering the temperature and RH data, it is likely one would conclude that room C was better than room D. However, the calculated functions clearly showed that room C had the higher estimated corrosion rate for ten of the twelve months assessed (one month being equal). The sculpture was displayed in room D. To assess the reliability of the calculations, a series of copper alloy corrosion rate measurements were run near the object during the six-month duration of the display. AirCorr loggers with pure copper and copper/tin CnSn8 sensors were deployed along with a Purafil Onguard 4000 logger with a pure copper sensor. The corrosion rate estimates were also repeated with temperature, RH, and pollution measurements taken in the six-month period. The results are shown in Figure 10.

The estimated values for the two six-month periods were slightly different in consecutive years, as would be expected, with January showing the largest difference, at almost 20%. The three measured corrosion rates showed the same trends and followed the estimated rates. The AirCorr copper showed a higher corrosion rate than the Onguard. This was, possibly due to the influence of particles on the copper corrosion rate. The Onguard system is essentially an extremely sensitive balance. As the copper corrodes, the coating mass increases, but the system has no way of differentiating from the mass gain caused by dust deposition. The sensor has a cover to reduce this and is deployed facing downwards, as per the manufacturer's recommendations. The AirCorr works on increased resistance as the metal track corrodes, and dust deposition does not interfere with the measurement but can increase the corrosion rate [42]. The copper tin alloy, as expected, showed significantly less corrosion than pure copper for all months. The function correctly predicted the corrosion rate trends, and the accuracy is probably sufficient for this use in preventive conservation. Historical copper alloys have a very wide range of compositions, and this has been shown to affect the corrosion rate. This emphasizes that the uncertainties in response functions in this area necessitate considering the results in terms of risk, and not as exact predictions. The statue was from a private lender and had not been analysed for composition.

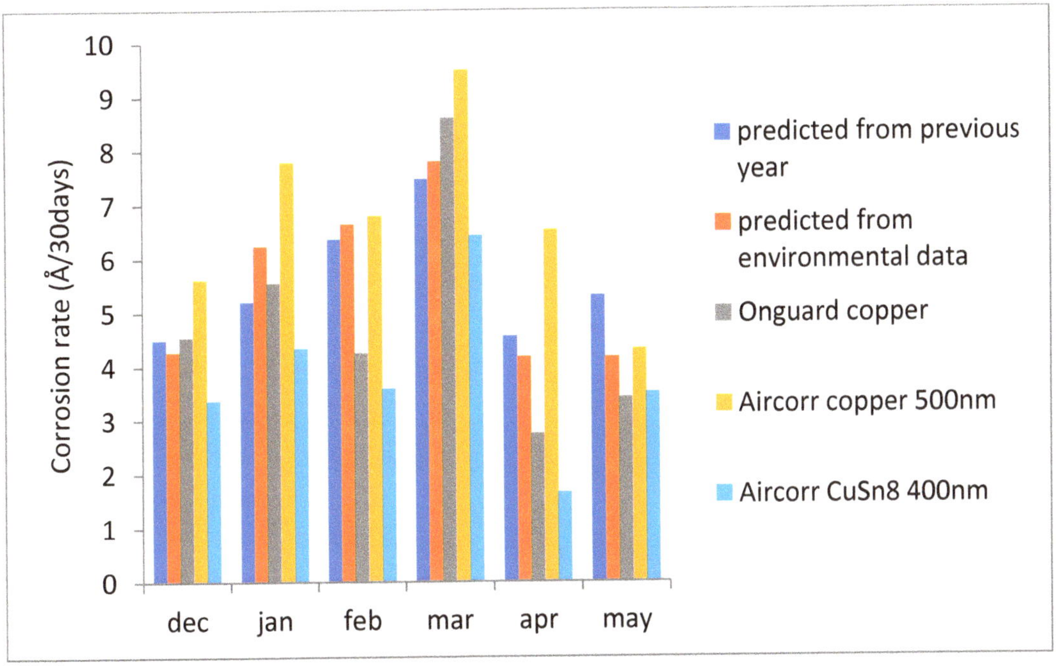

Figure 10. Corrosion rates over six-month periods, estimated from the function and measured for copper and copper tin alloy.

Another common preventive conservation task is predicting the response in new builds or refitted existing buildings. Predicted environmental data can be obtained from building modelling for new builds (modelling is becoming increasing common in the design phase) or from the previous monitoring and experience of the likely changes from the refitting of existing buildings. Pollution data can be calculated from nearby automated station results and ingress modelling, such as the IMPACT model [16,43]. English Heritage refitted a 1950s industrial building at Wrest Park as a collection store in 2011. The refit incorporated many features of the Danish passive storage concept [44]. No mechanical conditioning was installed in the larger-volume space in the new store. Two smaller-volume spaces were dehumidified to house more sensitive collections (Figure 11).

Figure 11. Wrest Park store.

Collections were planned to move from four buildings at two sites. The predicted environments were compared with the measured environments at the existing sites via four response functions: the multiassess functions for carbon steel, cast bronze, and Portland limestone; an internally derived function for sodium-sulfate-laden polychrome limestone [45], and a European standard for hygroscopic material [14]. These functions were selected from those available to cover the majority of the collections present. For example, very little copper, zinc, or cast iron was to be stored there. To validate the approach, annual corrosion rates were measured at one of the existing stores by exposing mild steel and copper tin alloy coupons to ISO 9223 [46]. After exposure, the coupons were chemically or electrochemically stripped, and the mass of metal lost was calculated [47]. Temperature and RH were measured (ACR SR002 loggers) and pollution measurements performed, as described previously. For the existing store, estimates were calculated for 2010-11 and compared to measured corrosion rates. Corrosion rates were also measured in 2011-12 to gauge year-to-year variation. The results are shown in Figure 12.

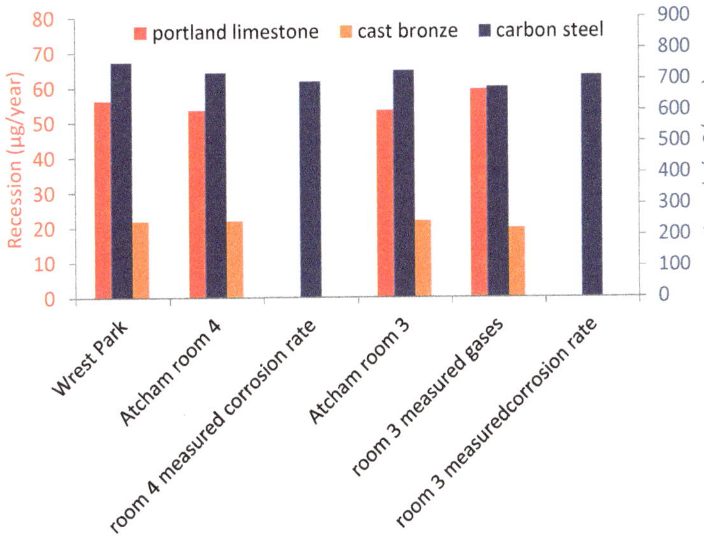

Figure 12. Estimated and measured corrosion rates at existing (Atcham) and proposed new (Wrest Park) stores.

The Wrest Park and Atcham estimated rates were similar. The predicted and measured corrosion rates were in reasonable agreement for 2010-11. There was some variation in the corrosion rate for 2011-12, but it was within 12% of the 2011-12 rates. The carbon steel corrosion rates were much higher than the copper tin alloy, and further measurements were only undertaken with carbon steel.

Prior to the move in 2012, a comparison was made between estimates based on IMPACT-modelled pollution data, estimates based on measured pollution data, and carbon steel corrosion rate measurements. Figure 13 shows the measured carbon steel and cast bronze rate, along with the estimated corrosion rate based on measured gas concentrations in one room.

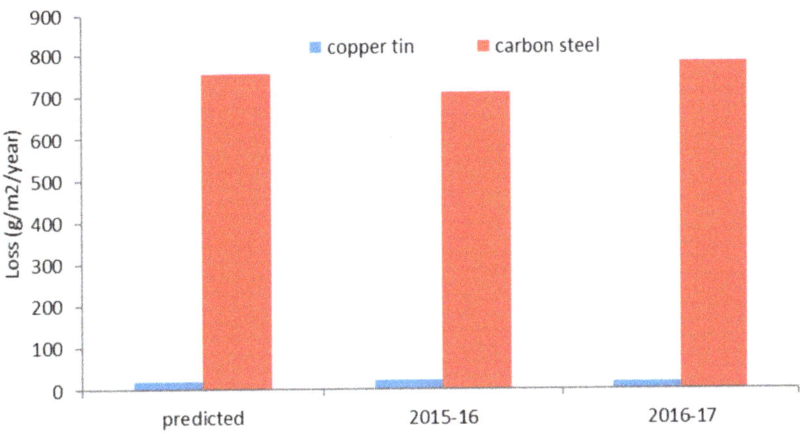

Figure 13. Predicted and measured cast bronze and carbon steel corrosion rates at Wrest Park store.

Good correlations were observed between the estimates and the measured corrosion rates. The new store showed comparable estimated corrosion rates to the existing store rooms, within the uncertainties of the method. This data helped the business case for the new store.

5. Sustainability

The use of response functions can aid sustainability in a number of ways. They can be used to inform an initial assessment of an environment for a collection and determine how much mitigation is required. This can involve microclimates such as showcases; full air conditioning; or less carbon-intensive room-conditioning solutions, conservation heating, dehumidification, or adaptive ventilation. Proposed approaches to reduce energy consumption and carbon footprints for existing air-conditioning systems include: seasonal shifts in set points; variable fan speeds; and reducing fresh air intake and periodic shut downs. These can produce short periods outside of the original temperature and RH ranges, the impact of which can be assessed with response functions. Reducing fan speeds and fresh air intake can increase pollutant levels, the impact of which can be assessed.

This can be combined with performance models of mitigation systems to generate estimated carbon footprints for different approaches. Performance models have been published for showcases and some dehumidifiers [48,49]. Building modelling frequently includes energy usage estimates for heating or air-conditioning systems. A model for managing silver tarnish has recently been developed [50]. This combines the results of a response function for the silver tarnish rate [16] with the perception level at which decision makers determine that silver needs cleaning to ascertain how frequently silver needs to be cleaned in a particular environment. Previous measurements are used to convert the response function rate in nm/year into the colorimetric co-ordinate b*, which best correlates with observers perception of early silver tarnish. The performance (expressed as the reduction in the tarnish rate) and carbon footprint of a number of different preventive conservation approaches are then calculated. The reference time period is set for the calculations. The embedded carbon of the materials used to clean the silver are calculated for the number of cleaning instances required in that period. For preventive conservation approaches, the embedded carbon and carbon footprint of the estimated energy and maintenance required are calculated. The model presently includes the following situations:

Room mitigation approaches—room filtration unit (Blue Clima), lacquering with Cannings Frigilene, a combination of both.

Display in showcases—showcase (based on air exchange rate), showcase filtration units (Camfil or Dynamx 5).

Cleaning approaches—prelim polish, silver dip (Cannings Frigilene removal).

Table 1 shows a set of data from an example calculated for Apsley House, London, UK.

The different approaches had a very wide range of carbon footprints. The model only considered the embedded carbon in the cleaning materials and did not include the scope-two emissions of running a conservation workshop and scope-three emissions of staff travel to work. The carbon footprint is only one factor feeding into decisions. Economics/resources are also clearly important and often the limiting factor. Each time a silver object is cleaned, there is a small loss of the original material, which will dramatically effect plated or finely detailed objects with repeated cleanings.

Table 1. Calculated silver tarnish rate and estimated carbon footprints for various preventive conservation approaches at Apsley House.

Time Period	20 Years
Environmental parameters (averages)	Temperature 18.2 °C, RH 42.4%, hydrogen sulfide 0.23 ppb, hydrogen chloride 0.02 ppb, nitrogen dioxide 14.17 ppb, ozone 1.32 ppb, sulfur dioxide 0.56 ppb
Tarnish rate estimated per year	110 nm

Table 1. *Cont.*

Time Period	**20 Years**						
b* rate	4.8/year						
Cleaning threshold b*	7.3						
Cleaning lifetime	2.28 years						
Silver surface area	10 m^2						
	No change	Room filtered	Lacquered	Existing showcase (1.23/day)	Plus Dynamx5 pump	New showcase (0.4/day)	Plus Dynamx5 pump
Embedded		223	3.71	0	42	90	132
Running		1430			372		124
Calculated cleaning requirement in period	8	5	1	3	0	1	0
Cleaning prelim	1.68	1.05	0.21	0.63		0.21	
Total carbon foot print (kg CO_2e)	1.68	1561.05	4.71	0.63	414	90.21	256

6. Conclusions

Damage functions clearly have a role in improving preventive conservation decisions by strengthening the evidence base. The very wide variety of cultural heritage material, the mainly unknown effects of aging and previous conservation treatments, and the different values ascribed to cultural heritage mean that care is required in interpreting these functions' results in this field. For organic materials, whose responses are predominantly determined by RH fluctuations, functions exist that can assess environmental data, determine the impact of short-term deviations from control limits, and provide an indication of when cracking is likely to have occurred in the recent past. For multi-parameter phenomena, such as mould growth, they rationalise a complex dataset into something that has been shown to reasonably approximate observed risk. The combination of temperature and RH thresholds and time dependence means that the visual assessment of such data is extremely challenging and prone to errors in intermediate situations. Such an approach is adequate when there is a very high mould risk, but many cultural heritage environments fall into the intermediate range. For metal corrosion and other pollutant-induced degradation processes, response functions allow an understanding of the synergistic effects taking place. Reliance on purely temperature and RH data can be misleading when comparing environments for these processes. The functions tested showed good correspondence to measured responses, at least in the instances tested.

To be widely utilised, beyond being representative of the collections investigated, a good user interface, allowing the easy input of environmental data, is essential. Comprehensive descriptions and, ideally, the underlying equations, assumptions, and potential issues are critical to build user confidence. Indications about the confidence and use of the results would also be extremely useful.

Response functions are well-suited to predicting environmentally induced effects. Their use allows new building/space specifications to extend beyond the rigid ranges and limits that are often applied presently, and they have been widely used in climate change studies. The more refined risk assessment of complex environments they engender can significantly benefit sustainability. Combined with control performance models and lifecycle data, they can be used to predict carbon and other sustainability measures for different approaches to preventive conservation.

Funding: This research received no external funding.

Data Availability Statement: All data are available from the corresponding author on request.

Acknowledgments: Acoustic emission equipment purchased with AHRC Capability for Collections Grant.

Conflicts of Interest: The authors declare no conflict of interest.

References

1. American Society of Heating. Refrigerating and Air-Conditioning Engineers. In *ASHRAE Handbook—HVAC Applications, Chap. 24: Museums, Galleries, Archives and Libraries (TC 9.8)*; ASHRAE: Atlanta, GA, USA, 2019.
2. *BS EN 16893*; Conservation of Cultural Heritage—Specifications for Location, Construction and Modification of Buildings or Rooms Intended for the Storage or Use of Heritage Collections. British Standards Institute: London, UK, 2018.
3. *BS 4971*; Conservation and Care of Archive and Library Collections. British Standards Institute: London, UK, 2017.
4. Guardian Newspaper. Available online: https://www.theguardian.com/world/2022/nov/27/new-berlin-art-museum-of-the-20th-century-climate (accessed on 18 January 2023).
5. Camuffo, D. *Microclimate for Cultural Heritage*; Elsevier BV: Milano, Italy, 2019.
6. Thickett, D.; Chisholm, R.; Lankester, P. Reactivity Monitoring Of Atmospheres. In *Metal 2013, Proceedings of the Interim Meeting of the ICOM-CC Metal Working Group, Edinburgh, Scotland, UK, 16–20 September 2013*; Hyslop, E., Gonzalez, V., Wilson, L.T., Eds.; Historic Scotland: Edinburgh, UK, 2013; pp. 129–135.
7. Strlič, M.; Thickett, D.; Taylor, J.; Cassar, M. Damage functions in heritage science. *Stud. Conserv.* **2013**, *58*, 80–87. [CrossRef]
8. Matthiesen, H.; Stemann-Petersen, K.S. A Fast and Non-destructive Method to Document and Quantify the Efficiency of Metals Conservation. In *Metal 2013, Proceedings of the Interim Meeting of the ICOM-CC Metal Working Group, Edinburgh, Scotland, UK, 16–20 September 2013*; Hyslop, E., Gonzalez, V., Troalen, L., Wilson, L., Eds.; Historic Scotland: Edinburgh, UK, 2013; pp. 175–180.
9. Watkinson, D.E.; Rimmer, M.B.; Emmerson, N.J. The Influence of Relative Humidity and Intrinsic Chloride on Post-excavation Corrosion Rates of Archaeological Wrought Iron. *Stud. Conserv.* **2018**, *64*, 456–471. [CrossRef]
10. Thickett, D. Post Excavation Changes and Preventive Conservation of Archaeological Iron. Ph.D. Dissertation, University of London, London, UK, 2012. Available online: www.english-heritage.org.uk/siteassets/home/learn/conservation/collections-advice--guidance/thickettthesisfinalversion.pdf (accessed on 21 February 2023).
11. Watkinson, D.; Lewis, M.T. Desiccated Storage of Chloride-contaminated Archaeological Iron Objects. *Stud. Conserv.* **2005**, *50*, 241–252. [CrossRef]
12. Feller, R.L. *Accelerated Aging: Photochemical and Thermal Aspects*; Research in Conservation 4; Getty Conservation Institute: Marina del Rey, CA, USA, 1994; Available online: http://hdl.handle.net/10020/gci_pubs/accelerated_aging (accessed on 29 January 2023).
13. Plazonic, I.; Bates, I.; Vukoje, M. Changes in Straw-Containing Laboratory Papers Caused by Accelerated Ageing. *Heritage* **2022**, *5*, 1836–1851. [CrossRef]
14. Nishimura, D.W. *Understanding Preservation Metrics*; IPI: Rochester, NY, USA, 2011.
15. Kucera, V. Final Report of Model for Multi-Pollutant Impact and Assessment of Threshold Levels for Cultural Heritage. Available online: www.corr-institute.se/MULTI-ASSESS (accessed on 23 January 2023).
16. Thickett, D.; Chisholm, R.; Lankester, P. Development of damage functions for copper, silver and enamels on copper. In *Climate for Collections, Standards and Uncertainties, Munich, 7 to 9 November 2012*; Ashley-Smith, J., Burmester, A., Eibl, M., Eds.; Archetype: London, UK, 2015; pp. 325–335.
17. Fenech, A. Lifetime of Colour Photographs in Mixed Archival Collections. Ph.D. Thesis, University College London, London, UK, 2011.
18. Pastorelli, G.; Cao, S.; Kralj Cigi, I.; Cucci, C.; Elnaggar, A.; Strlič, M. Development of dose-response functions for historic paper degradation using exposure to natural conditions and multivariate regression. *Polym. Degrad. Stab.* **2019**, *168*, 108944. [CrossRef]
19. Heri-e Heri-e. 2022. Available online: https://herie.pl (accessed on 18 February 2022).
20. Kupczak, A.; Jedrychowski, M.; Strojecki, M.; Krzemin, M.; Bratasz, L. HERIe: A Web-Based Decision-Supporting Tool for Assessing Risk of Physical Damage Using Various Failure Criteria. *Stud. Conserv.* **2018**, *63*, 151–155. [CrossRef]
21. Jakiela, S.; Bratasz, Ł.; Kozłowski, R. Numerical modelling of moisture movement and related stress field in lime wood subjected to changing climate conditions. *Wood Sci. Tech.* **2008**, *42*, 21–37. [CrossRef]
22. Michalski, S. Stuffing everything we know about mechanical properties into one collection simulation. In *Climate for Collections, Standards and Uncertainties, Munich, 7 to 9 November 2012*; Ashley-Smith, J., Burmester, A., Eibl, M., Eds.; Archetype: London, UK, 2015; pp. 349–362.
23. Bosco, E.; Suiker, A.S.J.; Fleck, N.A. Moisture-induced cracking in a flexural bilayer with application to historical paintings. *Theor. Appl. Fract. Mech.* **2021**, *112*, 102779. [CrossRef]
24. Eumelen, G.J.A.M.; Bosco, E.; Suiker, A.S.J.; Hermans, J.J.; Van Loon, A.; Keune, K.; Iedema, P.D. Computational modelling of metal soap formation in historical oil paintings: The influence of fatty acid concentration and nucleus geometry on the induced chemo-mechanical damage. *SN Appl. Sci.* **2020**, *2*, 1310. [CrossRef]

25. Gambarelli, S.; Ozbolt, J. 3D hygro-mechanical meso-scale model for wood. *Constr. Build. Mater.* **2021**, *311*, 125283. [CrossRef]
26. Godts, S.; Steiger, M.; Orr, S.A.; Stahlbuhk, A.; Desarnaud, J.; De Clercq, H.; Cnudde, V.; De Kock, T. Modeling Salt Behavior with ECOS/RUNSALT: Terminology, Methodology, Limitations, and Solutions. *Heritage* **2022**, *5*, 3648–3663. [CrossRef]
27. Richards, J.; Brimblecombe, P. The transfer of heritage modelling from research to practice. *Herit. Sci.* **2022**, *10*, 17. [CrossRef]
28. Michalski, S. The Ideal Climate, Risk Management, the ASHRAE Chapter, Proofed Fluctuations, and toward a Full Risk Analysis Model. *Environ. Sci.* **2014**.
29. *EN 15757*; Conservation of Cultural Heritage—Specifications for Temperature and Relative Humidity to Limit Climate-Induced Mechanical Damage in Organic Hygroscopic Materials. British Standards Institute: London, UK, 2012.
30. eClimateNotebook. Available online: https://www.imagepermanenceinstitute.org/environmental-management/eclimatenotebook (accessed on 3 August 2015).
31. Pretzel, B. Reasonable—Broadening acceptable climate parameters for furniture on open display. In Proceedings of the ICOM-CC 17th Triennial Conference Preprints, Melbourne, Australia, 15–19 September 2014; Bridgland, J., Ed.; International Council of Museums: Paris, France, 2014.
32. Blades, N. *Specification for Environmental Control*; National Trust: London, UK.
33. Frame, K.; Vlachou-Mogire, C.; Hallett, K.; Takami, M. Balancing Significance and Maintaining 'Sense of Place' in the Sustainable Display of Tudor Tapestries in the Great Hall, Hampton Court Palace. *Stud. Conserv.* **2018**, *63* (Suppl. S1), 87–93. [CrossRef]
34. Strojecki, M.; Łukomski, M.; Krzemień, L.; Sobczyk, J.; Bratasz, Ł. Acoustic Emission Monitoring of an Eighteenth-Century Wardrobe to Support a Strategy for Indoor Climate Management. *Stud. Conserv.* **2014**, *59*, 225–232. [CrossRef]
35. Thickett, D.; Vilde, V.; Lankester, P.; Richardsen, E. Using Science to Assess and Predict Object Response in Historic House Environments. In *Preventive Conservation in Historic Houses and Palace Museums, Versailles, 29th November to 1st December 2017*; Silvana Editoriale S.p.A.: Milan, Italy, 2017.
36. Wilk, D.; Bratasz, L.; Frczek, P.; Obarzanowski, M.; Klisińska-Kopacz, A.; Czop, J. Construction and use of microclimatic frames in the National Museum in Krakow. *Conserv. News* **2011**, 154–167.
37. Thickett, D. Sustainable Collections Environments. *Estud. Conserv. Restauro* **2019**, *11*, 93–103.
38. Thickett, D.; Lankester, P.; Pereira Pardo, L. Testing Damage Functions for Mould Growth. In Proceedings of the ICOM-CC 17th Triennial Conference Preprints, Melbourne, Australia, 15–19 September 2014; Bridgland, J., Ed.; International Council of Museums: Paris, France, 2014.
39. Vernon, W.H.J. Laboratory study of the atmospheric corrosion of metals. *Trans. Far. Soc.* **1935**, *31*, 1678–1700. [CrossRef]
40. Gibson, L.T.; Cooksey, B.G.; Littlejohn, D.; Tennent, N.H. A diffusion tube sampler for the determination of acetic acid and formic acid vapours in museum cabinets. *Anal. Chim. Act.* **1997**, *341*, 11–19. [CrossRef]
41. Ankersmit, H.A.; Carbo, A.D.; Tennent, N.H. Tarnishing of silver: Evaluation by colour measurements. In *Metal 2001, Proceedings of the International Conference on Metals Conservation, Santiago, Chile, 2–6 April 2001*; MacLeod, I.D., Theile, J.M., Degrigny, C., Eds.; Western Australian Museum: Freemantle, Australia, 2004; pp. 157–166.
42. Thickett, D.; Costa, V. The effect of particulate pollution on the corrosion of metals in heritage locations. In Proceedings of the ICOM-CC 17th Triennial Conference Preprints, Melbourne, Australia, 15–19 September 2014; Bridgland, J., Ed.; International Council of Museums: Paris, France, 2014.
43. Blades, N.; Kruppa, D.; Cassar, M. Development of a web based software tool predicting the occurrence and the effects of pollution inside museum buildings. In Proceedings of the ICOM Committee for Conservation, Preprints of the 13th Triennial Meeting, Rio de Janeiro, Brazil, 22–27 September 2002; Vontobel, R., Ed.; James and James: London, UK, 2003; pp. 9–15.
44. Xavier-Rowe, A.; Newman, C.; Stanley, B.; Thickett, D.; Pereira Pardo, L. A new beginning for English Heritage's archaeological and architectural stored collections. In Proceedings of the ICOM-CC 17th Triennial Conference Preprints, Melbourne, Australia, 15–19 September 2014; Bridgland, J., Ed.; International Council of Museums: Paris, France, 2014.
45. Thickett, D.; Stanley, B. Management of sodium sulfate damage to polychrome stone and buildings. In Proceedings of the SWBSS 2017—4th International Conference on Salt Weathering of Buildings and Stone Sculpture, Potsdam, Germany, 20–22 September 2017; Available online: https://www.saltwiki.net/index.php/SWBSS_2017 (accessed on 29 January 2023).
46. *ISO 9223*; Corrosion of Metals and Alloys—Corrosivity of Atmospheres. International Standards Institution: Geneva, Switzerland, 2012.
47. *ASTM G1 E1.1*; Standard Practice for Preparing, Cleaning and Evaluating Corrosion Test Specimens. American Society for the Testing of Materials: Conshohocken, PA, USA, 1999.
48. Thickett, D. Better use of showcases for preservation and sustainability. *Stud. Conserv.* **2022**, *67* (Suppl. S1), 267–276. [CrossRef]
49. Thickett, D. Specifying air exchange rates for showcases. In *Chemical Interactions between Cultural Artefacts and Indoor Environment*; Adriaens, M., Bioletti, S., Rabin, I., Eds.; ACCO: Leuven, Belgium, 2020; pp. 25–48.
50. Thickett, D.; Lankester, P.; Odlyha, M. Assessing and Predicting Sustainability for Maintaining Silver Collections. In Proceedings of the 19th Triennial Conference of ICOM-CC, Beijing, China, 17–21 May 2021. *Submitted.*

heritage

MDPI

Article

Painted Wood Climate Risk Analysis by the HERIe Model of Building Protection and Conservation Heating Scenarios in Norwegian Medieval Stone Churches

Terje Grøntoft [1,*] and Lena P. Stoveland [2]

1 NILU-Norwegian Institute for Air Research, Instituttvn 18, Box 100, 2027 Kjeller, Norway
2 NIKU-Norwegian Institute for Cultural Heritage Research, Storgata 2, 0155 Oslo, Norway
* Correspondence: teg@nilu.no

Abstract: HERIe was used to model the effect of changes to indoor climate on the risk of humidity-induced mechanical damage (cracking and plastic deformation) to wooden panels painted with stiff gesso in two Norwegian medieval stone churches: Kinn (mean relative humidity (RH, %) = 79%) on the humid west coast, and Ringsaker (mean RH = 49%) in the drier eastern part of the country. The risk involved in moving cultural heritage objects (paint on wood) between the churches and a conservation studio with more "ideal", stable conditions was also modeled. A hypothetical reduction in RH to ~65% and, proportionally, of the climate fluctuations in Kinn, and an increase in the RH in Ringsaker to a more stable value of ~63% via conservation heating, were found to improve (Kinn) and uphold (Ringsaker) the conformity to relevant standards and significantly reduce the risk of damage, except in the scenario of moving objects from Ringsaker to a conservation studio, when the risk would increase. The use of conservation heating could save ~50% of the heating cost. The estimated risk reductions may be less relevant for objects kept in situ, where cracks in the original paint and gesso have developed historically. They may be more relevant when moving original objects away from their proofed climate into a conservation studio for treatment.

Keywords: Norwegian medieval stone churches; cultural heritage degradation; polychrome painted wood; conservation treatment; HERIe climate risk modeling; humidity-induced mechanical damage; climate standard; conservation heating; energy saving

Citation: Grøntoft, T.; Stoveland, L.P. Painted Wood Climate Risk Analysis by the HERIe Model of Building Protection and Conservation Heating Scenarios in Norwegian Medieval Stone Churches. *Heritage* **2023**, *6*, 3089–3112. https://doi.org/10.3390/heritage6030165

Academic Editors: Peter Brimblecombe, Jenny Richards and Claudia Pelosi

Received: 13 January 2023
Revised: 22 February 2023
Accepted: 13 March 2023
Published: 15 March 2023

1. Introduction

As in the more famous wooden stave churches, Norway has several medieval stone churches that house religious objects of great cultural significance. Their indoor climates are characterized by large seasonal fluctuations and, often, extremely dry or moist conditions that are governed by the outdoor conditions or the heating regimes in the buildings [1]. Incorrect indoor relative humidity (RH, %) and temperature (T, °C) values and fluctuations, also termed the indoor climate in this work, can cause plastic deformation and cracking in polychrome painted wooden objects [2–4], followed by different and potentially accelerating damage processes such as cupping and the flaking of paint and ground layers. Changes to the historical climate, to which the objects are acclimatized [5], may thus result in reduced material value and increased conservation costs.

Energy saving is a requirement to mitigate greenhouse gas emissions. Heritage buildings, such as churches, often have high energy consumption and energy cost, and adopting energy-saving measures can help reduce the carbon footprint [6]. Recent international crises, with the accompanying dramatically increasing energy prices, are an additional concern. As the outdoor climate in Norway changes to become more humid, tempered, and fluctuating [7], the chances for change in the indoor climate in churches beyond the range of that experienced in historically proofed climates [5] may also increase. Significant increases in rainfall, leading to the occurrence of wood rot, have also been predicted [8] and were recorded in the recent past [9] in western Norway. However, only small changes to the humidity content of walls have been predicted [10]. It is essential that energy-saving measures in the churches should help reduce, or at the least not increase, the climate risks and conservation costs regarding building interiors and objects.

The aim of this work was to improve our understanding of the climate sensitivity of polychrome painted wooden objects in Norwegian stone churches by analyzing and modeling how changes in indoor RH and T might affect their preservation. The effect of measures to improve preservation climates and/or save energy on the risks of mechanical damage, plastic deformation, and cracking to painted wooden objects was assessed. Two churches, Kinn and Ringsaker (both dating from around 1150), which are located in very different climates and have different heating regimes, were selected for the study. Both house valuable polychrome wooden objects with past and/or present condition issues that can be related to unfavorable indoor climates [11–15]. The indoor climate data from the two churches were compared to the CEN 15757 standard and ASHREA guidelines (the term "guideline" is also used in this work as a generic term describing both formal "standards" (CEN EN 15757:2010) and other specific "guidelines" (ASHRAE)), and the HERIe [16] (https://HERIE.pl/, accessed on 12 January 2023) modeling tool was used to assess the risk of mechanical damage. To our knowledge, only a few exemplifying studies recording the application of HERIe have been published by others than the model developers [17].

The limitations of simplified modeling are well known. Accurate and precise modeling can only be performed if the model represents the phenomena of interest and possess good input data. HERIe simulates the worst-case scenarios of humidity-induced mechanical damage risks to paint and gesso on wooden panels, which may not represent the actual risk to historically acclimatized and naturally aged (and already cracked) objects [18]. HERIe is, however, a powerful tool in aiding the decision-making processes by estimating the risks involved when changing the indoor climate or moving objects between locations, for example, for treatment in a conservation studio [17]. It could be expected that the trend in the risk of damage to different geometries, depending on changes in the environment, resembles that of panels, but it was outside the scope of this work to assess such variations.

The HERIe model is briefly described herein to understand its use in this work. For a more detailed description, readers are referred to the published literature. Derivations that are particular to this work are described in detail. The current modeling did not consider the multitude of possible local objects and microclimate variations, combinations, and synergies. The evaluations were based on measurements of the local climate near the altarpieces in the Kinn and Ringsaker churches and, thus, do not represent the indoor climate of these buildings as a whole. The practical means and possibilities [19] of changing the buildings' indoor climates and of realizing the indoor climate scenarios were not evaluated. It has been noted that engagement with modeling in heritage science and practice has been limited [20]. The model application below is a contribution intended to redress this situation.

2. Locations and Objects

The Kinn and Ringsaker churches are located in the southern part of Norway. Kinn church is a (mainly) unheated stone building on a small island on the west coast. It is located north of a rising cliff and is directly exposed to the harsh weather conditions of the open North Sea in the west and north. Ringsaker church is a heated stone building, located inland east of the main Scandinavian mountain range (Figure 1). In the Köppen system, Kinn is in an oceanic climate and Ringsaker is in a subarctic climate [21,22]. In the following sections, the churches will mostly be referred to simply as "Kinn" and "Ringsaker".

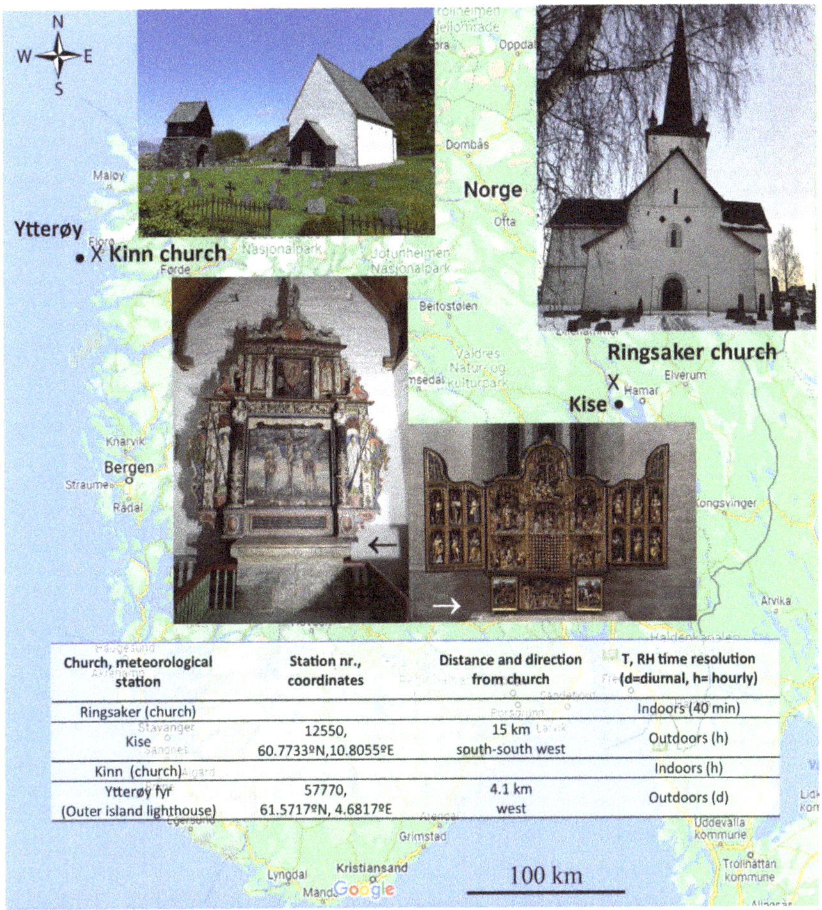

Church, meteorological station	Station nr., coordinates	Distance and direction from church	T, RH time resolution (d=diurnal, h= hourly)
Ringsaker (church)			Indoors (40 min)
Kise	12550, 60.7733ºN,10.8055ºE	15 km south-south west	Outdoors (h)
Kinn (church)			Indoors (h)
Ytterøy fyr (Outer island lighthouse)	57770, 61.5717ºN, 4.6817ºE	4.1 km west	Outdoors (d)

Figure 1. The Kinn and Ringsaker churches (X) and locations of the climate loggers (arrows) near their altars. The meteorological stations at Ytterøy, near Kinn, and Kise, near Ringsaker, are also marked. Photos: Kinn (left): Smestad, T.R., NIKU, 2020. Ringsaker (right): church: Jernæs, N.K., NIKU, 2019. Altar: Lindstad, B., 2020.

Kinn houses religious objects from the 16th century and later, including a painted wooden altarpiece carved in 1644, with an integrated central section from a late medieval triptych and painted elements from 1703 (Figure 1) [23]. The altarpiece's material composition, condition, and conservation treatments in 1971 and in 2004–2005 have been described elsewhere [14,15]. The sculptures and sculpted scenes in the altarpiece are made from oak, while lime and pine were used in other parts of the construction, such as the panel paintings. The altarpiece is probably painted with a thin oil-based paint on a relatively thick and porous chalk-glue ground. The objects have a long history of wear, cracking, flaking paint, and paint loss, undergoing restoration treatments before 1971 about which there exists relatively little information. Loose and flaking paint was consolidated with polyvinyl acetate (PVA) in 1971, with sturgeon glue in 2004, and with Lascaux medium for consolidation (MFK) in 2005. Newly flaking paint, paint loss, and possible activity from wood borers were observed on the altarpiece during a condition re-assessment in 2020 [15].

Although Kinn remains mainly unheated, it has ~15 electrical tube ovens of 1000 W output efficiency each, located below the seats. These are turned on during the occasional gatherings and events in the church, such as during a confirmation ceremony in May 2012 when all the ovens were turned to maximum, which gave a temperature varying between 8 and 12 °C [11]. In 2012, it was reported that a portable oil-filled electrical radiator was used for shorter periods for local heating behind the altar, in an attempt to reduce the observed moisture in this location. However, due to the very humid outdoor and indoor conditions in Kinn, it was found that any advantages to using heating to reduce the RH or surface moisture were very slight or non-existent. Heating a church that has remained largely unheated until this day is generally not recommended from a conservation perspective [24]. Heating could also affect the moisture content and moisture transport in the walls, which could potentially cause evaporation and increased RH [11]. Continuous heating would necessitate unwanted energy use and increased electrical bills. It seemed better that a reduction in the indoor humidity (RH and condensation) should, as far as possible, be via measures that reduce the moisture infiltration into the church through openings, for example, by doors and windows, and in the church fabric from the rain and the ground [25,26]. Occasional local heating might be more efficient if this moisture infiltration could be reduced. It has, however, been suggested that improved insulation, resulting in less infiltration, could reduce indoor air circulation and air movement and increase the risk of mold growth [11].

Ringsaker has a richly ornamented interior, with one of the most important and well-preserved altarpieces in Norway [27]. It was produced from oak wood in Antwerp in the early 16th century and consists of polychrome sculptures and sculpted elements and doors, with beautifully painted scenes [12]. It was painted with oil-based paint and has large areas of gilding, metal foils, and decorative punch marks. Detailed reports of its materials, construction, past treatment history before 1982 (including varnish removal, consolidation with wax, and revarnishing), its condition, and the indoor climate in the church have been provided [12,13]. Observations during the last investigations and a condition assessment in 2019 reported the shrinking of painted panels, craquelures on the paintings, some new loose and flaking paint on the sculptures, pieces of the sculptures that had fallen off, and signs of overcleaning, as well as layers of soil and dust. The need for structural repairs and the stabilization of mechanical elements such as doors, due to loads and movements, were also noted. The 2020 treatment included surface cleaning, the consolidation of loose paint on the sculptures (using Lascaux medium for consolidation), gluing the loose parts of the construction, retouching, and re-varnishing the paintings.

Ringsaker was unheated until wood stoves were installed in 1865 [12]. Electric tube stoves with two heating settings were in place along the north and south walls and beneath the seats before floor heating was installed in the 1960s. The sacristy has electrical panel stoves. The heating control is manually operated, with the aim of keeping a stable temperature of 18 °C. The few RH measurements reported from before 2019 were of a "variation in the RH of between 42 and 58%" in the winter of 1968, and of "a similar variation as in 1968" in the winter of 1982. Dry air due to the electric heating and RH fluctuations were reported to cause cracking and flaking paint.

Figure 2 shows damage to the painted wood of the altarpieces in Kinn and Ringsaker before the conservation treatments.

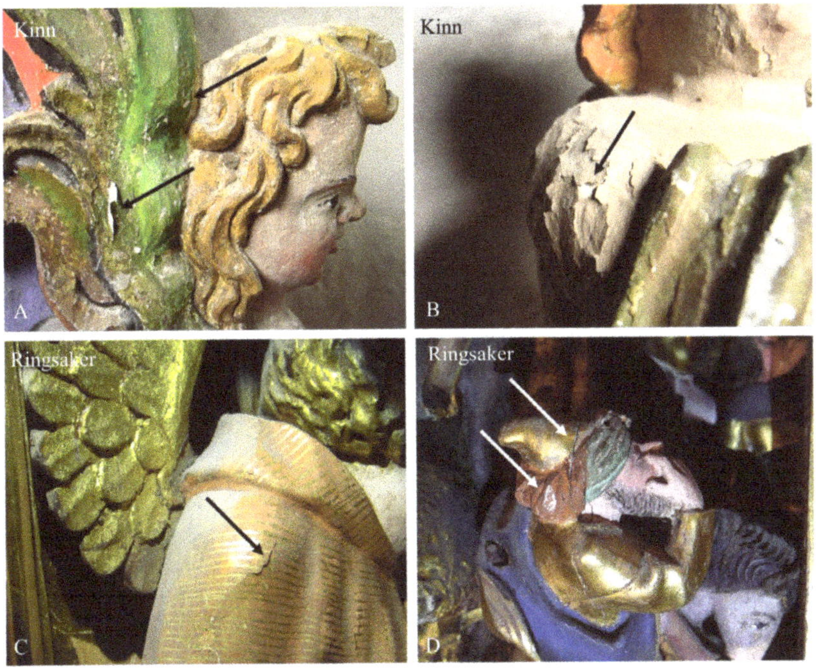

Figure 2. Details of loose, flaking paint and paint loss (**A**,**B**) of carved figures on the Kinn altarpiece before treatment in 2005. Details of cracked paint in the cape of a sculpture (**C**) and of cracks and small paint loss on the head and shoulders of a figure (**D**) of the Ringsaker altarpiece, before treatment in 2020. Photos: (**A**,**B**): Solstad, J., NIKU, 2005. (**C**,**D**): Olstad, T.M., NIKU, 2019.

3. Methods

3.1. Climate Measurements

The indoor climate conditions (RH and T) in the churches were measured during annual monitoring campaigns, using pre-calibrated Testo 177-H1 loggers with hourly resolution in Kinn in 2011–2012 [28], and Tinytag ultra2 loggers with 40-min resolution in Ringsaker in 2019–2020 [12,13]. The logger of the climate data for this work, in Kinn, was mounted on the front of the base of the altarpiece, between 0.1 and 1 m above the floor. The logger in Ringsaker was mounted on the side of the base of the altarpiece, ~1.5 m above the floor (Figure 1). The logger in Ringsaker was moved from the measurement location from 29 April to 25 May 2020. The RH and T values over this period were simply linearly interpolated, from the last value measured in the church at the end of April to the first value again measured in the church at the end of May. The interpolation was not expected to change the overall results of the study, although any variations of concern in this period will have been omitted (see Section 4, Results). The outdoor values of RH and T were collected from the nearest meteorological stations found to best represent the ambient conditions of the churches [29] (Figure 1).

3.2. Evaluation of Indoor Climate Adjustments in the Churches Using Conservation Guidelines

A description of the indoor climate guidelines is provided in Appendix A. The general climate risk in the churches was assessed by calculating the conformity to the CEN15757 standard [30] and ASHRAE guidelines [31] (in Excel). It was considered that the comparison of the climates in the churches with the CEN standard, which specifically address the risk of mechanical damage in organic hygroscopic materials, and the ASHRAE B class, which addresses temperature reduction in the winter, were of particular interest. The evaluation was made using a combination of the short-term, seasonal, and fixed limits of the guidelines/classes, calculated to always present the most stringent of the three limits via the RH and T time series, resulting in the least conformity. The conformity of the RH and T values with the limits was then calculated as the "percentage of time when the guideline limits were met", alternatively interpreted as the "percentage outside of limits" [32,33]. The guidelines provide, by their very definition, the limits of critical risk regarding damage at transgression. It is therefore important to avoid any transgression, and high conformity (near 100%) will still indicate a risk. However, the conformity, besides noting the risk at values of <100%, is also a measure that is usually expected to correlate with the frequency of critical transgressions, signifying the probable rate of the development of the resulting damage and, thus, condition at any future point in time (until "total damage" is assessed). The conformity comparison of time outside of different (guideline) limits is, in our view, a useful and simple additional measure to compare the variations (fluctuation amplitudes and durations) in RH and T time series.

HERIe modeling maps the risk of mechanical damage in terms of painted wooden panels in situ in the churches, after hypothesized changes in indoor climates, and when hypothetically moving a painted panel to a conservation studio. The indoor climate scenarios (see Section 3.4) used lower RH values in Kinn than might be obtained by building measures [26] in a situation with some, but probably still sparse, heating, and of a higher but more stable RH (<65%) in Ringsaker that could be obtained via conservation heating with a reduced temperature, and that could still be considered acceptable for preservation. Conservation heating, or humidistat-controlled heating, refers to the concept of heating a building to keep the relative humidity below given limits [34–38]. The energy consumption calculations were finally made as monthly heating degree days for comfort [39] and conservation heating.

3.3. Mold and the High RH Limit

The climate scenarios were of RH changes in the churches moving toward an (upper) RH limit of 65%, considering the mold risk, which has been a matter of major concern when suggesting limits. An RH limit (of 65%) has been adopted by the National Trust for its properties in England, Wales, and Northern Ireland, "considering primarily the risk of mould growth, with this threshold allowing for the difference between measurement of a bulk room environment and cooler microclimates close to external walls, as well as providing for sensor inaccuracies" [35]. The intention was to keep the coolest parts of a room below the more widely recognized 70 to 80% RH mold threshold. In Norwegian winters, which are colder than British winters, high RH levels close to, and condensation on, external walls would be more likely. In the recorded high humidity in Kinn (see Section 3.4), mold occurrence could reasonably be expected. However, no mold was reported on objects, room surfaces, or the external walls. It was reported in 2013 that this might be due to the building's natural ventilation [11]. Rot was later observed in 2016 in the building's structure, and some mold was recorded in the cellar [26]. The sea salt exposure of the church from the north Atlantic westerly winds [40,41], which would, in some amounts, be ventilated to the indoor air, may inhibit mold growth [42]. Salt exposure can, however, also damage wooden materials [43–45]. The amount of salt on the wood and painted surfaces in the church has not, as far as we know, been measured. No mold occurrence has been reported in Ringsaker [12,13]. A scenario change in Ringsaker to a more stable RH of <~65%

was not expected to increase the mold risk. Zero (mm) expected mycelium growth was calculated in Ringsaker over the period of the 2019–2020 in situ recorded and scenario (see Section 3.4) climate data, according to the method described in earlier publications [33,46].

3.4. Climate and Object Scenarios

In Kinn, a situation was hypothesized wherein building construction measures were possible and were effective in reducing the indoor humidity to RH scenario values, RH_{adj}, in steps of ~5% and below the set scenario RH maximum limits of $RH_{max,lim}$ = 85%, 80%, 75%, 70%, and 65%, assessed proportionally to the measured values over the whole measurement range, according to Equation (1):

$$RH_{adj} = RH_{measured} \cdot \frac{RH_{max,lim}/0.95}{RH_{max,measured}}. \quad (1)$$

It was assumed that it was inconvenient, or improbable, to adjust the highest peaks of the measured RH values throughout the year to below the limit values. The absolute maximum adjusted RH value was, therefore, set to a 5% higher value than the RH scenario maximum limit value. In that case, $RH_{measured} = RH_{max,measured}$, and the absolute maximum $RH_{adj} = RH_{max.lim}/0.95$. For example, the 65% scenario for the highest RH (peak), adjusted from the measured values in 2011–2012, would thus be $RH_{adj.max} = 65/0.95 = 68.4\%$.

In Ringsaker, a modified temperature should approach, as near as possible, a constant RH = 65% through the year, and be between a minimum temperature of 10 °C and a maximum of 25 °C, and was expected to result in conservation-heating saving. This modified temperature was calculated from the recorded RH and T values (in 2019–2020) and the calculated absolute humidity (AH, g/m^3) from the equation given by the authors of [47] (p. 51). The estimations were made by first calculating the absolute humidity in each time step (of 40 min) of the measurements. The RH was then fitted, again using the equation from [47] (p. 51) to be as close as possible to the set limit (of 65%) at that absolute humidity via a sequential numerical adjustment of the temperature from that measured. The fitting involved a changing of the temperature in each time step, proportionally to the respectively calculated distance of the adjusted RH (via Equation (2)) from the RH limit (= 65%), until the least possible sum of the estimated RH differences, from the RH limit throughout all the time steps of the annual measurement series, was obtained. In time steps wherein the maximum or minimum set temperature limits (25 °C and 10 °C) were reached, no further RH changes were allowed. A graphic representation of the scenarios is provided in Section 4.2.

The HERIe modeling process (Section 3.5) was performed for the measured and adjusted climates, and, in addition, for a scenario with the hypothetical removal of painted wooden panels from the churches in the winter (1 January) and summer (1 July) to a conservation studio with a stable indoor climate (RH = 50% and T = 20 °C), for a duration of one year. It must still be taken into consideration that maintaining a stable indoor climate (±5% RH) is not achievable in most conservation studios. The modeling was performed regarding mechanical damage under the scenarios and conditions listed in Table 1, which were found to represent possible climate modifications and resemble important painted wooden objects (Section 2) in the churches most closely. The only differences between the model characterization of the objects in the two churches were the thicker wooden panel and the additional modeling of lime and pine wood, compared with that of oak, in Kinn.

Table 1. Climate scenarios and modeling relevance for original wooden objects with gesso and paint (see Section 5, Discussion), along with modeling conditions in both Kinn and Ringsaker. RH_{av} and T_{av} are the average relative humidity and temperature of the data series.

Scenario	Kinn	Ringsaker	HERIe Risk Relevance
Recorded RH and T	RH_{max} = 96%, T_{max} = 20 °C RH_{av} = 79%, T_{av} = 9 °C	RH_{max} = 77%, T_{max} = 21 °C RH_{av} = 49%, T_{av} = 17 °C	Low relevance for original objects with proofed fluctuations Unknown relevance as a proxy for original objects that have recently undergone conservation treatments
Modified RH and T	RH reduction by building measures, $RH_{max\ limit}$(%)–RH_{av}(%): 85–74; 80–69; 75–65; 70–61 and 65–56, T_{av} = 9 °C	Conservation heating to RH_{max}–RH_{av}: ~65–63% and T_{min} = 10 °C (T_{av} = 13 °C)	Kinn: Low relevance for original objects if reductions in RH stay within proofed fluctuations Ringsaker: High relevance for original objects as the change in RH from proofed fluctuations Unknown relevance as a proxy for original objects that have recently undergone conservation treatments
Removal to conservation studio (RH = 50%, T = 20 °C)	From recorded and reduced RH (scenarios 1 and 2) in January and July	From recorded and adjusted RH (scenarios 1 and 2: T_{av} = 13 °C) in January and July	High relevance for original objects when moved outside proofed RH fluctuations to a conservation studio Moderate relevance for original objects when returning to the proofed climate from a conservation studio Unknown relevance as a proxy for original objects that have recently undergone conservation treatments
	HERIe modeling conditions		
	Damage		Mechanical
	Paint Material		Painting on wood
	Wood species		Oak. Kinn: also lime and pine
	Direction of cut		Tangential, radial
	Panel thickness (mm)		40 (Kinn), 20 (Ringsaker)
	Gesso		Stiff
	Water vapor transport		Through one side
	Long-term mean RH and T values		Scenario 2: From uploaded data Scenario 3: Set to the recorded or adjusted annual average scenario values, but set up to the available modeling RH maximum = 70%

In Kinn, the wooden panels and other wooden elements are probably both tangentially and radially cut (pers. comm with T.M. Olstad on 2 June 2022). In Ringsaker, they seem to be mostly radially cut. Modeling was conducted for both tangential and radial cuts. The tangential cut then represents a "worst case" scenario. Due to the slow humidity response of the oak and thick panels (40 mm), two years of data are required for modeling. As this was not available, data were obtained by duplicating the available annual measurements series to two years, which would then still represent the measured annual variation but not any real two-year period. The annual data series of seasonal (ASHRAE) and monthly (CEN15757) moving averages needed for the calculation of the ASHRAE guideline limits were also obtained from these extended data series, as recommended by the authors of [48]. The data series for modeling the move to a conservation studio (Scenario 3) was obtained by simply extending the two-year series to a three-year series by adding an annual data set of constant RH = 50% and T = 20 °C, from 1 January and 1 July in the second year. In these model climate data series, the dates of the approximate first year represent the measured values (see Section 3.1). Later dates that will still be used below to describe the data sets represent the simulated values.

3.5. The HERIe Modeling of Mechanical Damage Risk to Painted Wooden Panels

The HERIe model [14] of mechanical damage to the painted wooden panels provides a linear risk indication between 0 and 1, representing the change in the relative strains from values of ±0.002 to ±0.004, at the interface between the gesso and wood support of the painted panels. The positive values represent the per-mill elongation in tension and the negative values the per-mill compression. When the strain reaches ±0.002, plastic deformation of the modeled gesso begins. When the strain reaches ±0.004, it cracks. The maximum strain (tension or compression) calculated over the time series of the climate data input [17] determines the damage risk indices, RIs, reported in HERIe. The HERIe model can evaluate the risk of the deformation and cracking of an initially undamaged paint layer on soft or stiff gesso on 5- to 40-mm-thick oak, lime, poplar, or pine wood panels, with water vapor transport into the panel from one or two sides. The long-term climate variables (RH and T) that determine the zero strain, from which the fluctuation strains of the input climate are calculated in the model, are calculated as the average of the input data or are set by the user from a selection of values. It should be noted that the model overestimates the strain on changes from or especially toward very high humidity, but that this is of little significance below an RH = 70% (pers. comm. with an HERIe developer).mm

The "long-term mean RH and T values", applied as input to scenario 2 (Table 1) in the HERIe modeling process were "from uploaded data". The maximum strain and risk in the churches would then be established by fluctuations from the measured averages, due to either tension in the paint at high RH or compression at low RH. The scenario 3 HERIe modeling process was performed for the risk of the mechanical compression damage of gesso (and paint) when moving panels to the conservation studio (at RH = 50% and T = 20 °C), and the risk of tension damage when moving them back to the church. The model input for "the long-term mean RH and T values" when moving the panels to the studio was selected as the closest annual average scenario RH values of the data sets at intervals of 5% from RH = 50%, up to the available maximum model RH value for this input of 70%, and to the annual mean scenario temperature (excluding the time in the studio). The model input for "the long-term mean RH and T values" when moving the panel back from a conservation studio to the churches was set at the scenario values of the studio, although using T = 21 °C instead of 20 °C since the modeling input options were 19 °C and 21 °C. Thus, the annual change to the stable condition in the studio was assessed in the modeling to be a "long-duration change" from the climate in the church and not an annual period in a cycle.

The accessible modeling maximum for "the long-term mean RH and T values" of 70% RH was significantly lower than the recorded RH and high RH scenarios in Kinn. In nearly all these instances with an RH > 70%, the compression risks (RI) in the conservation studio predicted by HERIe were ~1 (the RI of the radial oak panels was 0.96). In scenarios of moving radial oak and pine wood panels back to the church from the studio, the RI was, however, modeled to be significantly lower than 1 at an RH between 70% and 80%. The model input of the "the long-term mean RH and T values" = 70% RH in these cases may have resulted in an underestimation of the risk, although, in such instances of changes to high humidity, it was also reported that the model overestimated the risk (as per the pers. comm. with one of the HERIe developers, https://herie.pl/, accessed on 12 January 2023). It seemed that this modeling analysis, with a comparison of the RH in the conservation studio (=50%) to a lower value than the mean recorded RH of the dataset (70% rather than the 70–80% of the dataset) was of little consequence for the overall risk assessment.

3.6. Energy Consumption Calculation by Heating Degree Days

The degree days for comfort heating, representing the heating needs in this respect, were calculated by subtracting the outdoor temperature from an indoor set point of 17 °C, as is customary in Norway [40]. The subtraction was made for each point in the time series of the data and was summarized over the months and year. The comfort heating was compared with the indoor measured situation by subtracting the indoor temperature from the outdoor measured temperature and then aggregating the degree days over the months and year. A comparison with the conservation heating scenario of reducing the temperature to a minimum of 10 °C, to approach an RH of 65%, was then made by subtracting this indoor (conservation heating) scenario temperature from the outdoor temperature and again summarizing the degree days over the months and year. The method of calculation of the conservation heating temperature modifications is described in Section 3.4.

4. Results

4.1. Conformity to Guidelines

The CEN 15757 standard and ASHRAE B class, which include seasonal relaxation, were thought to be the most interesting criteria to apply in the churches for this risk modelling. Figure 3 shows the measured data compared with the ASHRAE B class's combined fluctuating and fixed limits. The CEN 15757 variable limits are (not exactly) similar, but that standard does not include a combination with outer fixed limits, as illustrated in Figure 3.

The period of linear interpolation in May 2020 is seen in Figure 3B. The temperature in the church was relatively stable throughout the year, and a regular RH increase and regular fluctuations were observed in the spring. The reduced level of total recorded variability through the year due to the missing data will have reduced, somewhat, the 7th to 93rd percentile variable limit band width of the CEN 15757 standard. The 7th to 93rd percentile band of the data was estimated to be ~±5%, which is much lower than the bandwidth of ±10% that should, in this case (due to low variability), be used according to the standard (see Appendix A). The interpolation will thus not have affected the comparison with this standard. Table 2 reports the percentage conformity to the CEN 15757 standard and ASHRAE classes (see Appendix A, Table A1) of the measured data in the churches and the scenario 2 situation, with modifications to about 65% RH.

Table 2. Conformity (%) to the temperature and RH limits of the CEN 15757 standard and ASHRAE classes of measured data; scenario 2 data were modified to RH_{lim} = 65%.

CEN Standard and ASHRAE Classes	Kinn		Ringsaker	
	Temp [1]	RH	Temp [2]	RH
CEN 15757 [3]	**n.a.**	**98; 100**	**n.a.**	**98; 100**
AA	44	0; 52	100	32; 56
A1	44	0; 79	100	75; 61
A2	44	0; 90	100	70; 65
B [3]	**100**	**14; 99**	**100**	**96; 100**
C	100	36; 100	100	100; 100
D	100	36; 100	100	100; 100

[1] In Kinn there were no scenarios for changes in the temperature. [2] The temperature in the heated Ringsaker church conformed fully to all standard classes in both scenarios. [3] The CEN standard and ASHRAE B class thought most appropriate/interesting for application to the churches are in bold script. n.a. = not available.

Table 2 shows a near-full conformity of the recorded climates in the churches to the CEN15757 standard but a zero to low RH conformity to the ASHRAE classes, which all include fixed outer RH limits (see Appendix A) for Kinn, and a reduction in RH conformity regarding the more stringent ASHRAE classes (AA to B) for Ringsaker. This situation is illustrated in Figure 3 with the ASHREA B combination of fluctuating and fixed outer limits. The RH conformity in Kinn increased with the scenario reduction to RH ~65% and to near 100% for the ASHRAE B to D classes. For Ringsaker, an increase in the RH to a stable value of ~63% (RH_{lim} = 65%) increased the RH conformity to the most stringent ASHRAE AA class but decreased the conformity somewhat to the A1 and A2 classes. The reason for this is the fixed upper limit of RH_{lim} = 65% in these classes (see Appendix A). With the RH modification to ~63%, more of the data points come above this outer fixed limit. With the stringent constant limits of the AA class (±5%) the original variable situation through the year was, however, worse (32% fit) than if all the data through the year were closer to RH = 65%.

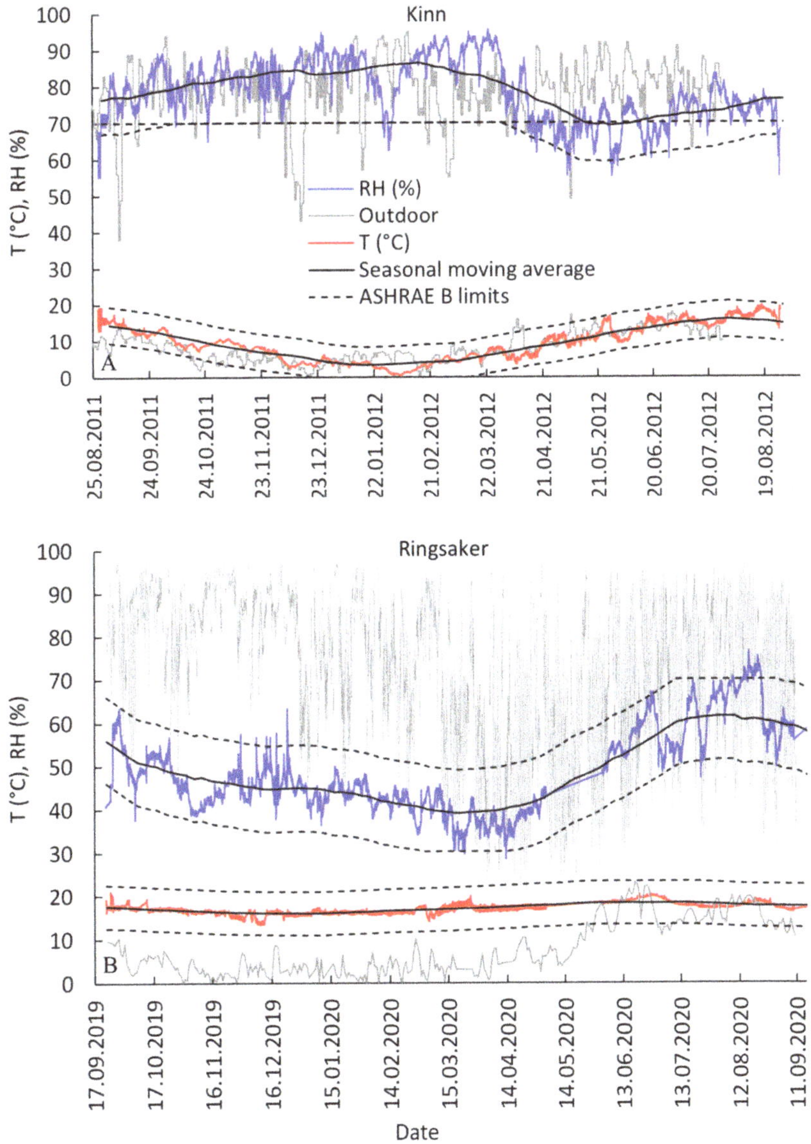

Figure 3. Comparison of the climate conditions in Kinn (**A**) and Ringsaker (**B**), with the ASHRAE B class's combined fluctuating and fixed outer limits. The outdoor RH and T are also shown.

4.2. Mechanical Damage Risk to Painted Wooden Panels

Figures 4 and 5 show the HERIe modeling results for the strain levels and risk of mechanical damage to the different and most relevant painted wood from various species, with the duplicated, bi-annual, RH and T values in Kinn and Ringsaker, in the measured situation (scenario 1), and in the scenario 2 situation, with the maximum modified indoor climate obtained by building modifications, in Kinn, to $RH_{max}(\%) - RH_{av}(\%)$ = 65–56, and by conservation heating in Ringsaker to RH_{max} ~65% and T_{min} = 10 °C (Table 1).

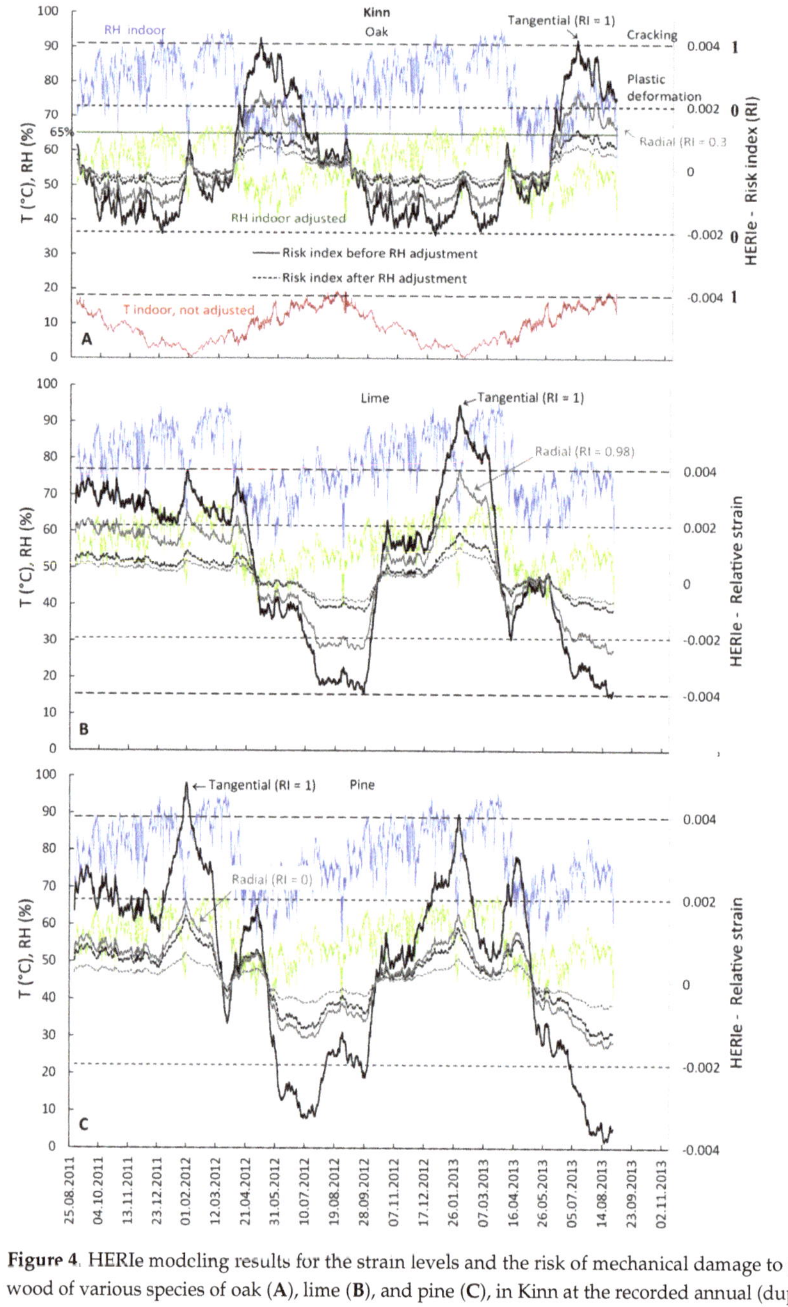

Figure 4. HERIe modeling results for the strain levels and the risk of mechanical damage to painted wood of various species of oak (**A**), lime (**B**), and pine (**C**), in Kinn at the recorded annual (duplicated to bi-annual) RH and T, and for the scenario of reduced RH (to < ~65%) according to building measures. The scenario temperature was not adjusted from that recorded. The descriptions of the curves are mainly given in the upper diagram (**A**).

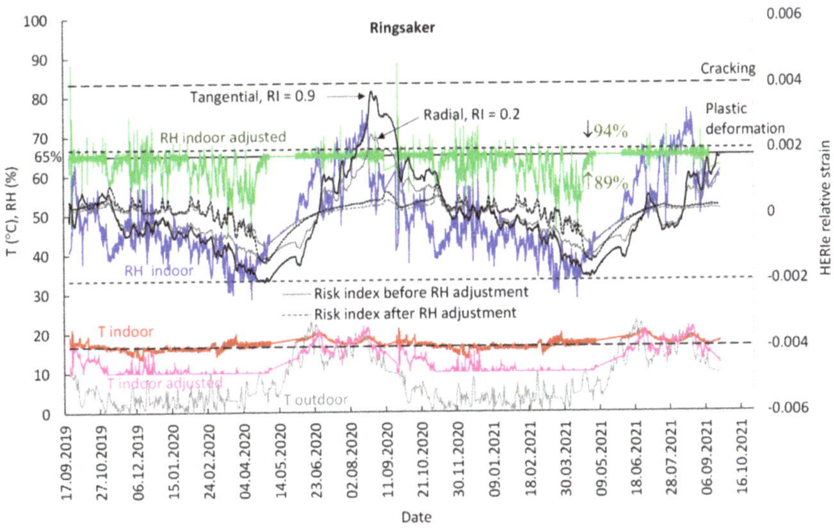

Figure 5. The HERIe modeling results of strain levels and the risk of mechanical damage to painted wooden panels in the Ringsaker church for the recorded annual (duplicated to bi-annual) RH and T measurements and for the hypothetical conservation heating scenario with an adjustment of the RH (to < ~65%) by reducing the temperature. The outdoor RH and T values are shown. The percentage of annual integrated fit of the RH values (to 65%) from the original measured RH values above (↓) and below (↑) the set conservation heating RH limit of ~65% are given.

It can be seen in Figure 5 that all the sharp RH peaks could not be fitted exactly using the sequential integration minimizing method. The percentage best fit, as shown in Figure 5 (as it would be for most data sets), was, therefore, somewhat less than 100%. During much of the autumn and winter in Ringsaker, the conservation heating would lower the temperature to the limit of 10 °C, resulting in an adjusted RH of lower than 65% and a somewhat lower percentage fit (89%) to the limit of RH = 65% from the lower (than 65%) measured RH values than from the higher (than 65%) measured RH values (that had a percentage fit of 94%, although also of much fewer data points). It should be taken into consideration that the adjusted RH curve in Figure 5 was obtained from modeling to investigate the energy-saving and object damage risks. The peaks are due to a lack of model fit and do not represent any real-world maxima and minima. This may, however, also reflect the practical difficulty in making precise RH modifications to limits via temperature adjustment.

Figures 6 and 7 show the HERIe modeling diagrams of the relative strains and risk indices of mechanical damage to painted oak wood panels in the tangential directions if the object was moved to a conservation studio in January (scenario 3, Table 1) from the recorded RH, and from the scenario when adjusted to below ~65% RH for Kinn and Ringsaker. The risk assessments for the recorded climate in the churches (i.e., from the recorded "long-term mean RH and T values") before moving a panel to a conservation studio are presented in Figures 3 and 4. It seems incorrect to present a risk assessment for this period, compared to the recorded "long-term mean RH and T values" (of T = 20 °C and RH = 50%) for the scenario 3 climate of the conservation studio; this period of the modeled risk assessment was removed in Figures 5 and 6. For paint on wooden panels that were cut in the radial direction, the shapes of the strain curves were similar but the risk was lower. Removal of the panel to a studio in July yielded different strain curves but a similar maximum high and low strain and risk.

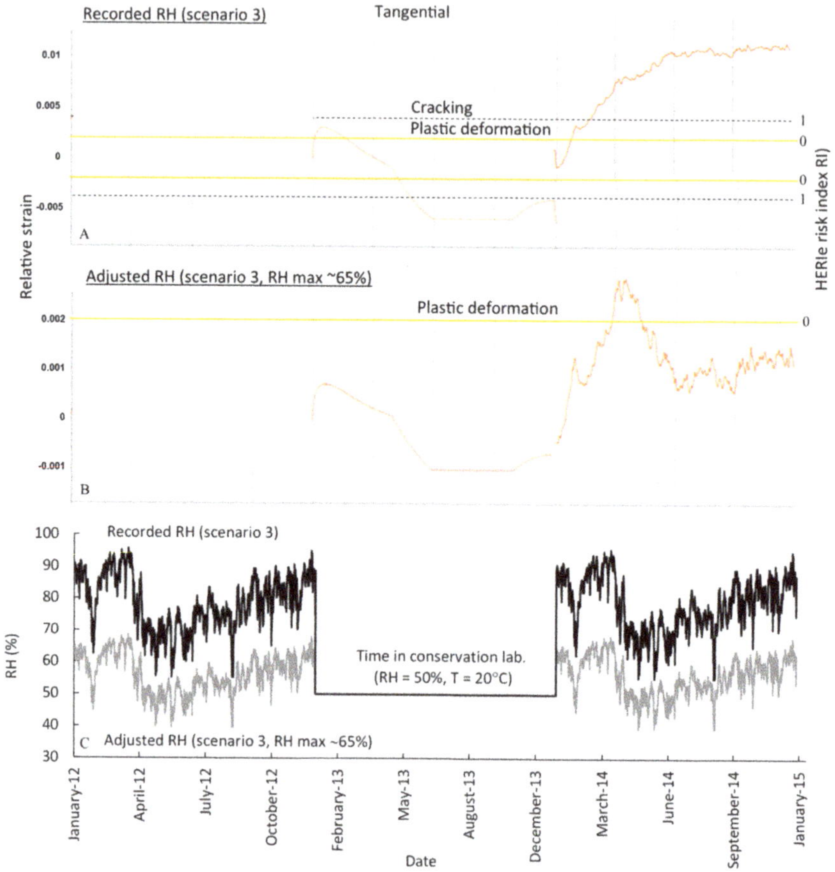

Figure 6. The HERIe modeling results of the strain levels and the risk of mechanical damage to painted oak cut in the tangential direction when it was moved to a conservation studio in January, from the recorded RH (**A**,**C**) and from an adjusted RH_{max} ~65% by the (hypothetical) building measures (**B**,**C**) for Kinn (scenario 3). An incongruity is seen at its removal back to a conservation studio (in January 2014), which was due to the combination of HERIe graphs for the Scenario 3 removal to and from the studio (for which the different input values for the "long-term mean relative humidity" were used).

It can be seen that the risk follows the RH fluctuations with some delay for the 40-mm-thick oak panel at Kinn (Figure 6), presumably due to the longer time needed for RH equilibration with the thicker oak panel at Kinn than at Ringsaker (Figure 7). With the panels from both Kinn and Ringsaker, the negative strain and, thus, the compression risk, developed to a constant level during their annual stays in a conservation studio. When the transport was from the adjusted low RH scenario for Kinn (RH_{av} = 56%, RH_{max} ~65%), or the recorded RH for Ringsaker (RH_{av} = 49%), HERIe predicted no risks in the studio (Figures 6B and 7A). When the transport was from the high RH situations in the churches (the recorded RH for Kinn was RH_{av} = 79%) and adjusted the RH for Ringsaker (RH_{av} = 63%), the constant compression strain of the paint layer in the studio represents a significant risk of cracking for Kinn and of plastic deformation close to cracking for Ringsaker (Figures 6A and 7B).

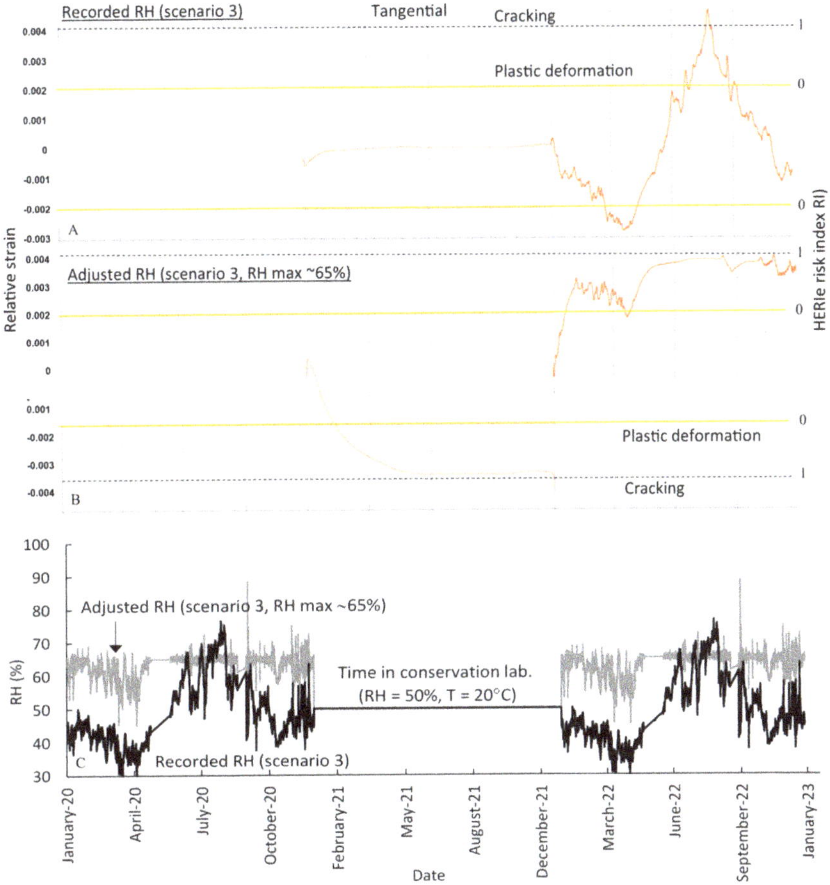

Figure 7. The HERIe modeling results of the strain levels and the risk of mechanical damage to painted oak cut in the tangential direction of the wood when moved to a conservation studio in January from the recorded RH (**A**,**C**) and an adjusted RH_{max} ~65% by conservation heating (**B**,**C**) at Ringsaker (scenario 3). An incongruity is seen at its return to the studio (in January 2022) due to the combination of HERIe graphs for the scenario 3 removal to and from the studio (for which different input values for the "long-term mean relative humidity" were used).

Some other variations in the curves in Figures 6 and 7 should be commented upon. On moving a painted panel from the higher humidity in the church to the dryer and warmer studio, HERIe predicted an instant increase in tension in the Kinn panel (Figure 6) that is much smaller in the Ringsaker panel when moving from the high RH scenario of ~65% (Figure 7B), before the slow drying and development of compression strain occur. This prediction of an initial increase in tension may be due to the instant effect of the higher temperature in the studio than in the churches. Due to the larger T increase in the studio from the higher RH in Kinn (ΔT = 11 °C, RH_{av} = 79%) than in Ringsaker (ΔT = 7 °C, RH_{av} = 63%), the duration of the period of constant compression strain in the studio is then shorter in Kinn than in Ringsaker.

The compression or tension risks in the scenario 2 and scenario 3 climate scenarios (Table 1, Figures 4 and 5), were assessed from the HERIe risk curves (as seen in the minimum and maximum RH scenarios in Figure 4 to Figure 7) and are shown in Figures 8 and 9.

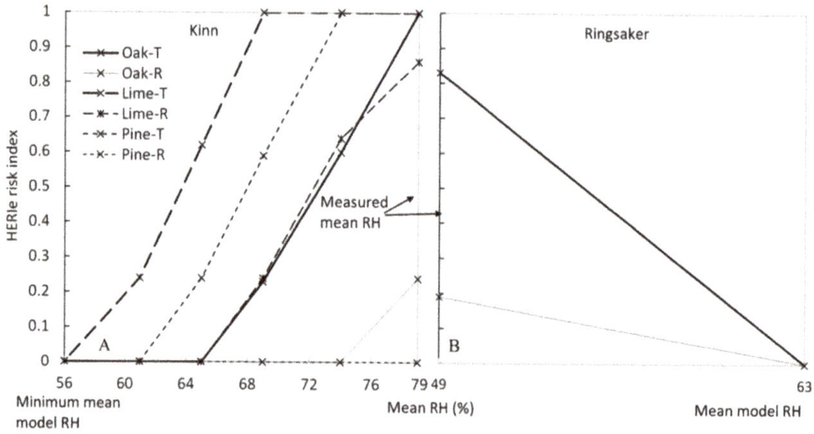

Figure 8. HERIe risk indices for the Scenario 2 (Table 1) reduction of RH to lower values for Kinn (**A**) and to a higher conservation heating value for Ringsaker (**B**). The crosses represent the modeling situations. All the curves reach from the lowest to the highest RH value in the diagrams and are partly overlapping.

Figure 8A shows that, according to HERIe, paint and stiff gesso on 40-mm-thick tangentially cut oak, pine, and lime panels in the very humid indoor climate of Kinn would probably crack within two years. If it was possible to reduce the mean RH to ~56%, along with, proportionally, the climate fluctuations, no deformations would probably happen. Between the 79% and 56% RH, there would be a near-linear reduction in the cracking and plastic deformation risk (from 1 to 0) over a 15% RH interval, with the reduction in risk starting for the oak panels already at ~79% RH, for pine at ~74%, and for lime at ~69% RH. The risk of damage to radially cut panels in the recorded climates in the churches was found to be less, at ~0.86 for lime, ~0.24 for oak, and zero for pine, implying some risk of plastic deformation, and decreased to zero with a similar near-linear rate to the tangential cut wood. Less (or no) new cracking might, however, be expected on old and already cracked panels. Figure 8B shows that the potential conservation heating in the Ringsaker church, with an increasing RH from an annual mean of 49% to a more stable mean of 63%, was found to reduce the risk of paint deformation on tangentially and radially cut 20-mm-thick painted oak panels in the church from risk indices of 0.83 and 0.19 (indicating a risk of plastic deformation) to zero risk.

The reductions in the RH for one year in Kinn, before moving the painted wooden panels to a conservation studio with a stable RH of 50% and T of 20 °C (scenario 3), were found to provide a similar reduction in risk in the studio (Figure 9A) and when moving the panels back to the church (Figure 9B) as when in situ (Figure 8). However, these figures include variations in the RH reductions needed in the church for an initial decrease in the risk and then in the rate of decrease, depending on wood species, the wood cutting direction, and the month of the year when reinstalling the panels in the church. The displacement of the July risk curves toward a lower RH and, thus, a higher risk compared to the January risk curves when moving the panels back to Kinn from the studio (Figure 9B) seemed to be due to a longer period of gradual increase in RH in July than in January, which resulted in the buildup of more tension strain before this was released by a sharp drop in RH (see the curves in Figure 6). In Ringsaker, the damage risk (RI) of moving a panel to a conservation studio after one year in the modified and more stable climate, with a higher (than recorded) mean RH = 63%, was found to increase when compared to moving it from the recorded mean RH = 49%, from 0.25 to 1 (with certain cracking) of the tangentially cut wood and from zero to 0.49 (with plastic deformation) for the radially cut wood (Figure 9C). The risk of moving the painted panel back to the conservation-heated church (RH_{av} = 63%) was found to be high (>1) but was slightly less than when moving it back to the recorded fluctuating climate situation (RH_{av} = 49%) (Figure 9D). As with the situation in situ (Figure 8), these indications of damage should be considered in the context of the historical condition (cracks) and proofed climates (see Section 5, Discussion).

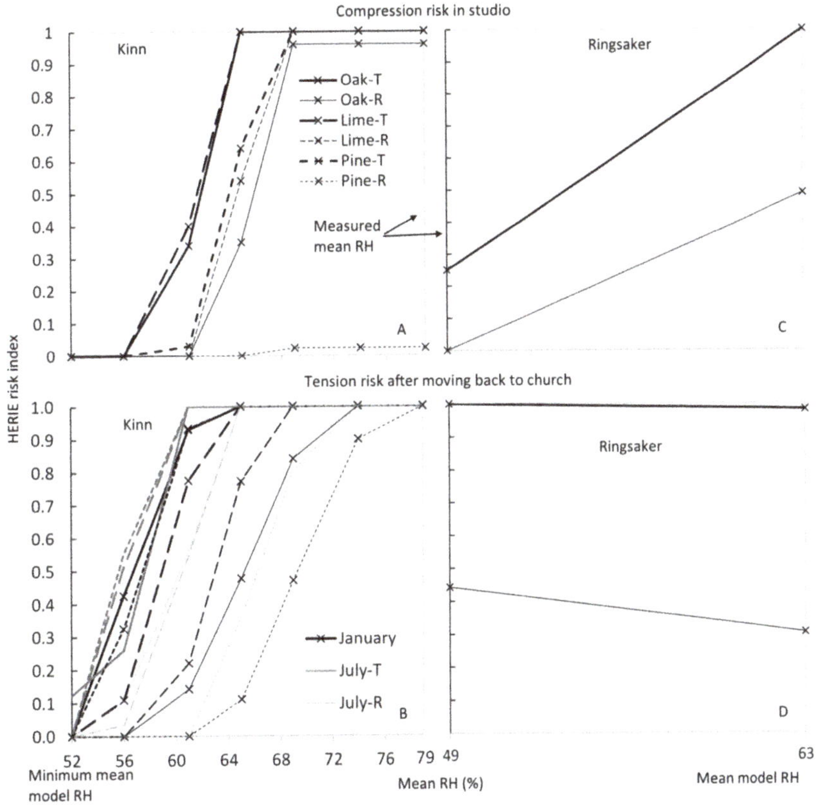

Figure 9. The HERIe compression risk indices when moving a painted panel to a conservation studio (RH = 50% at T = 20 °C) in January (**A**,**C**), and the tension risk on moving the panel back to the churches after one year (**B**,**D**), from the recorded and reduced scenario values of RH for Kinn and from the increased conservation heating RH scenario value for Ringsaker (scenario 3, Table 1). A difference in the risk in January and July was only found for the time when moving the panel back to Kinn (**B**). The crosses represent the modeling situations that were at the same RH in July as in January but, for clarity, are only given in January. T = gesso on tangentially cut wood. R = gesso on radially cut wood. All the curves (except for the tangentially cut oak in July) reach from the lowest to the highest RH value in the diagrams and are partly overlapping. The legend in (**A**) is applicable to all the diagrams.

4.3. Conservation Heating

Figure 10 shows the calculated heating degree days for Ringsaker in the recorded situation and for different heating scenarios.

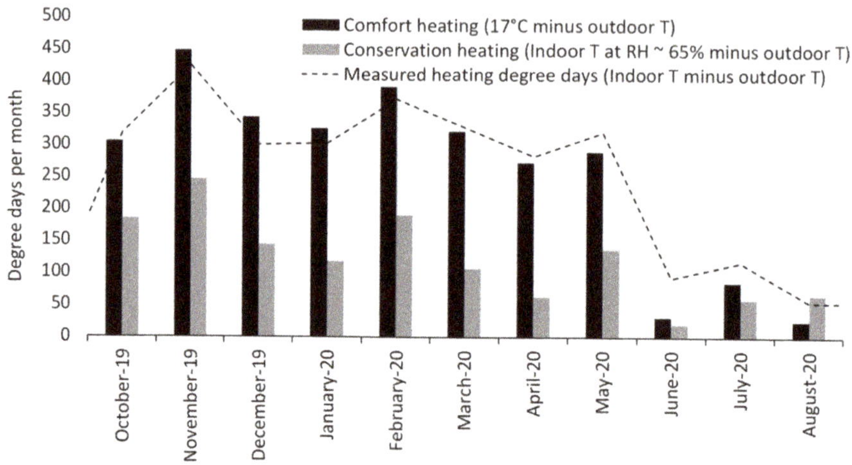

Figure 10. Monthly heating degree days for the different scenarios for Ringsaker.

The larger number of degree days for conservation heating than for comfort heating in August reflects the high humidity measured in this month (Figure 5) and, thus, a need to increase the temperature to a level above that needed for comfort heating, and above that of the measured situation, to reduce the RH to below 65%. In all the other months of the year, it would be possible to reduce the heating, if the scenario 2 conservation heating regime was accepted. If needed, the degree days of a regime with different set points (than RH_{max} ~65% and T_{min} = 10 °C) could, of course, be calculated. About 1400 annual heating degree days would be needed in this conservation heating regime. About 1600 heating degree days, or ~50% of the energy costs, would then be saved, compared to the approximately similar 3000 annual degree days of both the comfort heating scenario and the measured situation. This saving should be added to the saving in terms of conservation costs obtained by an expected reduction in the deformation and deterioration rate of conserved painted wooden objects.

5. Discussion

The near-full conformity of the recorded climates in the churches and of the scenarios of changed climates to RH ~65% to the variable limits of the CEN15757 standard and ASHRAE B class indicated that the measured fluctuations in the churches may not cause new critical mechanical damage. As, however, one critical fluctuation event can be sufficient to cause non-reversible mechanical deformations [33], this 98% conformity may still indicate a risk. The very high recorded RH for Kinn resulted in low conformity to all standard classes that include a maximum RH limit, raising concerns, especially about possible mold and rot, little of which was, however, observed in the church (see Section 3.3). The comparison with the standards was made to a single year with the available data. The variation in the historical fluctuations has certainly been larger and the damage and deterioration that was noted in the conservation reports may have occurred in worse situations or for reasons that were related, for example, to the specific properties of original painting materials and methods or to past conservation treatments.

In situations with intermittent heating, it may be an issue for conservation if significantly more fluctuations would be recorded with a higher resolution in terms of the measurements than one hour, and a different evaluation would then be obtained by the guidelines. However, most materials do not respond to fluctuations immediately. It has been found that fluctuations of a duration of less than one hour do not affect most museum objects [49] (p. 626). In a museum with a controlled indoor environment (~22.5% < RH < ~42.5%), RH measurements along with hourly resolution were reported to give the same evaluation via guidelines (CEN

15757 and ASHRAE) for measurements with a higher time resolution than one hour [50]. Based on the situation in the churches, with heating aiming to keep a stable temperature for Ringsaker, and the observed nominal effect of the infrequent intermittent heating on the high RH in Kinn, a higher measurements resolution than was used in this study would probably not significantly affect the results. This possibility, however, cannot be excluded.

The HERIe modeling process represents well-defined situations that are explored experimentally in the laboratory. The modeling is more applicable to "worst case" scenarios [17] involving original undamaged gesso layers, in situations where a climate-induced mechanical damage mechanism is dominating. This highlights the fact that the cumulative deterioration process of an aged painted object is influenced by its condition and environmental history. Most objects with a gesso/wood interface that are subjected to fluctuating climates are likely to have developed multiple small cracks (known as craquelure). The cracks release the tension and reduce the chance of further deformation and cracking [16]. The development of cracking is a complex matter involving the different dimensions of the cracking patterns and paint layer(s) [51]. The damage mechanisms and risks to a painting with a fully developed craquelure pattern are very different from that of a recently painted object. Cracking is not expected to happen in the future if the climate (RH and T) fluctuations (periods, amplitudes, and frequency) are within those of the proofed historical climate [5]. The risks to original or old painted wooden objects in the churches would thus appear if there were climate variations in situ, or, in the case of moving objects to a different location, changed beyond the historical variations. There may have been periodic changes in the historical climates that affected the condition of the objects, due, for example, to the introduction of changing heating regimes. This could, however, not be determined from the historical records. It has also been stated that a "conservation treatment erases proofed fluctuations" [5], and thus irreversible changes could occur to a newly treated object located in the "proofed climate" due to processes that will, however, be more or less different than those for the original paint and gesso in the HERIe model [14].

There was, thus, considerable uncertainty in how well the modeling of the risk of damage represented the risk of the original conserved painted wooden objects in Kinn and Ringsaker, and, thus, the model's relevance. The mechanical damage risks indicated by the analysis from the HERIe modeling in situ in both churches, and in the scenario where RH values were over about 70% in Kinn, may be of less concern for the original/old paint and gesso or previous conservation treatments that have been acclimatized to the churches' environments. This could be interpreted as a reason for the apparent lower risk assessment by the CEN 15757 standard and ASHRAE B class than for the modeling analysis. It is not always possible to identify the exact causes of observed damage to aged cultural heritage objects that are subject to concomitant deterioration processes and need treatment. New cracking and flaking paint and ground layers after conservation treatments might relate not only to movement in the paint-gesso-wood interfaces due to climate fluctuations but also to such factors as the initial adherence of the paint and ground, the curing of the paint, and the influence of the surrounding atmosphere and environment. Due to these uncertainties, the HERIe modeling results are presented only as an indication of the risk to newly treated objects.

The high risks, which are probably more directly relevant, indicated whether moving a panel/object, from the high humidity recorded in Kinn (RH_{av} = 79%, T_{av} = 9 °C) and the conservation heating regime in Ringsaker (RH_{av} = 63%, T_{av} = 13 °C) to a (yet unrealistically stable) conservation studio (RH = 50%, T = 20 °C) and then back to the churches, could perhaps imply that the compression of the gesso and paint due to the drying and shrinking of the wood in the studio would produce additional cracking to that already present in the proofed climate, and that moving the item back to the church could worsen its condition. The modeling analysis indicated that a reduction to below 60–65% RH in Kinn would significantly reduce these risks, with a decreasing risk also for plastic deformations, along with the further lowering of the RH (toward 50–55%). The modelling indicated that to reduce the tension risk of cracking a reduction to below 60% of RH would only be necessary in the situation of moving painted panels back to Kinn from a studio in July. These modelling assessments do not evaluate the appropriateness of such changes or consider the effect of a lower RH on the building. Although, in the case of Ringsaker, it was found that the conservation heating regime (at a stable RH of ~63%) would reduce the risk of cracking and plastic deformation, the risk of moving a panel/object from the conservation heating regime to a conservation studio (and back to the church after one year) could be significant and higher than if moving the panel/object from the recorded climate in the church, due to an increased compression risk in the studio (Figure 9C). The high indicated risk when moving a panel/object back to Ringsaker seems of less concern for the original paint and gesso since the scenario RH and T values in the conservation studio were near to the average of the recorded and proofed climate in the church; this risk was indicated to be slightly less when moving back to the conservation heating scenario than to the recorded climate (Figure 9D). The risk of new damage may be higher, although uncertain, when moving an object that has been treated in a conservation studio back to either the recorded fluctuating climate or to a more stable conservation heating scenario.

Thus, the HERIe risk indications seem mainly to be the drying-out of the wood and the compression of the original paint and gesso in a studio, with the possible strains in objects newly treated in situ, and tensions put a strain on objects treated in a conservation studio when they are returned to the churches. Lastly, it could be noted as a general observation about heritage model applications that it can be challenging for non-developers (users) to understand all the conditions of modeling. The clear explanations on the HERIe web pages were helpful but, still, the HERIe model is based on the application of complex physics and time dependencies in the data, which can be difficult to fully understand.

6. Conclusions

The conformity to relevant climate guidelines (CEN 15757 and ASHRAE B) was improved in several scenarios of reduced indoor RH in the church at Kinn (from RH_{av} = 79%) on the west coast of Norway, and was upheld in a conservation heating scenario (RH_{av} = 63%) that was found to save about 50% of the heating cost in the drier Ringsaker church (RH_{av} = 49%), which was east of the southern Norwegian mountains. Analysis via HERIe (https://HERIE.pl/, accessed on 12 January 2023) modeling indicated a significant reduction in the risk of mechanical damage (the cracking and plastic deformation) of painted wooden panels with stiff gesso in Kinn, in the context of a scenario with a mean RH of ~70% and no risk below a mean RH of ~56%, and also with no risk in the scenario of stabilized conservation heating climate in Ringsaker (as would also be expected if stabilizing the humidity in the church around the recorded mean RH of 49%). A lowering of the RH in Kinn to below 60–65% would significantly reduce the risk of damage when moving panels out of the church to drier and warmer indoor climates and then back to the church. The risk of damage when moving panels from the scenario of conservation heating climate in Ringsaker was, however, found to increase, compared to the recorded climate. Although there were considerable uncertainties related to the representation in the modeling of the complex historical deterioration of objects in the churches, the damage mechanisms, and the environmental influences, the results offer clear indications that modification of the RH in the churches could have substantial benefits in reducing the risk of mechanical damage to the painted wooden objects. This reduction in risk would be most significant for those objects with original paint and gesso that might be moved from (and back to) the churches, for example, because of conservation treatment.

The analysis shows the usefulness not only of the HERIe modeling tool but also that its representation of old, painted, and often conserved, wood objects is not straightforward. HERIe provides important warnings about the worst-case critical risks to case panels with original paint and gesso, in the current case, which may, however, not represent the situation of many objects in situ. To predict the deterioration that commonly leads to the conservation of painted wooden heritage objects it seems that other descriptive methods, possibly more specific models of different paint deterioration mechanisms and, probably, more general models of environmental doses and the effects on object changes are also needed.

Author Contributions: Conceptualization, methodology, visualization, investigation, and writing—review and editing, T.G. and L.P.S.; formal analysis, data curation, and original draft preparation, T.G. All authors have read and agreed to the published version of the manuscript.

Funding: This research received no external funding.

Data Availability Statement: The data are available as figures in the manuscript. Underlying data files can be obtained from the corresponding author.

Acknowledgments: We want to thank Kjersti Marie Ellewsen (NIKU) and Tone Marie Olstad (formerly at NIKU) for providing the indoor climate data and information about the churches, objects, and their condition used in this work. We also wish to thank Łukasz Berger in the HERIe team for his responses to our questions about HERIe modeling conditions.

Conflicts of Interest: The authors declare no conflict of interest.

Appendix A

The CEN15757 standard recommends RH limits that are equal to the 93rd and 7th percentile of the measured fluctuations from the monthly moving average, but of ±10% RH from the monthly moving average if these percentile values are less than 10%. The ASHRAE guideline classes, AA, A, B, C, and D, are briefly explained as follows [5]: A1 (formerly A) allows for the seasonal relaxation of RH ±10% and T of up 5 °C and down 10°C, compared to the customary strict AA controls of ±5% RH and ±2 °C variation from the annual average and only a seasonal T adjustment of ±5 °C. A2 allows the double extent of short-term RH fluctuations, compared to A1 (10% vs. 5%), but there is then no seasonal adjustment, which should provide the same protection against mechanical damage as under A1. Conformity to the A classes signifies a "small risk of mechanical damage to high vulnerability artifacts, no mechanical risk to most artifacts, paintings, photographs, and books". The A (A1 and A2) relaxation of the limits was not meant to be seen as a larger change from the "no mechanical risk" situation of the AA class, similar to that of B–D, which are assigned to provide guidance for smaller museums or more vulnerable buildings. B addresses those situations where very low winter temperatures are preferable to very low RH. C addresses the need to control just the RH between 25% and 75%, to avoid the rapidly increasing risks at lower and higher values.

Table A1. Climate parameters and allowed fluctuations and limits in standards [31].

CEN Standard and ASHRAE Classes	Short Term (Hour, Day)		Seasonal		Outer Fixed Limits, Low–High	
	RH (%)	T (°C)	RH (%)	T (°C)	RH (%)	T (°C)
CEN 15757	7th and 93rd percentile of annual fluctuations from the monthly moving mean but maximum ± 10%	n.a.	n.a.	n.a.	n.a.	n.a.
AA	±5	±2	no	±5	35–65	10–25
A1	±5	±2	±10	−10–+5	35–65	10–25
A2	±10	±2	no	−10–+5	35–65	10–25
B	±10	±5	±=10	−20–+10	30–70	n.a.–30
C	n.a.	n.a.	n.a.	n.a.	25–75 *	n.a.–40
D	n.a.	n.a.	n.a.	n.a.	–75 *	n.a.

* Not continually above 65% RH for longer than number of days needed for visible mould growth, given from reported Table [31].

References

1. Olstad, T.M.; Stein, M. Saving art by saving energy. *Niku Temah.* **1996**, *2*, 1–20.
2. Michalsk, S. Agent of Deterioration: Incorrect Relative Humidity. 2021. Available online: https://www.canada.ca/en/conservation-institute/services/agents-deterioration/humidity.html (accessed on 5 November 2022).
3. Łukomski, M. Painted wood. What makes the paint crack? *J. Cult. Herit.* **2013**, *13S*, S90–S93. [CrossRef]
4. Bratasz, L. Allowable microclimatic variations for painted wood. *Stud. Conserv.* **2013**, *58*, 65–79. [CrossRef]
5. Michalski, S. The ideal climate, risk management, the ASHRAE chapter, proofed fluctuations, and towards a full risk analysis model. *Expert. Roundtable Sustain. Clim. Manag. Strateg.* **2007**, 1–19. Available online: https://citeseerx.ist.psu.edu/document?repid=rep1&type=pdf&doi=f00555145ab24e771bb117f43e37d84f9a325992 (accessed on 14 March 2023).
6. Sesana, E.; Bertolin, C.; Gagnon, A.S.; Hughes, J.J. Mitigating Climate Change in the Cultural Built Heritage Sector. *Climate* **2019**, *7*, 90. [CrossRef]
7. Hanssen-Bauer, I.; Førland, E.J.; Haddeland, I.; Hisdal, H.; Mayer, S.; Nesje, A.; Nilsen, J.E.Ø.; Sandven, S.; Sandø, A.B.; Sorteberg, A.; et al. (Eds.) Climate in Norway 2100—A knowledge base for climate adaptation. *NCCS Rep.* **2017**, *1*, 2017.
8. Lisø, K.R.; Hygen, H.O.; Kvande, T.; Thue, J.V. Decay potential in wood structures using climate data. *Build. Res. Inf.* **2006**, *34*, 546–551. [CrossRef]
9. Grøntoft, T. Observed Recent Change in Climate and Potential for Decay of Norwegian Wood Structures. *Climate* **2019**, *7*, 33. [CrossRef]
10. Sabbioni, C.; Brimblecombe, P.; Cassar, M. (Eds.) *The Atlas of Climate Change Impact on European Cultural Heritage: Scientific Analysis and Management Strategies, Anthem Environmental Studies*; Anthem Press: London, UK, 2010.
11. Ellewsen, K.M. Rapport Klimadata Kinn Kirke September 2011–September 2012. Internal report—The Norwegian Directorate for Cultural Heritage, Oslo, Norway, 2014. Non-Published Report. Available on request from The Norwegian Directorate for Cultural Heritage. Available online: https://www.riksantikvaren.no/en/ (accessed on 14 March 2023). (In Norwegian).
12. Olstad, T.M.; Jernæs, N.K.; Mengshoel, K.; Smestad, T.R.; Vestvik, I.; Spaarschuh, C.; Wedvik, B.; Kaun, S. Ringsaker kirkes gamle herlighet. Undersøkelser og dokumentasjon av alterskapet i A97 Ringsaker kirke, Innlandet. NIKU commissioned report 5/2020. Norwegian Institute for Cultural Heritage Research, Oslo, Norway, 2020. Available on request from The Norwegian Institute for Cultural Heritage Research. Available online: https://www.niku.no/en/ (accessed on 14 March 2023). (In Norwegian).
13. Olstad, T.M.; Berg, D.; Smestad, T.R.; Spaarschuh, C.; Vestvik, I.; Wedvik, B. Ringsaker Kirkes Alterskap. Konserveringsprosjekt 2020. NIKU Oppdragsrapport 3/2021. 2021. Available online: https://ra.brage.unit.no/ra-xmlui/handle/11250/2753430 (accessed on 14 March 2023). (In Norwegian).
14. Solstad, J.; Kirke, K. Behandling av Altertavle, Epitafium og Helgenfigurer NIKU Rapport Kunst og Inventar nr 12/2006. 2006. (In Norwegian). Available online: https://ra.brage.unit.no/ra-xmlui/bitstream/handle/11250/176684/Kirkekunst_Kinn_NIKURapportKunst_12_2006.pdf?sequence=1 (accessed on 14 March 2023).
15. Smestad, T.R. Tilstandsundersøkelse og metodeutvikling, altertavle og epitafiet i Kinn kirke. NIKU Commissioned Report 149/2020. Norwegian Institute for Cultural Heritage Research, Oslo, Norway, 2020. Available on request from The Norwegian Institute for Cultural Heritage Research. Available online: https://www.niku.no/en/ (accessed on 14 March 2023). (In Norwegian).
16. HERIE2.0. Quantitative Assessment of Risks to Heritage Assets. 2020. Available online: https://HERIE.pl/ (accessed on 14 March 2023).
17. Taylor, J.; Łukomski, M.; Bratasz, Ł. Increasing evidence-based decision-making for loan agreements. In *Transcending Boundaries: Integrated Approaches to Conservation, Proceedings of the ICOM-CC 19th Triennial Conference Preprints, Beijing, China, 17–21 May 2021*; Bridgland, J., Ed.; International Council of Museums: Paris, France, 2021. Available online: https://www.icom-cc-publications-online.org/4355/Increasing-evidence-based-decision-making-for-loan-agreements (accessed on 8 February 2023).
18. Bratasz, L.; Akoglu, K.G.; Kékicheff, P. Fracture saturation in paintings makes them less vulnerable to environmental variations in museums. *Herit. Sci.* **2020**, *8*, 11. [CrossRef]
19. *EN 16883: 2017*; Conservation of Cultural Heritage—Guidelines for Improving the Energy Performance of Historic Buildings. European Committee for Standardization: Brussels, Belgium, 2017.
20. Richards, J.; Brimblecombe, P. The transfer of heritage modelling from research to practice. *Herit. Sci.* **2022**, *10*, 17. [CrossRef]
21. Wikipedia. 2022. Available online: https://en.wikipedia.org/wiki/Climate_of_Norway (accessed on 29 August 2022).
22. Kottek, M.; Grieser, J.; Beck, C.; Rudolf, B.; Rubel, F. World Map of the Köppen-Geiger climate classification updated. *Meteorol. Z.* **2006**, *15*, 259–263. [CrossRef] [PubMed]
23. Daly, A.; Olstad, T.M. Fragile fragments—A new provenance for the late medieval triptych in Kinn Church, Norway. *MoK Medd. Om Konserv.* **2022**, *1*, 49–67. Available online: https://niku.brage.unit.no/niku-xmlui/handle/11250/3048386 (accessed on 14 March 2023).
24. Olstad, T.M.; Haugen, A. Kirker og oppvarming—Hva skjer? *MoK Medd. Om Konserv.* **2012**, *1*, 21–29. Available online: https://mok.scholasticahq.com/api/v1/articles/36366-kirker-og-oppvarming-hva-skjer.pdf (accessed on 14 March 2023). (In Norwegian).
25. Hoem, S.; Marthinsen, E.; Haugen, A.M. Kinn Kyrkje, Flora Kommune, Tilstandsanalyse NS-3424 Nivå 2. NS-EN 16096. Forsvarsbygg. 2016. Available online: https://www.mercell.com/m/file/GetFile.ashx?id=81680337&version=0 (accessed on 14 March 2023). (In Norwegian).

26. Hoem, S.; Marthinsen, E.; Haugen, A.M. Faktaark Tilstandsanalyse Kinn Kyrkje. *Forsvarsbygg* **2018**. (In Norwegian). Available online: https://www.forsvarsbygg.no/no/radgivingstjenester/vern-av-kulturminner/referanseprosjekter/tilstandsanalyser/tilstandsanalyse-kinn-kyrkje-i-flora-kommune/ (accessed on 14 March 2023).
27. Olstad, T.M. *Alterskapet i Ringsaker Kirke—Et Uendret Klenodium? Ringsaker Kike–Landets Sognekirke*; Hauglid, K., Stige, M., Bø, R.M., Eds.; Instituttet for Sammenlignende Kulturforskning. Ringsaker Kirkes Venner. Novus forlag: Oslo, Norway, 2021; pp. 257–281. (In Norwegian)
28. Ellewsen, K.M. Notat—Klimamålinger i Kinn Kirke 2011−2012. Saksnummer: 07/02104. Internal Short Report ("Note")—The Norwegian Directorate for Cultural Heritage. Oslo, Norway, 2014. Non-Published Report. Available on Request from The Norwegian Directorate for Cultural Heritage. (In Norwegian). Available online: https://www.riksantikvaren.no/en/ (accessed on 14 March 2023).
29. Norsk Klimaservicesenter. 2021. Available online: https://seklima.met.no/ (accessed on 14 March 2023).
30. *CEN/TC 346: EN 15757*; Conservation of Cultural Property—Specifications for Temperature and Relative Humidity to Limit Climate-Induced Mechanical Damaging Organic Hygroscopic Materials. European Committee for Standardization: Brussels, Belgium, 2010.
31. ASHRAE. Museums, Galleries, Archives and Libraries. In *ASHRAE Handbook: Heating, Ventilating, and Air-Conditioning Applic., SI Edition*; American Society of Heating, Refrigerating and Air-Conditioning Engineers, Inc.: Peachtree Corners, GA, USA, 2019; pp. 1–46.
32. Martens, M.H.J. Climate Risk Assessment in Museums: Degradation Risks Determined from Temperature and Relative Humidity Data. Ph.D. Thesis, Eindhoven University of Technology, North Brabant, The Netherlands, 2012. Available online: https://pure.tue.nl/ws/files/3542048/729797.pdf (accessed on 14 March 2023).
33. Huijbregts, Z.; Kramer, R.P.; Martens, M.H.J.; van Schijndel, A.W.M.; Schellen, H.L. A proposed method to assess the damage risk of future climate change to museum objects in historic buildings. *Build. Environ.* **2012**, *55*, 43–56. [CrossRef]
34. Broström, T.; Vyhlídal, T.; Simeunovic, G.; Larsen, P.K.; Zítek, P. Evaluation of different approaches of microclimate control in cultural heritage buildings. *Clim. Collect. Stand. Uncertain* **2013**, *7*, 105–106. Available online: https://www.climateforculture.eu/index.php?inhalt=download&file=pages/user/downloads/publications/Climate_for_Collections.pdf (accessed on 14 March 2023).
35. Blades, N.; Rice, K. Conservation heating and energy efficiency at the national trust: Theory and practice. In *Developments in Climate Control of Historic Buildings*; Kilian, R., Vyhlídal, T., Broström, T., Eds.; Linderhof Palace: Stuttgart, Germany, 2010. Available online: https://www.climateforculture.eu/index.php?inhalt=download&file=pages/user/downloads/publications/2010_DevelopmentsClimateControl.pdf (accessed on 14 March 2023).
36. Blades, N.; Poupard, S.; Barber, L. Analysing the energy consumption of conservation heating systems at the National Trust. *J. Inst. Conserv.* **2011**, *34*, 16–27. [CrossRef]
37. *EN 15759-1:2011*; Conservation of Cultural Property—Indoor Climate—Part 1: Guidelines for Heating Churches, Chapels, and Other Places of Worship. European Committee for Standardization: Brussels, Belgium, 2011.
38. Schellen, H.L.; Neuhaus, E. Conservation heating in a historical building: Results from an experimental and simulation study. In *Developments in Climate Control of Historic Buildings*; Kilian, R., Vyhlídal, T., Broström, T., Eds.; Fraunhofer IRB Verlag: Stuttgart, Germany, 2011.
39. Pöyry Management Consulting AS. Evaluering av Modeller for Klimajustering av Energibruk. Norges Vassdrags-og Energidirektorat, Report no. 7 2014. (In Norwegian). Available online: https://publikasjoner.nve.no/rapport/2014/rapport2014_07.pdf (accessed on 14 March 2023).
40. Grøntoft, T.; Svenningsen, G. Windborne Sea Salt Aerosol Fluxes and Deposition Inland from Ocean Shorelines—Measurements and Modelling of Climate Change Effects. In *Climate Change and Cultural Heritage, Proceedings of the Ravello International Workshop, 14–16 May 2009, and Strasbourg European Master-Doctorate Course, 7–11 September 2009, Ravello, Italy and Strasbourg, France*; Lefèvre, R.-A., Sabbioni, C., Eds.; Edipuglia: Bari, Italy, 2010.
41. Grøntoft, T. Beregning av Korrosjonsklasse fra Miljø-Parametere i Fitjar Lokasjon (59°56′11.5″N 5°19′58.4″Ø). NILU Rapport 12/2022. 2022. Available online: https://nilu.brage.unit.no/nilu-xmlui/handle/11250/2996943 (accessed on 14 March 2023). (In Norwegian).
42. Higashijima, K.; Hori, C.; Igarashi, K.; Enomae, T.; Isogai, A. First aid for flood-damaged paper using saltwater: The inhibiting effect of saltwater on mold growth. *Stud. Conserv.* **2012**, *57*, 164–171. [CrossRef]
43. Kirker, G.; Glaeser, J. Salt damage to wood "fuzzy wood" often confused with fungal decay. *Pile Driv.* **2011**, *Q3*, 85–86.
44. Blanchette, R.; Held, B.; Farrell, R. Defibration of wood in the expedition huts of Antarctica: An unusual deterioration process occurring in the polar environment. *Polar Rec.* **2002**, *38*, 313–322. [CrossRef]
45. Klüppel, A.; Mai, C. Effect of seawater wetting on the weathering of wood. *Eur. J. Wood Wood Prod.* **2018**, *76*, 1029–1035. [CrossRef]
46. Sedlbauer, K. *Prediction of Mould Fungus Formation on the Surface of and Inside Building Components*; Fraunhofer Institute for Building Physics: Stuttgart, Germany, 2001.
47. Camuffo, D. Microclimate for cultural heritage. In *Measurement, Risk Assessment, Conservation, Restoration, and Maintenance of Indoor and Outdoor Monuments*, 3rd ed.; Elsevier: Amsterdam, The Netherlands; Oxford, UK; Cambridge, MA, USA, 2019.
48. Camuffo, D.; Della Valle, A.; Becherini, F. The European Standard EN 15757 Concerning Specifications for Relative Humidity: Suggested Improvements for Its Revision. *Atmosphere* **2022**, *13*, 1344. [CrossRef]

49. Michalski, S. Relative humidity: A discussion of correct/incorrect values. In Proceedings of the ICOM-CC 10th Triennial Meeting Preprints, Washington, DC, USA, 22–27 August 1993; Bridgland, J., Ed.; International Council of Museums: Paris, France, 1993; pp. 624–629.
50. García-Diego, F.-J.; Verticchio, E.; Beltrán, P.; Siani, A.M. Assessment of the Minimum Sampling Frequency to Avoid Measurement Redundancy in Microclimate Field Surveys in Museum Buildings. *Sensors* **2016**, *16*, 1291. [CrossRef] [PubMed]
51. Jamalabadi, M.Y.A.; Zabari, N.; Bratasz, L. Three-dimensional numerical and experimental study of fracture saturation in panel paintings. *Wood Sci. Technol.* **2021**, *55*, 1555–1576. [CrossRef]

heritage

Article

Quantifying Housekeeping Challenge and Conservation Need

Helen Lloyd [1,2]

1 Trusted Conservators, London SW11 5SB, UK; helen.lloydtc@gmail.com
2 Formerly National Trust, Heelis, Kemble Drive, Swindon SN2 2NA, UK

Abstract: This simple model, developed by conservators, assists in the challenge of making preventive conservation, housekeeping, and care of historic interiors and collections, physically and economically sustainable, in historic houses welcoming increasing volumes of visitors (a primary source of dust). It introduces objectivity into conservation advice and management decisions: how many collections care staff should each historic house ideally employ, and how large an annual budget is required to fund the non-wage costs of routine preventive and interventive conservation? Are staffing structures rational and consistent, and tailored to the individual and developing circumstances of multiple properties? Eight qualitative and quantitative criteria are each given a score from 1–4 in relation to their data ranges. The total scores for each property are converted to percentages, correlated with staffing structures, and used to estimate the requirements for daily, weekly and annual housekeeping and conservation cleaning. Selected data are used to measure housekeeping performance against weekly targets, and to rationalize the distribution of financial resources for preventive conservation and maintenance. The model can be adapted for use in any museum or heritage building which needs to assess and quantify the routine care of interiors and collections on open display to visitors.

Keywords: preventive conservation; housekeeping; cleaning routines; staffing; maintenance needs; resource distribution; sustainability

Citation: Lloyd, H. Quantifying Housekeeping Challenge and Conservation Need. *Heritage* **2023**, 6, 3757–3776. https://doi.org/10.3390/heritage6040199

Academic Editor: Fernanda Prestileo

Received: 27 March 2023
Accepted: 9 April 2023
Published: 19 April 2023

1. Introduction

This novel, simple but effective model, using only an Excel spreadsheet, was developed by conservators working with environmental scientists, for use in historic buildings in the UK, as a means of promoting preventive conservation and housekeeping standards, and ensuring that the care of interiors and collections remains both physically and economically sustainable. It was initially devised in response to the challenge of determining appropriate staffing levels in properties which were welcoming ever-increasing numbers of visitors, and where staff needed sufficient time, skills and resources to care for historic interiors and collections. The model evolved over several years, becoming more sophisticated with time and external advice, and is easily adapted for use in other heritage buildings, museums and galleries with collections on open display.

The practice of preventive conservation was introduced to historic houses in the late 1970s [1]. Based on traditional housekeeping practices documented since the eighteenth century, these were updated with conservation science used in national museums and galleries. In the 1990s, the preventive conservation framework devised by the Canadian Conservation Institute [2,3] was adopted, and further developments in collective knowledge, skills and experience led to the publication of a much-expanded 'National Trust Manual of Housekeeping—care and conservation of collections in historic houses' [4].

In 2000, a research partnership was formed between English Heritage, Historic Royal Palaces, the National Trust and the University of East Anglia Environmental Sciences, initially funded by the Leverhulme Trust, to investigate 'Controls on Irreversible Soiling' in historic buildings open to visitors [5,6]; this work built on earlier studies of coarse particulate soiling in museums [7–9]. The research studied the effect of visitors on dust in historic collections [10], and the consequent impacts of dust on the care and management of collections [11], dust in historic libraries [12], staff and visitor perceptions of dust [13–17], the economics of dust [18], and causes of cementation [19,20]. The work concluded with guidance on low-technology methods of monitoring dust deposition [21] and a dust atlas enabling the identification of coarse dust particles [22].

Research outcomes provided scientific evidence to support housekeeping policies and practices, adding authority to the associated advice and guidance. Conservators are now better equipped to make rational assessments of the cleaning time for historic interiors of all shapes and sizes, ranging from small closets to long galleries; these spaces display collections of diverse fragility and significance, from robust brown furniture to decaying silk curtains, and from unique tapestries and state beds to assorted ephemera.

Heritage organisations have always been aware of the need to balance access and preservation. For example, John Bailey, chairman of the National Trust in 1923-31, said: "Preservation may always permit of access, while without preservation access becomes for ever impossible" [23]. In 2005, its responsibilities for care of land, buildings and nature were redefined: "Conservation is the careful management of change; it is about revealing and sharing the significance of places and ensuring that their special qualities are protected, enhanced, enjoyed and understood by present and future generations" [24]. This twenty-first century definition enshrines the Trust's dual purposes (of preservation and public benefit), supports the ongoing need to achieve a sustainable balance between preservation and access, and emphasises the positive contribution which conservation makes to enabling access.

However, since the 1970s, the volume of visitors desiring access to heritage sites, both outdoors and indoors, has grown steadily (Figure 1). Over 40 years, access to historic interiors has increased dramatically—not only in the numbers of people, but also in the hours of opening—thus reducing the time available for collections care, and exposing furnishings to more hours of daylight, and greater rates of dust deposition [25].

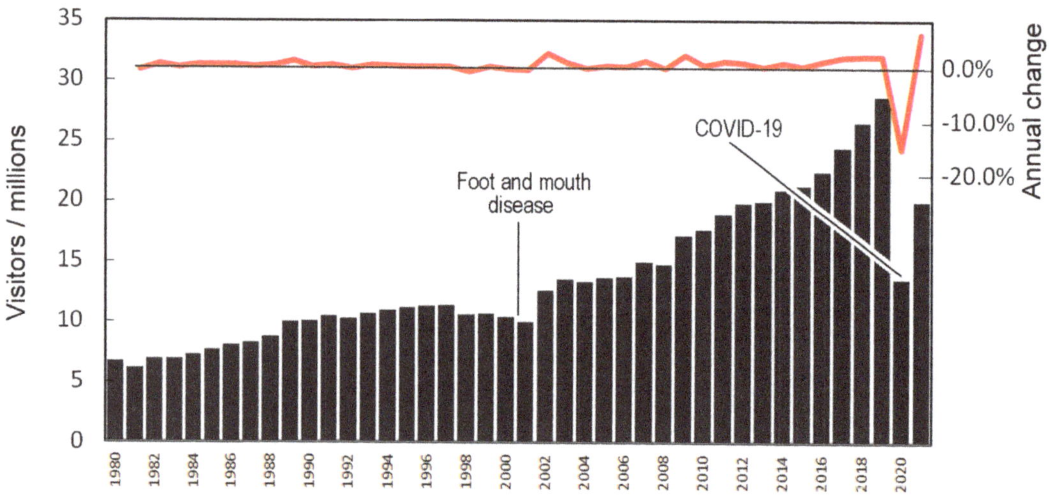

Figure 1. Annual visitors to pay-for-entry National Trust properties, houses and gardens, 1980-2022, with annual increases rising more rapidly since 2001 (lower numbers in 2000 and 2020-21 when properties were closed, due to Foot and Mouth Disease and COVID-19). The red line shows the annual growth rate, with the fine black line as zero.

For example, from 1980–2020, National Trust properties as a whole experienced a five-fold increase in visitor numbers. Houses, at first open to the public for six months in spring and summer, for five days a week, four hours a day, and closed for six months in autumn and winter, could expect to welcome 20,000–50,000 visitors each year. In the twenty-first century, many houses are open for 10–12 months a year, from February to November and over Christmas/New Year, for 6–7 days a week and 6–7 h a day; they now anticipate between 50,000 and 150,000 annual visitors.

This increase prompted the senior staff responsible for gardens, houses and visitor services to collaborate with property managers to devise a method of assessing the impact of public access on the management of each property. Together they created processes by which the impacts of changing access and the accompanying need for conservation resources could be measured objectively, thus supporting a sustainable balance between staffing, visitors, conservation and costs. Since 2005 a 'Conservation for Access' toolkit has been used by properties which plan to change their hours, days and months of access [26,27]. In essence, the toolkit prompts property managers and regional consultants to assess the current levels of staffing, and hours committed to housekeeping and/or gardening, and promotes the cost-benefit of increasing these resources in response to higher visitor numbers.

Worksheets have been used to compare the current numbers of visitors and hours of access with the capacity of each space to accommodate more; additionally, to measure the peaks and troughs of visitor entry flow and assess the effect of overcrowding on visitors' experience and enjoyment, as well as the physical impact on the building/landscape and fragile materials/plants. In interiors, the toolkit assesses the condition and fragility of the collections, and the sensitivity of materials to light exposure, together with the costs of protection measures, and of maintaining the standards of housekeeping and preventive conservation. And last, but not least, the toolkit compares the costs of changing access with the potential for increased revenue, and estimates the budgetary requirements to make conservation sustainable. The flowchart decision tree enables property staff to determine which parts of the 'Conservation for Access' toolkit it will be most useful for them to complete [28,29] (Figure 2).

It is relatively straightforward to count the number of visitors who can be accommodated in each space at one time, without risk of physical damage to furnishings or detriment to their experience; then an entry flow rate can be calculated, which ensures this capacity is not exceeded [29]. It is also possible to calculate the annual hours of light exposure, using blue wool dosimeters and data loggers [30].

However, it is less easy to assess how many housekeeping hours are required daily to remove the dust deposited on the furnishings by yesterday's visitors. The rate of deposition relates directly to the number of people and their proximity to the collections, but the frequency of cleaning varies with the fragility of materials and their proximity to visitors [10].

A simple model was needed, which could account for the significance and fragility of each house, the diversity of its collections and complexity of their needs for care, the size of its inventory and library, the number of spaces accessible to visitors, the total annual visitors and hours of access, and weekly hours needed to clean its spaces and collections on open display. Ideally, the model would also help to estimate the non-wage costs of preventive conservation in historic properties; this calculation would set realistic targets for annual budgets at each place, and support a proportionate distribution of available resources between properties.

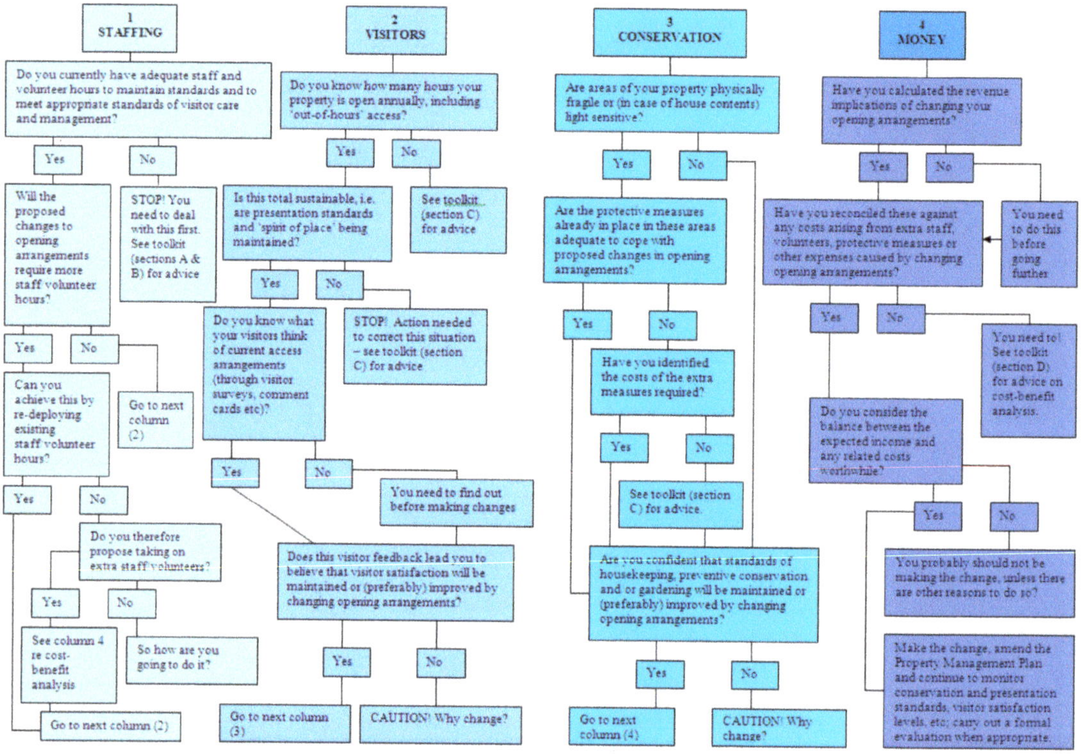

Figure 2. The Dimensions model evolved to support the 'Conservation for Access' toolkit, a multidisciplinary collaborative process designed between 2000–2005 to determine a balance between the amount of physical access each property could tolerate, and the resources necessary for conservation to sustain that level of access. This decision tree helps to identify where proposed changes to access may require additional resources. Reprinted with permission from Ref. [4]. 2023, National Trust.

Over two decades, an operations model evolved, the quality and quantity of data improved, calculations became more sophisticated, and further uses were developed in response to conservation management challenges. This article describes its uses at many stages of evolution from simple to more complex; the data is, therefore, occasionally inconsistent between the tables, which illustrate calculations made at different times for a variety of purposes and properties.

The model sets out to quantify the 'dimensions' of the housekeeping challenge at each property, and is thus known as the Dimensions model. Its purpose is to introduce both facts and fairness into conservation recommendations and management decisions: how many collections care staff should each historic house employ, and how large an annual budget is required to fund non-wage costs of preventive and interventive in situ conservation? How could central advice on staffing structures be made consistent and rational, yet tailored to the individual and developing circumstances of each property?

The Dimensions model has also been used to monitor and rate conservation performance, for example, how well house teams are meeting their target of cleaning hours per week, as part of a wider property conservation performance review [31]. A further use has been to estimate the scale of budget needed by each property for preventive conservation equipment and materials, and to allocate fairly between properties the resources for routine maintenance and cyclical in situ treatments by freelance conservators [32].

2. Method

Every historic property is unique, and offers a different amount of access, to a variable number of interiors with more or less dense displays of collections, some fragile and others robust. The aim was simply to find a way to understand, quantify and rationalise the factors which influence resource needs for the diverse interiors and collections found in a national collection of 360 historic properties, of which 160 contain libraries.

The model evolved as an iterative process over two decades, and was tested at every stage by national and regional conservators, validating the outcomes against their practical experience of working with staff in a wide range of historic houses. Initially, the method was largely intuitive and experimental, as there was no existing internal or external benchmark against which the process could be measured, nor another model to follow.

Eight criteria for measurement were selected:

- four qualitative: significance, fragility, complexity of interiors and collections, and average of the significance and size of library;
- four quantitative: size of inventory, number of historic spaces, annual visitors, and the hours of access.

Initially, the ranges for each criterium were estimated, and each property given rough scores of 1 to 4, where 1 was low and 4 was high. The 1–4 scoring was chosen to encourage decisiveness in allocating qualitative scores, and to prevent overloading of the central value. A larger number would have made choices difficult, and an uneven number of, say 1–5, could have produced too many scores of 3, on the cusp between high and low.

Qualitative assessments were made by comparing individual properties with their neighbours; quantitative data for each criterion was obtainable from individual properties, but not yet centrally collated. Regional conservators obtained the data for properties in each region, and their qualitative scores were moderated by a national conservator (familiar with all the properties), to ensure equity within and between the diverse regional portfolios of properties.

2.1. Criteria and Simple Methods of Scoring

- Significance:

  At the outset, there was no external or internal process for assessing significance, such as the methodologies subsequently published [33,34]. A national curator collaborated with a national conservator to define a simple ranking from 1 (low/local significance) to 4 (high national/international significance) (Table 1). Using their respective professional knowledge of individual property histories and characteristics, they agreed on a score for every property listed in the members' Handbook [35]. Today, significance is more formally evaluated by regional and national curators, and individual scores are adjusted from time to time, in response to the development of house interiors and collections.

- Fragility:

  In the absence of a formal method of assessing fragility, regional conservators ranked each property and collection in their portfolio from 1 to 4, where 1 is robust (no structural weakness, displays brown furniture and replica textiles), and 4 is most fragile (interiors, furnishings, textiles and collections highly vulnerable to damage from key agents of deterioration (e.g., physical forces, vibration and abrasion, mishandling and cleaning, exposure to light, and incorrect relative humidity)) [2,3] (for definitions see Table 1).

- Complexity:

  Regional conservators ranked each collection from 1–4, according to the diversity of its furnishings; 1 indicates basic materials (brown furniture, and modern textiles) needing only basic care and cleaning, whereas 4 indicates the broadest range of materials (including historic textiles, decorative surfaces, natural history specimens, photographs and early plastics), thus requiring routine care by house teams with greater knowledge, skills and experience in preventive conservation. However, complexity can evolve, as new collections are acquired and other items are returned to lenders or deaccessioned. (For definitions, see Table 1).

Table 1. Revised ranges, showing how the scores were adjusted to represent the full extent of the data initially collected.

Criteria/Scores	4	3	2	1	Data Source
Significance	Highest/ International	High/ National	Medium/ Regional	Low/ Local	Consult regional/national curators and conservators
Fragility	Most Fragile	Fragile	Moderately Fragile	Least Fragile	
Complexity	All Materials	Broad Range	Narrow Range	Basic Materials	
Inventory size	15 – >20 k	10–15 k	5–10 k	<5 k	Incl. loans; excl. books, specialist collections
Library significance	Highest/ International	High/ National	Medium/ Regional	Low/ Local	Consult Libraries Curator
Library size	9 – >12 k	6–9 k	3–6 k	<3 k	Unknown volumes = titles + 25%
Historic spaces in daily care	36 – >48	24–35	12–23	<12	Cleaning units, incl. outdoor collections (excl. stores)
Annual visitors (000 s)	120 – >160 k	80–120 k	40–80 k	<40 k	Visitors to house only
Annual open hours	1500 – >2000	1000–1500	500–1000	<500	Incl. events held on closed days

- Size of inventory:

  Pre-digitisation, the number of items in each collection was roughly assessed by the total quantity of index cards, overlooking pairs or sets of items represented by a single card. The scores were allocated to four ranges: 1 = less than 1000, 2 = 1000–3000, 3 = 3000–5000, 4 = 5000+. Post-digitisation and population of a national Collections Management System (CMS), more accurate totals gradually became apparent, requiring an adjustment to the ranges (up to 50,000). Today, 1 unit = 12,000 items (log score).

- Significance and size of library (averaged):

  In many libraries the cleaning of books is undertaken by trained volunteers; therefore, this is not a routine responsibility of house teams. To ensure that library scores are not disproportionate to regular staff responsibilities for cleaning interiors and other collections, the model uses an average of the scores for significance and size. The significance of each library had been assessed by the national libraries curator, using scores from 1–5; as only a handful of 160 libraries merited the top score, scores of 5 were reduced to 4*. In each library, the number of titles was known but one title could have multiple volumes; therefore, 'total titles' under-represented the labour of cleaning individual books. Where the quantity of volumes was unknown, it could be estimated by increasing the number of titles by 25% (with the average uplift calculated by a libraries conservator). Post-digitisation of records, the range of library sizes (originally assessed as 1–5000) was adjusted to accommodate libraries of 16,000 volumes. Today, 1 unit = 4000 volumes (log score).

- Historic spaces:

  Initially, the counting of rooms/spaces on the visitor route included every lobby, corridor, passage, flight of stairs, landing and closet—irrespective of the individual size or density of furnishings. This rough assessment is still valid, but has been refined to reflect the disparity in room sizes and cleaning requirements; spaces are now defined as 'cleaning units', as discussed in 2.4 below. In the model, 1 = 12 rooms/spaces or cleaning units, and scores are calculated to two decimal points.

- Visitor numbers:

  Annual visitors to a house can be difficult to quantify; each place counts admissions to the property as a whole (including garden, park and other attractions within the pay barrier). Visitors to the house are rarely recorded, but house staff may provide rough estimates, for example 75% of total property admissions. At the outset, visitor numbers ranged from 1–90,000 but the growing popularity of historic house visits demanded an adjustment. Today, 1 = 40,000 visitors (with the score calculated to two decimal points, e.g., 70,000 visitors = 1.75).

- Hours of access:

  Annual hours when a house is open to visitors can be relatively simple to calculate where a property is open 4–6 h per day, 5 days per week, 6–7 months per year; however, when the hours and days vary with the seasons, and where access in winter is restricted to discrete parts of each building, calculations become more complex (a model for simplifying the calculation would be useful). At the outset, the scores spanned a range of 1–1200 h, but were subsequently adjusted to accommodate public access up to 2000 h per annum. In the model, 1 = 500 h (log score).

After adding the scores for each property, and converting the total to a percentage of the maximum possible score, the properties can then be ranked in descending order of percentage score, revealing the diversity of the housekeeping challenge presented by each property, from the largest to smallest.

2.2. Fine-Tuning the Ranges and Calculations

The initial simple scoring system was useful, and offers a good starting point when adapting the model for use in other circumstances (e.g., museums, galleries, churches, private historic houses and other heritage buildings). However, initial results demonstrated that many quantitative criteria were under-represented; to achieve results more evenly distributed across a nationwide portfolio of collections, the ranges needed adjustment. With the full scope of quantitative data now evident, it was decided to abandon simple 1–4 scores, in favour of accurate calculations to two decimal points, as a proportion of the maximum for each range. Table 1 tabulates the revised ranges.

2.3. Fine-Tuning the Structure of the Spreadsheet (Quality and Quantity)

The Excel spreadsheet is now subdivided into several sections, with the first two summarizing quality and quantity. Some quantitative factors, such as inventory and library sizes, and the annual hours of access, have extensive ranges by comparison with the 1–4 scale adopted for qualitative and other factors. A logarithmic calculation now represents these larger ranges, using scores from 0–4, comparable with the other factors.

2.4. Fine-Tuning Historic Spaces as Cleaning Units

The sizes of rooms in historic houses vary widely, from an intimate closet to a long gallery. The density and fragility of each room's furnishings influence the extent of daily cleaning required, in response to the number of visitors and their proximity to the collections, both factors which influence the rate of coarse dust deposition [10–12].

Conservators generally recommend that daily cleaning in historic interiors is restricted to vacuuming of every robust floor surface on which visitors have walked, and removing dust particles from robust horizontal surfaces within sight of the visitor route. Weekly cleaning routines tackle 2–3 rooms each week in rotation, cleaning vertical surfaces up to head height, and horizontal surfaces further from the visitor route, wherever particles are visibly accumulating and undermining standards of presentation [36,37].

Annual 'spring' or 'deep' cleaning includes inspection, monitoring and, where necessary, cleaning of all surfaces—from ceilings and cornices via the walls and windows to the floor—also paying attention to surfaces usually concealed behind paintings, wall hangings and furniture, and beneath carpets. The purpose is to remove accumulations of dust and reduce the risk of particles becoming cemented to surfaces and thus requiring more interventive—and expensive—removal by specialist conservators using more sophisticated skills [19,20,32]. Annual deep cleaning also includes detailed cleaning (e.g., inside, outside, behind and beneath) of individual portable items of furniture, upholstery, ceramics, metalwork and decorative objects; this latter work may be undertaken at any time of year, often in front of visitors, to engage their interest in traditional housekeeping methods, and in the benefits of using preventive conservation techniques and materials to improve collections care.

Figure 3a–c illustrates how resources for cleaning are allocated in proportion to the distribution of furnishings and extent of public access, in Mr Bankes' Bedroom at Kingston Lacy in Dorset:

Figure 3. Generic cleaning routines: (**a**) visitor access is restricted to the drugget (green area), using rope and stanchion barriers (not illustrated) along the inner edge. Daily vacuuming focusses on floors accessed by visitors, and removing dust from robust polished surfaces close to the visitor route (green); (**b**) weekly/monthly cleaning focusses on horizontal and vertical surfaces up to head height (pink area), and within sight of the visitor route; (**c**) annual inspection, monitoring and deep cleaning is carried out from ceiling to floor (blue area), including all surfaces inaccessible while rooms are open to visitors. Detailed cleaning of individual portable items is continued throughout the year, for visitors to see. © National Trust Images/James Mortimer.

Initially, a 'rule of thumb' was used to assess the cleaning time needed for each historic space. Experience of cleaning heavily-visited houses over three decades (prior to the COVID SARS-2 pandemic in 2020) suggested that, on average, 20 min/day are required to clean those parts of each room accessed daily by visitors; an equal amount of time is needed weekly on closed days to clean areas not accessed by visitors, but where coarse dust will accumulate and become visible, if not occasionally removed. These 20 min are counted 'from door to door', in other words from the moment a room is entered (including opening shutters and blinds and plugging in vacuum cleaners) to the moment when a room is vacated (cleaning completed, equipment and tools collected, shutters and blinds closed). For the 'rule of thumb' to work, every accessible space must be counted, irrespective of size and density of furnishings—including every porch, lobby, closet, landing, flight of stairs, corridor and passage along the visitor route [37].

It is important to emphasise here that the average of 20 min cleaning per space is indicative, not an absolute; the model can be adjusted to suit different circumstances, for example where more or less access—temporal and/or spatial—is offered to a building, its interiors, and its collections. Pictures hanging in galleries, or collections protected within glazed cabinets in museums, might require a different allocation of time per space;

similarly in cathedrals and churches where access for worship and tourism follows a variable weekly pattern; and in private houses where the needs of the residents parallel those of the collections and the visitors. The number of visitors per day will also influence the rate of dust deposition and frequency of need for cleaning [10]. In every case, it is helpful for the house team to undertake a time-and-motion study, testing the allocation of time per space to verify that it has not been over- or under-estimated.

Following the research into 'Controls on irreversible soiling' [5,6,10–22], and a greater understanding of the sources and distribution of coarse dust in response to the number of visitors and their proximity to collections, the cleaning schedules were refined. Cleaning resources were allocated more precisely, in proportion to the size of each room, the area accessible to visitors, and the fragility and density of its furnishings and collections. Staffing resources for routine housekeeping are now counted as 'cleaning units', with each unit requiring just 20 min of one person's input per day.

An Excel worksheet, from the 'Conservation for Access' toolkit, is used by house teams and conservators to fine tune the 'rule of thumb' assessment, allocating more precise timings to each individual space [26]. For example, a closet might require a 0.25 cleaning unit per day (5 min), whereas 9 units (3 h) might be needed to vacuum a long gallery whose area is equivalent to 9 large rooms on an adjacent floor. A large, densely-furnished room with a drugget walkway over a fragile carpet and barrier ropes to prevent visitors straying, might require 2 units (40 min) per day, whereas a sparsely furnished hall of a similar size, allowing unrestricted visitor access, might need 4–6 units of time to vacuum the wood or stone floor and remove dust from robust furniture (80–120 min).

The resulting daily total is the indicative time needed for cleaning the house, but it is only sufficient if every member of the team is present every day. There are many reasons for absence—holidays, sickness, meetings, training, etc—and it is essential to recognise that many houses now open to visitors on six or seven days per week, whereas staff are generally employed for only five days. To create enough flexibility in the weekly rota to compensate for staff absences, the cleaning time per day should be increased to an average of 30 min per room or cleaning unit; this generates a more realistic and sustainable outcome. This 'optimum' allowance of time (30 min) enables the construction of a staffing rota to deliver sufficient cleaning time every day, and, therefore, ensures sustainable housekeeping practices for the long term. See the Supplementary for sample staff rotas for details of these calculations, and sample staff rotas for five- and seven-day working weeks (Table S1a–c). Where houses are closed to visitors during the winter months, cleaning hours can be consolidated into a five-day working week.

3. Results

The length and breadth of the Excel spreadsheet used by the Dimensions model makes illustration in its entirety difficult; therefore, the text describes the calculations used in each section of the model.

3.1. *Quality*

The first section of the model calculates the Quality factor. The working assumption is that each factor has roughly the same weighting, to avoid biasing one factor over another. The calculation takes the scores for the significance, fragility and complexity of interiors and collections, to which it adds an average of the two scores for the size and significance of the library. It calculates the number of volumes in libraries where only the number of titles is known, and the logarithmic adjustment to fit within the 0–4 scale. The calculation can be summarised as:

$$Quality = \overset{\text{Interiors and Collections}}{(significance + fragility + complexity)} + \overset{\text{Library}}{\frac{(significance + log\ of\ volumes)}{2}}\ /4$$

Table 2 illustrates the calculation of quality factors for six historic buildings from the fourteenth to nineteenth centuries, housing collections from the seventeenth to twentieth centuries, with and without libraries, with some located in towns and others in the countryside.

Table 2. Determining the Quality factor for a selection of six historic properties, each with interiors and collections of varying significance, fragility and complexity of materials, and most with libraries of individual significance and size.

Historic Property	Interiors and Collections			Library					Quality Factor
Description	Significance	Fragility	Complexity	Significance	Titles	Volumes (or titles + 25%)	Log of volumes	Significance and size (avg)	Sub Total/4
Cottage ornee and collections, 18c	4	4	4	3	1056	1264	3.10	3.05	3.76
Mansion, late 18c, early 19c collections	4	3	3	4	4506	5632	3.75	3.92	4.23
Farmhouse and collections, 17–19c	4	3	3	4	4000	5000	3.70	3.90	3.97
Town house, 18c, 20c collections	3	4	3	1	-	797	2.90	1.63	3.41
Terraced house and collections, early 20c	3	2	2	2	3974	4968	3.70	2.57	2.89
Homestead, 17c, mid-18c collections	2	2	2	1	12	20	1.30	1.10	2.28

3.2. Quantity

The second section of the model provides the Quantity factor. Again, the working assumption is that each factor has roughly the same weighting, so that one is not biased over another. This section calculates the quantity of collections (via inventory or CMS digital records), accessible spaces (counted as cleaning units), annual visitors and the hours of access. The calculation can be summarised as:

$$Quantity = (log\ of\ CMS\ records + cleaning\ units + visitors + log\ of\ access\ hours)\ /\ 4$$

The data in Table 3 represents the same six historic buildings as in Table 2, of which one property is large with a substantial collection, two are medium with a modest collection, and three are smaller with fewer furnishings. Visitor numbers vary greatly, from >100,000 to <2000, and the annual hours of access range from almost 2000 to less than 500.

3.3. Linking Dimensions Data to Staffing Structures and Pay Ranges

The subtotals for quality and quantity are averaged and converted to a percentage of the maximum achievable. These percentages are ranked in descending order and, starting from the highest, a different level of staff seniority (reflecting their qualifications and experience) is allocated to each 10%, to indicate a recommended staffing structure for full-time house roles.

For example, the resulting structure may include, where merited, a qualified and experienced museum professional as the head of department for the house (two grades), with 1–2 full-time supporting staff (e.g., house operations manager, and house team leader—two grades), as necessary to provide 24/7 security cover, oversee staff rosters, supervise collections care, and manage public access. Where scores drop below 40%, the interiors and collections may merit only a part-time conservation-trained cleaner, and below 30%, only a seasonal or contract cleaner (conservation training not essential). Table S2 provides the data and staffing structure for six historic collections.

In other heritage organisations, a similar process of relating Dimensions scores to salary grades can be adopted, but a different allocation of percentages might be needed, according to the local circumstances.

Table 3. Determining the Quantity factor for six historic buildings offering varying amounts of access (hours, visitors, and spaces) to disparate numbers of collection items.

Property	Inventory		Historic Spaces		Visitors to House		Hours of Access		Quantity Factor
Description	CMS records	Log of records	Cleaning units	Score per 12 units	Number (or % of property total)	Score per 40,000 visitors	Annual total	Log of access hours	Sub Total/4
Mansion, late-18c, early 19c collections	9103	3.96	33	2.75	103,000	2.58	1983	4.77	3.51
Cottage ornee and collections, 18c	4240	3.63	18	1.50	50,000	1.25	1614	4.69	2.77
Homestead, 17c, mid-18c collections	1441	3.16	19	1.58	3274	0.08	515	4.19	2.25
Terraced house and collections, early 20c	1270	3.10	9	0.75	9818	0.25	1092	4.52	2.15
Town house, 18c, 20c photographic Collections	4636	3.67	8	0.67	1800	0.05	405	4.08	2.12
Farmhouse and collections, 17–19c	78	1.89	8	0.67	15,943	0.40	488	4.17	1.78

The resulting staffing recommendations are indicative and proportionate to the housekeeping challenges quantified in the Dimensions model. The number of full-time supporting roles required at each property is validated by the regional conservator, using their experience of how well a house team is currently performing in the collections care, and what improvements in resources might be needed. Decisions on staffing are based on many factors, and each property manager must reconcile conservation advice with the availability of property resources, and the staffing needs of other departments on the property (for example, challenges presented by the garden may outweigh those of the house, or vice versa).

3.4. Calculating Cleaning Hours

The next section of the Dimensions model calculates how many cleaning hours should be achieved per house per week, in response to the days of access, and how many part-time team members are needed, with each achieving 20 h of cleaning per week, but employed for 22.5 h (i.e., 0.65 full-time equivalent (FTE)) to allow 30 min per day for a refreshment break and to maintain cleaning equipment and tools.

As discussed in 2.4 above, if every member of the team were on duty every day, only the indicative number of roles would be needed. However, staff work a 5-day week, and cover may be needed for 6–7-days of visitor access, as well as holidays, sickness and absence for meetings and training; therefore, the optimum recommendation is more realistic and sustainable. Where possible, rotas are designed to include working on one or two closed days per week to undertake the weekly/occasional tasks which are difficult to achieve while visitors are present; where a house is open 7 days, teams work additional hours each week, equivalent to an 8th day.

The quality factor is used to guide decisions on whether a house requires indicative or optimum hours of cleaning—the higher the quality factor, the greater the need for optimum hours. The calculations can be summarised as:

$$\textit{Indicative team: } (\textit{cleaning units}/3) \times (\textit{access days} + 1)/20\ \text{h} = \textit{number of roles}$$

$$\textit{Optimum team: } (\textit{cleaning units}/2) \times (\textit{access days} + 1)/20\ \text{h} = \textit{number of roles}$$

Figure 4 shows the variable number of part-time conservation cleaning roles, commensurate with the dimensions of the housekeeping challenge presented by each property. The results range from the need for a team of 6.6 part-time conservation-trained roles at a large mansion with high quality collections, open to visitors 7 days per week, to a few hours of weekly cleaning at a small property with robust collections. Table S3 provides the data used in the calculation.

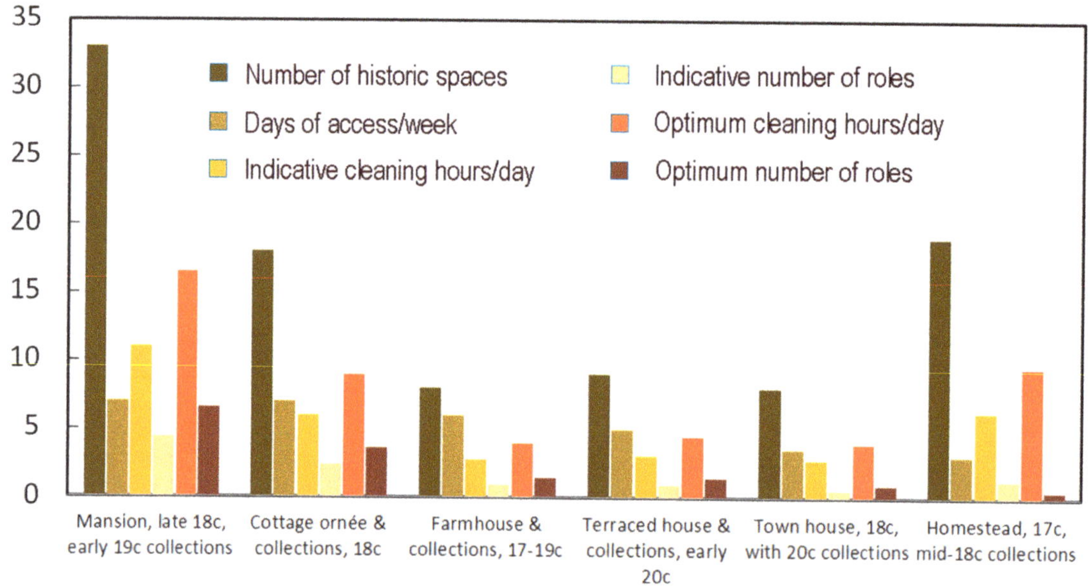

Figure 4. Calculation of cleaning hours and roles for six houses of different periods and size, displaying varying quantities of historic collections, from fragile to robust; showing the number of spaces and weekly days of access, indicative cleaning required (hours/day) and indicative number of roles, the optimum cleaning hours/day and the optimum roles. The 17c homestead is open only 3 days/week; therefore, it only requires indicative roles, whereas the *Cottage ornée* with a similar number of historic spaces is open 7 days/week and requires the optimum number of roles.

3.5. Winter Access and Deep Cleaning

A further section of the Dimensions model is used to estimate the number of weekly hours needed to undertake annual 'deep cleaning'. Smaller properties may close for several months during the winter, and undertake this work using the same number of daily cleaning hours as calculated for the open season; the weekly hours of cleaning are re-allocated between spaces, so that rooms densely furnished with fragile materials receive more detailed attention to deep cleaning than the rooms sparsely furnished with robust materials (which would have benefitted from more frequent cleaning during the open season). However, to accommodate Christmas and New Year festivities, and school half-term holidays in February, larger properties may open intermittently or consistently throughout the winter (i.e., 363 days per year); this leaves a house team with the double challenge of preparing the house daily for visitor access, while at the same time undertaking the annual deep clean.

Two calculations are made: one for properties offering little or no visitor access between November and March, where a team can concentrate on deep cleaning; the other estimating how much additional time is needed to prepare a house for visitor access in winter, while also embarking on a programme of deep cleaning the ceilings, walls and floors, and collection items inaccessible during public access. Deep cleaning involves a significant disruption to each room in turn, moving furniture and carpets so that tower scaffolds and stepladders can be used; to make this activity safe for staff and visitors, one room at a time may need to be closed to public access but, where possible and safe to do so, the conservation work is demonstrated to visitors. The calculations can be summarized as:

For limited winter access + deep cleaning:

$$\textit{Indicative team: } (\textit{cleaning units}/3) \times (\textit{winter access weeks})/20\ \text{h} = \textit{number of roles}$$

$$\textit{Optimum team: } (\textit{cleaning units}/2) \times (\textit{winter access weeks})/20\ \text{h} = \textit{number of roles}$$

For year-round access + deep cleaning:

$$\textit{Indicative team: } (\textit{cleaning units}/3) \times (1\ \textit{year} + \%\ \textit{winter weeks})/20\ \text{h} = \textit{number of roles}$$

$$\textit{Optimum team: } (\textit{cleaning units}/2) \times (1\ \textit{year} + \%\ \textit{winter weeks})/20\ \text{h} = \textit{number of roles}$$

Small houses with few and/or robust collections may not need many, if any, additional hours for deep cleaning; however, highly significant houses with larger fragile collections may need to employ additional roles, or supervise the work of volunteers providing additional support, in order to complete deep cleaning simultaneously with visitor access. The optimum weekly cleaning hours become the target for measuring the conservation performance (CP), as discussed in 4.1 below. Figure 5 shows the results for the same six properties, and Table S4 provides the data used in the calculation.

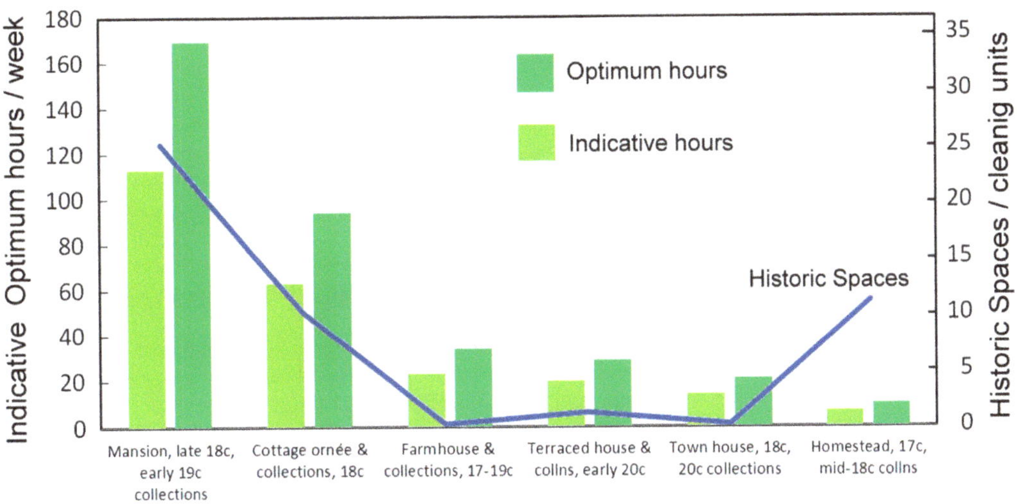

Figure 5. Calculation of additional cleaning hours needed per week during the winter months, in historic buildings where visitor access is year-round, and daily preparation for winter access and deep cleaning of the shell of each room are undertaken simultaneously. The 17c homestead is closed to visitors; therefore, although of a similar size to the *cottage ornée*, it requires fewer cleaning hours in winter.

4. Discussion

Over the two decades during which the Dimensions model has gradually evolved, national conservators have occasionally been challenged by management colleagues to devise ways in which property performance could be objectively measured and national resources equitably distributed. The Dimensions model has proved a useful starting point for developing new calculations, as it contains core data which enable distinctions to be made between the quality and quantity of the housekeeping challenge faced by each individual property, enabling national resources to be apportioned fairly between them.

4.1. Conservation Performance Assessment

The Dimensions model can be used to set, monitor and rate the conservation performance (CP), for example, how well house teams are meeting their target of housekeeping and cleaning hours per week. This assessment could form part of a periodic review, designed to measure improvements in conservation across every department on an estate including the archaeology, buildings, gardens, parks, woodlands and farmland [28].

For this assessment, properties can be divided into four bands, namely, A-D, in relation to their percentage score for Quality and Quantity; the targets for each band can vary according to the risks associated with a failure to achieve sufficient hours of conservation cleaning. Taking the optimum cleaning hours per week as the target for each large and significant collection (and in small robust properties, taking the indicative hours), the percentage of hours achieved can be measured, and the property's performance graded accordingly. Table S5a,b illustrates a set of bands, success measures, and ratings of performance against relevant targets.

If higher percentage targets are set for significant and fragile collections, and lower percentage targets for smaller, more robust collections, small houses which meet most of their target may achieve a greater measure of success than larger houses which fail to reach their target. This weighting of the targets and success measures is logical, as the conservation consequences of failure to meet the relevant target are more serious for highly significant, fragile and complex collections, especially those offering the longest hours of access to the largest numbers of visitors and/or accumulating the highest rates of deposition of coarse dust.

4.2. Estimating and Distributing National Resources to Meet Preventive Conservation and Specialist Maintenance Needs

The Dimensions model can be used to inform the re-balancing of historical inequities in individual property endowment funds, and channel additional resources to those properties unable to meet their routine conservation and maintenance needs. (Through this process, no property already consistently spending more on conservation than its basic need loses any funds, but those properties historically disadvantaged are eligible for additional central support.) Each property is assessed for its ability to fund its maintenance needs from its own resources (i.e., investment and rental income; visitor admissions and member visit credits; and retail and catering profits); where necessary, an amount of top-up funding is agreed on to support those needs.

The model supports properties in anticipating and making provision for the non-wage costs of preventive conservation, such as equipment, materials and fuel costs for conservation heating; it also helps to distribute top-up funding for routine maintenance and in situ treatments by freelance conservators. Two calculations are required, as follows.

4.2.1. 'Basic' Conservation and Maintenance Need

Heads of profession and national specialists assess the basic conservation and maintenance needs for their area of responsibility, whether they be buildings, countryside, collections and interiors, gardens or interpretation. The 'basic need' for each component of a property is consistently defined as:

$$\textit{Distribution parameter/property} \times \textit{Rate £/unit}$$

Calculation of the 'maintenance need' for collections and interiors is more complex than for other components of an estate. It cannot be simply reduced to, say, GBP 1 per item on each property's inventory, as this favours numerically large collections of robust items (e.g., brown furniture) usually needing less attention, while neglecting the fragility and complexity of significant items requiring intermittent specialist care (e.g., historic textiles).

To determine the scale of the national budget required annually for basic collections conservation needs (i.e., specialist care and maintenance), a rough calculation first takes the average maintenance cost per item for the most numerous collection items (e.g., textiles, costumes, and books) and the average maintenance cost for all other material items, multiplied by the average quality and quantity factors across all properties. This calculation (made in 2014) encompasses data from 348 properties with collections, 870,456 digital CMS records, and a total of 660 Quality and Quantity factors, averaging 1.896 each. It indicates that the national objective should be to allocate c. GBP 3m per year to conservation and maintenance needs (Table 4).

Table 4. A rough calculation made in 2014 to determine that the national conservation 'need' budget for collections should be c. GBP 3 million per year (in addition to the objective to spend GBP 3 million per annum on preventive conservation, as described in 4.2 above).

Maintenance Costs	Textile	Costume	Book	Other Item	Avg Cost	Annual 'Need'
Avg cost per item (2014):	GBP 15.00	GBP 5.00	GBP 0.10	GBP 1.90	GBP 2.46	
Total cost excl. Q factors:	GBP 880 k	GBP 158 k	GBP 21 k	GBP 1084 k		GBP 2143 k
Q factored to 1:	GBP 1504 k	GBP 256 k	GBP 34.5 k	GBP 1770 k		GBP 3565 k
Avg factored cost:	GBP 25.62	GBP 8.13	GBP 0.16	GBP 3.10	GBP 4.10	

4.2.2. Preventive Conservation Equipment, Materials and Fuel

A questionnaire was used to gather information from 30 properties about the quantities and average lifespans of preventive conservation and housekeeping equipment in use (e.g., environmental control, protective materials for access/display and storage, emergency salvage equipment, cleaning equipment and materials, and personal protective equipment) (Table 5).

Table 5. Items of preventive conservation equipment and materials for which each house needs an annual budget, proportionate to the quantity of each item, its anticipated lifespan, and the number of team members.

Environment	Protection: Access	Protection: Storage	Salvage Store	Cleaning: Equipment	Cleaning: Consumables
Blinds	Entrance matting	Dust covers	Tools	Vacuum cleaners	Brushes
UV filter	Floor coverings	Racking	Equipment	Floor polishers	Dusters
Data loggers	Rope barriers	Boxes	Crates	Work lights	Polishes
Dehumidifiers	Stanchions	Rollers		Tower scaffold	
Pest control	Display stands	Dustsheets		Ladders	
Environmental monitoring and control software		Acid-free tissue	Personal protective equipment		Personal protective equipment

These costs (updated periodically using the UK Retail Price Index) are divided by the equipment's lifespan (in years), and multiplied by the quantity at each property, to estimate an annual figure sufficient to replace these resources as they become exhausted. The data is collated in an Excel spreadsheet within the 'Conservation for Access' toolkit, to support property teams in calculating appropriate budgets for preventive conservation equipment. This process is described in other publications [18,26–28,32].

For financial planning purposes, specialist conservation advisers suggest the average frequency of need for condition surveys and STC maintenance of each category of collection material, together with an average cost per item. Three examples of the frequencies and costs of STC maintenance are shown in Figure 6a–c below, and more detail is published in [32].

Figure 6. (**a–c**): Examples of average costs per item for short-term cyclical maintenance by specialist conservators. (a) Wall paintings: average GBP 5000 each, every 50 years. (**b**) Paintings: Average GBP 200 each, every 25 years. (**c**) Upholstered seat furniture: average GBP 400 each, every 10 years. © National Trust Images/Andreas von Einsiedel and James Dobson.

Together, these calculations indicate that national objectives should be to spend GBP 6 m per year on meeting STC conservation needs—GBP 3 m on preventive conservation equipment and GBP 3 m on the maintenance of collections by specialist conservators. The Dimensions model is used to indicate roughly how much each property should spend per year, as its contribution to achieving the national objective to spend GBP 3 m on preventive conservation. It also offers a mechanism for allocating central top-up resources to under-endowed properties, as percentage shares of GBP 3 m, in proportion to the quality and quantity factors of each collection.

The calculation can be summarised as:

Conservation need = quality (collections + library) × cleaning units, expressed as % of national total

The same % of £3m is the property target for its annual preventive conservation equipment budget.

Table S6 provides the data and calculation for the six properties discussed above. More recent outcomes (2019) for six different properties, using updated scores for significance, are illustrated in Figure 7, with the underlying data and calculations in Table S7.

The calculations raise awareness of the individual property responsibilities for identifying annual budgets for maintenance and conservation; however, the national top-up funds remain insufficient to bridge all the historic gaps in resources, leaving some properties still challenged to generate sufficient income to meet their basic preventive conservation equipment and maintenance needs. Regional conservators monitor the expenditure at each place, and use the Dimensions model to promote the need for funding commensurate with the quality of each collection, and the quantity of interiors in which each collection is displayed, with each space requiring a regular renewal of room-scale preventive measures and protective materials, as well as specialist maintenance.

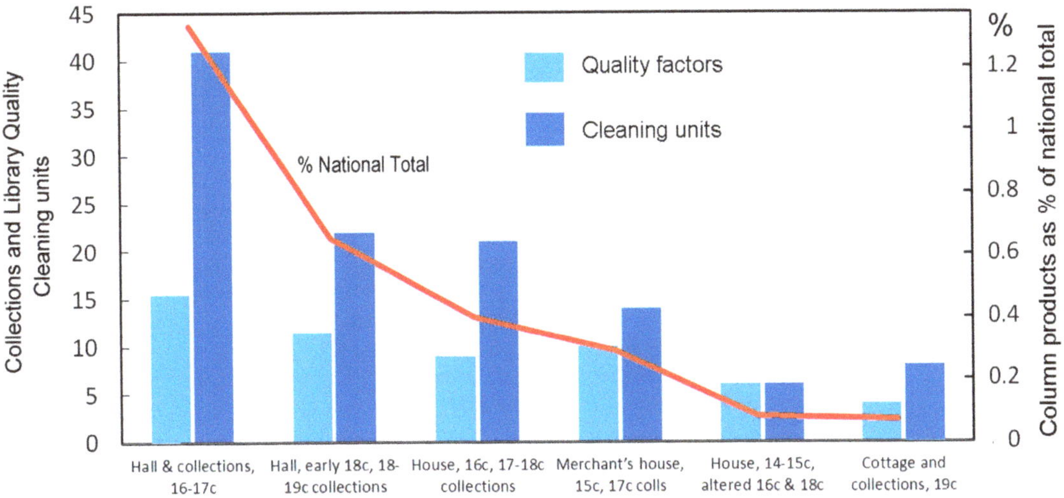

Figure 7. Extract from 2019 calculation to inform the distribution of the national GBP 3 m conservation and maintenance 'need' budget, in proportion to the quality of collections, and quantity of interior spaces open to visitors. See Table S7 for the underlying data, using updated qualitative scores for significance, fragility and complexity for six properties (in addition to those discussed above).

4.3. Staff Resources and Job Evaluation

In a large national organisation, where the local management is devolved to regions and properties, there are advantages to having one simple system which can be applied consistently at a local level, as well as supporting a central overview of the national resource needs. The transparency of the data underpinning the calculations enables everyone to understand the full extent of the differences between properties and their needs for resources. The simplicity of the model enables it to be readily understood by non-specialists, whether they be local house staff, consultants in other professional disciplines, or national heads of profession who collaborate to develop national strategy and operational policy, influence the distribution of funds, and support professional development.

Using the model, advice on staffing structures and cleaning hours can be offered by email or telephone, without the need for time-consuming travel to each property, provided that care is first taken to update the qualitative and quantitative data relevant to each property. Discussion with each property manager of the core ingredients in the calculations increases their understanding of the scale of the challenge at their property and how it differs from its regional neighbours, but how it might compare more directly with a property in another region, with which they may not be familiar. This understanding reduces the likelihood of competition for resources between dissimilar neighbouring properties; it also helps property heads of department (who aspire to equal salary grades) to recognise that the challenges in the house may merit leadership of a greater/lesser calibre than the garden, countryside, visitor services, retail or catering.

When applications for re-grading of individual staff roles are received, the Dimensions model is effective in quantifying the scale of challenge at that property, and guiding a central discussion on whether the property merits a more senior role, or whether staff should be encouraged to consider a career move to a property offering greater challenges and opportunities for promotion.

5. Conclusions

The Dimensions model contributes to the achievement of a sustainable balance between access and conservation. It is effective in helping conservators and property managers to ensure that each historic house is aware of its needs for conservation housekeeping staff and financial resources, commensurate with the quality and quantity of its interiors and collections, and with the extent of access offered to its visitors. The model has proven effective in practice, and its outputs reflect conservators' professional assessments of the individual property needs for staff and financial resources. The regular refreshing of data ensures that the results of calculations correlate with the practical experience of supporting house teams and property portfolios. The model also establishes benchmarks against which resources and performance can be monitored and measured. For national consultants, it provides reassurance that their professional opinion is underpinned by hard facts, and that their advice is consistent over time and between properties.

The model is effective in making multiple uses of one large set of data; this simplifies the updating of core information, while also supporting the evolution and adaptation of the model for exploratory calculations in response to new management challenges. It enables the monitoring of how well resources are used; it helps to identify gaps, and supports advocacy for how those gaps might be filled, whether by employing more staff, increasing the skills of existing staff, seeking external funds, or prompting a fairer allocation of the existing resources.

To be effective, the criterion chosen for use in the model must be relevant to the local circumstances, and be able to be tested and validated, so that conservators, property managers and their teams have confidence in the outcomes. The assessments of cleaning units, and the allocations of indicative/optimum timings to each unit, are not absolutes; these figures should always be tested and validated, and may need adjustment before use in different contexts. The methodology described here had worked well for more than two decades but, following the COVID-19 pandemic, some parameters were adjusted to reflect the changing circumstances: reduced visitor numbers (generating less coarse dust), reduced access (fewer rooms open for fewer days/weeks/months), and/or more diverse visitor offerings, including pre-booked guided tours, events, and demonstrations of conservation in action. These parameters remain under review as visitor access at properties gradually returns to pre-pandemic levels.

More sophisticated formulae could be developed, but those described and illustrated here require only GCSE mathematics and an elementary use of an Excel spreadsheet; therefore, they are within the capability of many potential users. To make the calculations more mathematically sound would require the factors to be individually weighted, necessitating a justification and validation of the weightings. Future refinements remain possible and, in the course of writing this paper, two recent publications were drawn to the attention of the author, suggesting different approaches for future investigation [38,39].

Last but not least, it is important to emphasise that, to remain effective, models need to be supported, so that their data is consistently refreshed, and their uses can be continuously fine-tuned, improved and developed to offer robust and sustainable solutions to conservation management challenges.

Supplementary Materials: The following supporting information can be downloaded at: https://www.mdpi.com/article/10.3390/heritage6040199/s1.

Funding: This research received no external funding.

Data Availability Statement: No new data were created or analysed in this study. Data sharing is not applicable to this article.

Acknowledgments: The author owes a huge debt of gratitude to Peter Brimblecombe for leading the collaborative research on 'Controls on irreversible soiling', advice on fine-tuning the model, and long-distance support while this paper was written. At the National Trust, to Katy Lithgow, former head conservator, enduring thanks for her advocacy for conservation and maintenance resources; to the late Stephen Page, management consultant, for supporting use of the Dimensions model to achieve a realistic distribution of resources proportionate to property needs; and to National Trust people business partners for welcoming the model as a guide to job evaluation. To Cara Wallace and Claire Hollinghurst, business support coordinators, and Penny Robbins, loyal volunteer, thanks for their tenacity and patience while repeatedly refreshing data in the model. To Nigel Blades, conservation scientist, for helpful comments on the first draft, and to Hannah Harte, head of conservation, for permission to publish this work.

Conflicts of Interest: The author declares no conflict of interest.

References

1. Sandwith, H.; Stainton, S. (Eds.) *The National Trust Manual of Housekeeping*; Allen Lane: London, UK, 1984.
2. Ten Agents of Deterioration. Available online: https://www.canada.ca/en/conservation-institute/services/agents-deterioration/physical-forces.html (accessed on 8 April 2023).
3. Framework for Preserving Heritage Collections. Available online: https://publications.gc.ca/collections/collection_2021/pch/CH57-4-29-2021-eng.pdf (accessed on 22 January 2023).
4. National Trust. *National Trust Manual of Housekeeping, Care and Conservation of Collections in Historic Houses*, revised ed.; The National Trust: Swindon, UK, 2011.
5. Lloyd, H.M.; Brimblecombe, P. *Controls on Irreversible Soiling, 2000–2003*; The Leverhulme Trust: London, UK, 2005.
6. Tackling Dust in Historic Houses. Available online: https://www.nationaltrust.org.uk/our-cause/history-heritage/tackling-dust-in-historic-houses (accessed on 8 April 2023).
7. Yoon, Y.H.; Brimblecombe, P. Contribution of dust at floor level to particle deposit within the Sainsbury Centre for Visual Arts. *Stud. Conserv.* **2000**, *45*, 127–137.
8. Yoon, Y.H.; Brimblecombe, P. Clothing as a source of fibres within museums. *J. Cult. Herit.* **2000**, *1*, 445–454. [CrossRef]
9. Yoon, Y.H.; Brimblecombe, P. The Distribution of Soiling by Coarse Particulate Matter in the Museum Environment. *Indoor Air* **2001**, *11*, 232–240. [CrossRef] [PubMed]
10. Lloyd, H.; Lithgow, K.; Brimblecombe, P.; Yoon, Y.H.; Frame, K.; Knight, B. The effects of visitor activity on dust in historic collections. *Conservator* **2002**, *26*, 27–84. [CrossRef]
11. Lithgow, K.; Brimblecombe, P.; Lloyd, H.; Thickett, D.; Yoon, Y.H. Managing dust in historic houses—A visitor/conservator interface. In *Our Cultural Past—Your Future Proceedings of ICOM Committee for Conservation 14th Triennial Meeting, The Hague, The Netherlands, 12–16 September 2005*; James & James/Earthscan: London, UK, 2005; pp. 662–669.
12. Lloyd, H.; Bendix, C.; Brimblecombe, P.; Thickett, D. Dust in historic libraries. In *Museum Microclimates, Contributions to the Conference in Copenhagen, 19–23 November 2007*; National Museum of Denmark: Copenhagen, Denmark, 2007; Available online: https://www.conservationphysics.org/mm/musmic/musmic150.pdf (accessed on 22 January 2023).
13. Lithgow, K.; Brimblecombe, P.; Knight, B.; Julien, S. Visitor perceptions of dustiness. In *Indoor Air Quality in Museums and Historic Properties Proceedings of the Indoor Air Pollution Working Group, Norwich, UK, 28–29 April 2003*; University of East Anglia: Norwich, UK, 2003; Available online: http://www.iaq.dk/iap/iaq2003/2003_05.htm (accessed on 22 January 2023).
14. Lithgow, K.; Brimblecombe, P. Dust, the visitors' point of view. *Views* **2003**, *39*, 47–49.
15. Lithgow, K.; Brimblecombe, P. Conservation assistants and room stewards focus on dust. *Views* **2004**, *41*, 22–25.
16. Lloyd, H.; Brimblecombe, P. Focussing on dust. *Views* **2003**, *39*, 49–52.
17. Dillon, C.; Golfomitsou, S.; Lithgow, K.; Ravaioli, F.; Storey, C.; Tully, C.; McArthur, G. A clean sheet: A bottom-up and mixed-methods approach to understanding visitors' perceptions of dust, dirt and cleaning. In Proceedings of the ICOM-CC 18th Triennial Conference Preprints, Copenhagen, Denmark, 4–8 September 2017; International Council of Museums: Paris, France, 2017.
18. Lloyd, H.; Brimblecombe, P.; Lithgow, K. Economics of dust. *Stud. Conserv.* **2007**, *52*, 135–146. [CrossRef]
19. Brimblecombe, P.; Thickett, D.; Yoon, Y.H. The cementation of coarse dust to indoor surfaces. *J. Cult. Herit.* **2009**, *10*, 410–414. [CrossRef]
20. Tarnowski, A.; McNamara, C.; Bearce, K.; Mitchell, R. Sticky microbes and dust on objects in historic houses. In *Objects Specialty Group Post-Prints Proceedings of the AIC 32nd Annual Meeting, Portland, OR, USA, 9–14 June 2004*; American Institute for the Conservation of Historic and Artistic Works: Washington, DC, USA, 2004; Volume 11, pp. 11–28. Available online: http://resources.conservation-us.org/osg-postprints/wp-content/uploads/sites/8/2015/02/osg011-02.pdf (accessed on 22 January 2023).
21. Lloyd, H.; Grossi, C.M.; Brimblecombe, P. Low-technology dust monitoring for historic collections. *J. Inst. Conserv.* **2011**, *34*, 104–114. [CrossRef]
22. Brimblecombe, P.; Grossi, C.M. The Identification of Dust in Historic Houses. Available online: https://connectingtocollections.org/the-identification-of-dust-in-historic-houses/ (accessed on 27 March 2023).
23. Waterson, J.M.W. *The National Trust: The First Hundred Years*; BBC Books and National Trust (Enterprises) Limited: London, UK, 1999; p. 73.
24. Lithgow, K. Sustainable decision making: Change in National Trust collections conservation. *J. Inst. Conserv.* **2011**, *34*, 128–142. [CrossRef]
25. Staniforth, S.; Lloyd, H. Use it or lose it: The opportunities and challenges of bringing historic places to life. *Stud. Conserv.* **2012**, *57* (Suppl. S1), S286–S294. [CrossRef]
26. Lloyd, H.; Lithgow, K. Conservation for access: A toolkit to promote sustainability. In *Conservation and Access, IIC London Congress Preprints*; Saunders, D., Townsend, J.H., Woodcock, S., Eds.; IIC: London, UK, 2008; p. 263.

27. Lloyd, H. Opening Historic Houses. In *The National Trust Manual of Housekeeping: Care and Conservation of Collections in Historic Houses*, revised ed.; The National Trust: Swindon, UK, 2011; pp. 671–685.
28. Lithgow, K.; Staniforth, S.; Etheridge, P. Prioritizing access in the conservation of National Trust collections. *Stud. Conserv.* **2008**, *53* (Suppl. S1), 178–185. [CrossRef]
29. Lithgow, K.; Lloyd, H.; Tyler-Jones, M. Conservation for access redux: Narrative, visitor flow and conservation. In *The Artifact, Its Context, and Their Narrative: Multidisciplinary Conservation in Historic House Museums Proceedings of the ICOM–DEMHIST and ICOM-CC Working Groups, Los Angeles, CA, USA, 6–9 November 2012*; Seymour, K., Sawicki, M., Eds.; The Getty Research Institute: Los Angeles, CA, USA, 2012; Available online: https://icom-demhist.org/wp-content/uploads/2019/05/Conference-Proceedings.-Los-Angeles-2012.pdf (accessed on 8 April 2023).
30. Bullock, L.; Saunders, D. Measurement of cumulative exposure using Blue Wool standards. In Proceedings of the ICOM-Committee for Conservation 12th Triennial Meeting, Lyon, France, 29 August–3 September 1999; James & James (Science Publishers) Ltd.: London, UK, 1999.
31. Lithgow, K. Delivering the National Trust's preservation purpose: Mission, strategy and structure. *Insight—Non-Destr. Test. Cond. Monit.* **2020**, *62*, 152–159. [CrossRef]
32. Lithgow, K.; Lloyd, H. Direct preventive conservation: Using information from the past to prevent small issues in the present from becoming bigger problems in the future. In Proceedings of the ICOM Committee for Conservation 18th Triennial Conference, Copenhagen, Denmark, 4–8 September 2017; International Council of Museums: Paris, France, 2017. Available online: https://www.icom-cc-publications-online.org/1622/Direct-preventive-conservation--Using-information-from-the-past-to-prevent-small-issues-in-the-present-from-becoming-bigger-problems-in-the-future (accessed on 22 January 2023).
33. Russell, R.; Winkworth, K. *Significance 2.0: A Guide to Assessing the Significance of Collections*; Collections Council of Australia Ltd (now Australian Government Office of the Arts): Canberra, Australia, 2009. Available online: https://www.arts.gov.au/sites/default/files/documents/significance20.pdf (accessed on 8 April 2023).
34. Reed, C. Reviewing Significance 3.0: A Framework for Assessing Museum, Archive and Library Collections' Significance, Management and Use. 2018. Available online: https://collectionstrust.org.uk/resource/reviewing-significance-3-0/ (accessed on 8 April 2023).
35. National Trust. *The National Trust Handbook for Members and Visitors*; National Trust: Swindon, UK, published annually; Available online: https://www.nationaltrust.org.uk/membership/our-members-handbook (accessed on 8 April 2023).
36. Lloyd, H.; Lithgow, K. Planning and managing housekeeping. In *The National Trust Manual of Housekeeping: Care and Conservation of Collections in Historic Houses*, revised ed.; The National Trust: Swindon, UK, 2011; pp. 114–123.
37. Lloyd, H.; Doyal, S. Staffing historic houses. In *The National Trust Manual of Housekeeping: Care and Conservation of Collections in Historic Houses*, revised ed.; The National Trust: Swindon, UK, 2011; pp. 686–695.
38. D'Ayala, D.F.; Copping, A. Ayala, D.F.; Copping, A. A conceptual model for Multihazard assessment of the vulnerability of historic buildings. In *Structural Analysis of Historical Constructions*; Lourenco, P.B., Roca, P., Modena, C., Agrawal, S., Eds.; Macmillan: New Delhi, India, 2006; ISBN 972-8692-27-7.
39. Ashrafi, B.; Kloos, M.; Neugebauer, C. Heritage Impact Assessment, beyond an Assessment Tool: A comparative analysis of urban development impact on visual integrity in four UNESCO World Heritage Properties. *J. Cult. Herit.* **2021**, *47*, 199–207. [CrossRef]

MDPI
St. Alban-Anlage 66
4052 Basel
Switzerland
www.mdpi.com

Heritage Editorial Office
E-mail: heritage@mdpi.com
www.mdpi.com/journal/heritage

www.ingramcontent.com/pod-product-compliance
Lightning Source LLC
Chambersburg PA
CBHW061437210926
51850CB00008B/1539
9783036586540